Global Perspectives on Health Geography

Series editor

Valorie Crooks, Department of Geography, Simon Fraser University, Burnaby, BC, Canada

Global Perspectives on Health Geography showcases cutting-edge health geography research that addresses pressing, contemporary aspects of the health-place interface. The bi-directional influence between health and place has been acknowledged for centuries, and understanding traditional and contemporary aspects of this connection is at the core of the discipline of health geography. Health geographers, for example, have: shown the complex ways in which places influence and directly impact our health; documented how and why we seek specific spaces to improve our wellbeing; and revealed how policies and practices across multiple scales affect health care delivery and receipt.

The series publishes a comprehensive portfolio of monographs and edited volumes that document the latest research in this important discipline. Proposals are accepted across a broad and ever-developing swath of topics as diverse as the discipline of health geography itself, including transnational health mobilities, experiential accounts of health and wellbeing, global-local health policies and practices, mHealth, environmental health (in)equity, theoretical approaches, and emerging spatial technologies as they relate to health and health services. Volumes in this series draw forth new methods, ways of thinking, and approaches to examining spatial and place-based aspects of health and health care across scales. They also weave together connections between health geography and other health and social science disciplines, and in doing so highlight the importance of spatial thinking.

Dr. Valorie Crooks (Simon Fraser University, crooks@sfu.ca) is the Series Editor of Global Perspectives on Health Geography. An author/editor questionnaire and book proposal form can be obtained from Publishing Editor Zachary Romano (zachary.romano@springer.com).

More information about this series at http://www.springer.com/series/15801

Sarah Atkinson • Rachel Hunt
Editors

GeoHumanities and Health

Springer

Editors
Sarah Atkinson
Department of Geography
Durham University
Durham, UK

Rachel Hunt
School of Geosciences
University of Edinburgh
Edinburgh, UK

ISSN 2522-8005 ISSN 2522-8013 (electronic)
Global Perspectives on Health Geography
ISBN 978-3-030-21405-0 ISBN 978-3-030-21406-7 (eBook)
https://doi.org/10.1007/978-3-030-21406-7

This Springer imprint is published by the registered company Springer Nature Switzerland AG
The registered company address is: Gewerbestrasse 11, 6330 Cham, Switzerland

Acknowledgements

The project of putting this book together would never have progressed beyond a passing thought were it not for the energy of Professor Valorie Crooks from Simon Fraser University in setting up this book series and her enthusiasm and encouragement to take on this collection.

We have benefited from windows of research time in which to get the job done allowed by our respective universities of Durham and Edinburgh and from the stimulating intellectual environments of Durham's Institute of Medical Humanities and Department of Geography and Edinburgh's School of GeoSciences. The Durham Institute of Medical Humanities is supported by the Wellcome Trust grant number WT209513. We have also enjoyed unstinting support at home, and our thanks go to David, Doug, Rosie, Merry, and Joe.

We particularly want to thank Sarah de Leeuw for allowing us to publish two poems from her exciting new collection, *Outside America*, and Faber and Faber Ltd. for granting permission for the reproduction of lines from Alice Oswald's poem, *The Thing in the Gap-Stone Stile*.

Finally, we have enjoyed stellar support throughout the process, from proposal to publication, from the team at Springer, Zachary Romano, Aaron Schiller, Silembarasan Panneerselvam and Gopalraj Chitra; it has been a real pleasure to work with you all.

Contents

Contributors

Editors

Sarah Atkinson Durham University, Institute of Medical Humanities, Department of Geography, Durham, UK

Sarah Atkinson is Professor of Geography and Medical Humanities at Durham University. Her work engages key concepts shaping contemporary practices of health and medicine through critical approaches in a humanities-facing social science. These include wellbeing, care, physical movement and experience. In addition, she is Associate Director of the Durham Institute for Medical Humanities and Associate Editor of the *Edinburgh Companion to the Critical Medical Humanities*.

Rachel Hunt University of Edinburgh, School of GeoSciences, Edinburgh, UK

Rachel Hunt is a Lecturer in GeoHumanities in the School of GeoSciences at the University of Edinburgh. Her work lies at the intersection of cultural, historical and rural geographies with a focus upon the self-landscape encounter. Her key research interests fall into three related areas: cultural geographies of land and landscape, rural lives and leisure and links between landscape experience and wellbeing.

Poetry

Sarah de Leeuw University of Northern British Columbia, School of Population and Public Health, Prince George, BC, Canada

An award-winning Creative Writer (poetry/literary non-fiction) and Canada Research Chair in Humanities and Health Inequities, Sarah de Leeuw's activism, writing, scholarship and teaching focus both on unsettling geographies of power and on the role of humanities in making biomedical and health sciences more socially accountable.

Essays

Gavin J. Andrews McMaster University, Department of Health, Aging and Society, Hamilton, ON, Canada

Gavin J. Andrews is Professor in the Department of Health, Aging and Society at McMaster University, Canada, with a background in Geography. His empirical interests include ageing, holistic medicine, healthcare work, fitness, health histories and popular music. Much of his work is positional and considers the development and progress of health geography. In recent years, he has become interested in the potential of post-humanist and non-representational theory to convey the processual nature and immediacy of health and wellbeing.

Rachel Bar Canada's National Ballet School and Ryerson University, Toronto, ON, Canada

Rachel Bar is a graduate of Canada's National Ballet School's Professional Ballet Program. She danced professionally with the English National Ballet and the Israel Ballet before pursuing academia. She is currently completing her PhD in Psychology as a Vanier Scholar, at Ryerson University, in Toronto, Canada. Her research explores the benefits of dance for older adult populations and the utility of arts-based knowledge translation of health research. She also manages health and research initiatives at Canada's National Ballet School (NBS) and is part of the team developing and researching NBS' dance initiatives for older adults.

Sarah Bell University of Exeter, European Centre for Environment and Human Health, Exeter, UK

Sarah Bell is a Lecturer in Health Geography, whose research focuses on the complex intersections between human health, wellbeing and the interlinked physical, social and cultural environments in which people live, work and move. She has recently completed a research fellowship, 'Sensing Nature' (www.sensing-nature.com), exploring how people living with varied forms and severities of sight impairment describe their experiences with(in) diverse types of nature through the life course.

Hannah Bradby Uppsala Universitet, Department of Sociology, Uppsala, Sweden

Hannah Bradby was born in Paisley (Scotland) because there was no space in the maternity hospitals in Glasgow that month. She mostly went to school in Kent (England), while her father ferry-commuted to work in Normandy (France). She is currently Professor in the Sociology Department at Uppsala University (Sweden). Her research was published as a novella *Skinfull* (Onlywomen Press, 2007) which included the funny, scandalous and paradoxical stories that could not be included in an academic monograph. The fictionalised version showed how young women confronted and resolved, or at least survived, the everyday contradictions of diverse inner-city living. She has also published plenty of empirically justified scholarly research of the more traditional variety.

Bentley Crudgington University of Manchester, School of Medical Science and the Animal Research Nexus, Manchester, UK

Bentley Crudgington directs public engagement on the Animal Research Nexus and is based at the University of Manchester with disciplinary expertise in Biology and Virology. As a Public Engagement Practitioner, Bentley has devised, implemented and delivered award-winning projects across art, science and technology for Wellcome, Imperial College London, Science Museum and the British Council. His role is to develop, drive and facilitate innovative and exciting public engagement on the animal research nexus. They have particular interests in queer and monstrous things, performative roles, living technologies, creation and destruction of unique ecosystems within biomedical facilities, human/nonhuman bond and the role of touch in human/nonhuman communication.

Gail Davies University of Exeter, Department of Geography, Exeter, UK

Gail is Professor in Human Geography at the University of Exeter. Her work is situated at the intersection of human geography and animal and science and technology studies. Her research seeks to understand the changing spaces of laboratory animal research, support decision-making in complex science policy contexts and develop innovative public engagements with science. She has previously published on natural history film-making practices, public attitudes to xenotransplantation, international scientific collaborations using mutant mice and interdisciplinary methods for engaging different knowledges. She is currently involved in collaborative research on the animal research nexus. Her contribution to this interdisciplinary programme involves working with Rich Gorman on understanding the changing patient expectations and engagements with animal research and discovering new possibilities in public engagement from Bentley Crudgington. She was member of the UK Animals in Science Committee from 2013 to 2019 and chaired the 2017 review of harm-benefit analysis in UK animal research.

Cameron Duff RMIT University, School of Management, Melbourne, VIC, Australia

Cameron Duff is a Senior Lecturer at the Centre for People, Organisation and Work in the School of Management at RMIT University, Melbourne, Australia. His research explores the role of social innovation in responding to complex health and social problems in urban settings. He has explored these themes in qualitative studies of precarious urban lives in Australia and Canada with a focus on problems of housing insecurity, addiction and mental illness. His first book, *Assemblages of Health: Deleuze's Empiricism and the Ethology of Life,* was recently published by Springer.

Ronan Foley Maynooth University, Department of Geography, Maynooth, Ireland

Ronan Foley is a Senior Lecturer in Health Geography and GIS at Maynooth University, Ireland, with specialist expertise in therapeutic landscapes and geospatial planning within health and social care environments. He has worked on a range of research and consultancy projects allied to health, social and economic data analysis in both the UK and Ireland. He is the current Editor of the academic journal *Irish Geography* and was an Erskine Fellow at the University of Canterbury (NZ) in 2015. His current research focuses broadly on the relationships between water, health and place, including two authored/coedited books, *Healing Waters* (2010) and *Blue Space, Health, and Wellbeing: Hydrophilia Unbounded* (2019), as well as journal articles on holy wells, spas, social and cultural histories of swimming and 'blue space'. He was the co-convener, with Prof. Thomas Kistemann (WHO/Bonn), of a special issue on the topic of healthy blue space for *Health & Place* (2015). He was the PI on an EPA-funded project on GreenBlue Infrastructure and Health with the UCD and EMRA in 2017–2018 and is collaborating on a number of water/health projects with colleagues within Ireland as well as internationally with the Universities of Exeter and Seville.

Clélia Gasquet-Blanchard École des Hautes Études en Santé Publique (EHESP), Saint-Denis, France

Clélia Gasquet-Blanchard is a Lecturer in Geography at the EHESP (French School of Public Health). Currently affiliated to the Department of Human and Social Sciences and Health Behaviour, she also attached to the UMR CNRS ESO 6590 (National Centre of Scientific Research/Research Unit 'ESO: Space and Society') at the University of Rennes 2. Her research aims at understanding health inequalities, considering social and spatial processes related to pregnant women. Her research projects target especially on migration issues, access to healthcare system and social relationships.

Richard Gorman University of Exeter, Department of Geography, Exeter, UK

Richard Gorman is a Postdoctoral Research Fellow within the University of Exeter's Department of Geography. His research is particularly interested in practices of health, care and medicine and how these intersect with human-animal relations and more-than-human worlds. In previous research, he has explored care farming and animal-assisted therapies, alongside the ideas of therapeutic spaces. His current work with Gail Davies, as part of the animal research nexus, explores the changing interfaces and relationships between people affected by different health conditions and laboratory animal research.

Kristofer Hansson Lunds Universitet, Institutionen för Kulturvetenskaper, Lund, Sweden

Kristofer Hansson is Associate Professor in Ethnology at the Department of Arts and Cultural Sciences, Lund University. He did his PhD studies at Vårdalinstitutet, the Swedish Institute for Health Sciences. His research foci are cultural analysis of medical practice in healthcare and biomedical research and phenomenological analysis of living with disability and illness. In recent years, much of his research has related to explaining those everyday processes that underpin social and cultural inertia in the societal work for accessibility in society and healthcare.

Rachel Herron Brandon University, Department of Geography, Brandon, MB, Canada

Rachel Herron is an Associate Professor in the Department of Geography at Brandon University and a Canada Research Chair in Rural and Remote Mental Health. Her current research examines the vulnerability and complexity of care relationships, social inclusion, and meaningful engagement for people with dementia and the diversity of lived experiences of rural mental health.

Anne-Cécile Hoyez Université de Rennes, L'Unité Mixte de Recherche - Espaces et Sociétés (UMR-ESO), Rennes, France

Anne-Cécile Hoyez is a Research Fellow at the CNRS (French national research centre) and based in the research unit 'ESO: Space and Societies' in the University of Rennes 2, France. As Geographer, specialised in health, she has been working on a geography of health with a social geography background. She is involved in ongoing reflections on the scope of health geography and immigration and works mainly on the health practices of immigrant populations in France.

François Lepage Photographer and Painter, Paris, France

François Lepage, after graduating in literature, has been working for several years in Africa and North America. From 2007 to 2017, he has been a Collaborator of the Sipa Press photojournalism agency, before he integrated the Studio Hans Lucas. Different major projects have marked his carrier: a series of photo report on exportation cultures, on French protected areas and on exile (which is described in the present book). He has also been involved in various expeditions in Austral territories and Antarctica, for which he has published three books and won different prizes.

Pia Kontos Professor, Toronto Rehabilitation Institute – University Health Network, Canada

Pia Kontos is a Senior Scientist at Toronto Rehab-University Health Network and an Associate Professor in the Dalla Lana School of Public Health at the University of Toronto. Her research focuses on the transformation of dementia care, so it is more humanistic and socially just. This involves the critical and qualitative exploration of the structural and relational dimensions of stigma, the creation of research-based dramas to effect personal and organizational change and the use of the arts (e.g. dance, improvisational play) to enrich the lives of people living with dementia.

Cheryl McGeachan University of Glasgow, School of GeoSciences, Glasgow, UK

Cheryl McGeachan is Lecturer in the School of Geographical and Earth Sciences, University of Glasgow. Her research interests include the history of psychiatry and forensic science and the experiences of mental (ill)health, biography and life writing, outsider art and art therapy.

Verena Menec University of Manitoba, Rady Faculty of Health Sciences, Winnipeg, MB, Canada

Verena Menec is a Professor in the Department of Community Health Sciences at the University of Manitoba. One of her research interests is social isolation and loneliness. Another area of interest is what factors make communities 'age-friendly' so that people can age in good health, be safe and continue to participate in the society and how communities can go about becoming more age-friendly.

Neil Overy Freelance Researcher and Photographer, Cape Town Area, South Africa

Neil Overy is a Historian (doctorate from London University), Writer and Professional Photographer specialising in environmentally themed images. He has worked in the non-profit sector in Southern Africa on issues as diverse as grand corruption to problems of sanitation. In recent years, after completing an MPhil in Environmental Humanities South at the University of Cape Town, he has proactively broadened his research expertise to include environmental issues, especially those that intersect with issues of social justice. Examples of his research and photography can be found at www.neilovery.com.

Hester Parr University of Glasgow, School of GeoSciences, Glasgow, UK

Hester Parr is Professor of Human Geography, University of Glasgow. She works on cultural geographies of mental health and illness via a range of qualitative methods, artistic practices and collaborative relationships.

Mark Skinner Trent University, Centre for Aging and Society, Peterborough, ON, Canada

Mark Skinner is Professor of Geography at Trent University, where he holds the Canada Research Chair in Rural Aging, Health and Social Care, and is Founding Director of the Trent Centre for Aging & Society. His research examines how rural people and places are responding to the challenges and opportunities of population ageing, particularly the evolving role of the voluntary sector and volunteers in supporting older people and sustaining ageing communities.

Michael Koon Boon Tan Nanyang Technological University, School of Art, Design and Media, Singapore, Singapore

Michael Koon Boon Tan is Assistant Professor in the School of Art, Design and Media, as well as a Programme Faculty in the MSc in Applied Gerontology at Nanyang Technological University, Singapore. As a leading Advocate for arts and health development in Singapore, his research, art practice and teaching to date explore the roles of creative practices in shaping culture of care, health and wellbeing in the context of ageing and chronic illness. His work has appeared in some journals, including *Journal of Arts and health* and *Journal of Applied Arts and Health*, and has also influenced arts and cultural policies as well as long-term care service development in Singapore. He holds a PhD (Durham University), a Masters in Photography and Urban Cultures (Goldsmiths College, University of London) and a Bachelor of Fine Arts (The School of the Art Institute of Chicago).

Carla Tsampiras University of Cape Town, Primary Health Care Directorate, Faculty of Health Sciences, Cape Town, South Africa

Carla Tsampiras is a Senior Lecturer in the Primary Health Care Directorate at the University of Cape Town. She is a Feminist, Vegan and Historian interested in social histories of health and healing, particularly early histories of AIDS and HIV. Her work is concerned with understanding how constructions of 'race', gender, sexuality and class – and the underlying powers and privileges associated with them – influence understandings of, and responses to, health concerns. Other areas of interest include gender, violence and slavery in South Africa, global crises and health and medical humanities.

Oonagh Walsh Glasgow Caledonian University, Glasgow School for Business and Society, Glasgow, UK

Oonagh Walsh is a History Professor in the Glasgow School for Business and Society at Glasgow Caledonian University, Scotland. A County Galway Native, she was educated at Trinity College, Dublin University, and Nottingham University and has held academic appointments at the University of Southampton (New College), Aberdeen University and University College Cork. Her research interests lie in medical history, especially mental health history, and epigenetic change linked to the Great Famine in Ireland (1845–1851).

Viv Wilson Journalist, Photographic Archivist, Local Historian, Bridge View, Teignmouth, UK

Viv Wilson, MBE, is a Journalist, Historian, Musician and Advocate who often works with older people in the community. She founded the Wilson archive and has published a number of books on the history of her home town (including *Images of England: Teignmouth* in 1997; *Teignmouth at War, 1939 to 1945* in 2000; *Teignmouth: Frith's Photographic Town Memories* in 2002; and *Teignmouth Through Time* in 2009). Over the years, she has produced and delivered numerous shows, films and performances for the general public.

List of Boxes

List of Figures

List of Tables

Chapter 1
GeoHumanities and Health

Rachel Hunt and Sarah Atkinson

And I certainly intended
anyone to be almost
abstracted on a gap-stone between fields.
(From The Thing in the Gap-Stone Stile by Alice Oswald 1996)

Abstract This is an introduction to the themes and contributions in the book. The essay discusses the potential and characteristics of the hybrid space of the GeoHumanities and the placing of health with this space. We consider the ways in which openings and opportunities may arise for successfully interdisciplinary and transdisciplinary scholarship related to medicine and health. We take a thematic approach in structuring the collection, through *bodies*, *voices* and *practice*, and chart the ways in which the contributing authors have conceived of GeoHumanities and Health.

The GeoHumanities

The decade of the 2010s has seen the emergence, recognition and expansion in the use of the term GeoHumanities within the spectrum of subcategories of geographical engagements. The term GeoHumanities encompasses scholarly practices that sit at the intersection of geography, a broadly conceived humanities and the creative arts. The hybrid quality of the activities celebrated under this term emerges from a

R. Hunt (✉)
School of GeoSciences, University of Edinburgh, Edinburgh, UK
e-mail: rachel.hunt@ed.ac.uk

S. Atkinson
Institute for Medical Humanities and Department of Geography, Durham University, Durham, UK
e-mail: s.j.atkinson@durham.ac.uk

S. Atkinson, R. Hunt (eds.), *GeoHumanities and Health*, Global Perspectives on Health Geography, https://doi.org/10.1007/978-3-030-21406-7_1

number of influences but unifies their diversity with a focus on space and place in a kaleidoscope of work that is both startling and comforting (Richardson 2011). There are several substantively defined areas of GeoHumanities already evident in the contemporary landscape of geography including creative, digital, environmental and historical. In this collection, we add to these the GeoHumanities of medicine and health. These engagements with the borderlands of ontological and epistemological foundations have inspired new ways of interrogating the world and our place within it and encourage future scholars to step outwith disciplinary bounds. Pick your metaphor for the potential of such hybridity, of border crossing or edgelands, of kaleidoscope or mosaic, and of dialogue or umbrella. In our cover image, we propose an addition to the panoply of metaphors, drawing on British poet Alice Oswald, of *The Thing in the Gap-Stone Stile* (1996). Oswald's *Thing in the Oswald's poem* echoes contemporary social and spatial engagements with the materiality of both organic and inorganic life but also enlivens space itself as the book is a gap connecting domains otherwise divided, including pasts, presents and futures, and bearing witness to the beings, doings, movements and patterns around and through its own in-betweenness (Oswald 1996).

In the first half of this introduction, we focus upon the potential of the GeoHumanities and seek to play with questions of what they are, where they have come from and what they might become. We then explore the placing of health within the GeoHumanities by assessing the present state of medicine, health and the humanities and by considering the ways in which openings and opportunities may arise for successful interdisciplinary scholarship. In the second half of the chapter, we turn to address the content of this collection. Taking a thematic approach to the delivery of an overview, we chart ways in which our authors have conceived of GeoHumanities and Health.

This recent manifestation of humanities facing engagement is formally marked by the publications, both in 2011, of *Envisioning Landscapes, Making Worlds: Geography and the Humanities* (Daniels et al.) and *GeoHumanities* (Dear et al.), and, the subsequent launch, in 2015, of the journal *GeoHumanities* (Cresswell and Dixon 2015). Yet this present wealth of GeoHumanities work has a past composed of at least four interrelated pathways. A longer historical view stresses the entwining of geographical sensibilities with the humanities from the discipline's earliest formulations, not least as shaped by Alexander von Humboldt and his insistence on the centrality of both the humanities and the sciences to the discipline's core business (Bunkše 1981; Marston and de Leeuw 2013). An origin story of the GeoHumanities picks up pace with the major influence of humanist geography during the 1970s and 1980s which, while attending to space and place, drew from and was, in turn, used by the humanities more broadly (see Bunkše 1990; Buttimer 1976, 1990; Entrikin 1976; Pocock 1981a, 1981b, 1988; Porteous 1985; Tuan 1974). Cutting across both these influences is geography's long tradition in attending to and valuing historically engaged research, comprising a third principal connection to the broader humanities (Livingstone 1992; Withers 2001). Finally, the contemporary GeoHumanities are indebted to the spatial turn informing theories within the humanities and social science which emerged over 30 years ago. Authors such as

Pierre Bourdieu in social science, Edward Said in literature, Judith Butler in gender studies or Bruno Latour and Michel Foucault in philosophy all looked at the importance of space in the creation of social life.

With such a rich pedigree, it is possible to question what is new about the GeoHumanities. Some have argued this new label is merely the latest incantation of the 'new' cultural geography given the lack of a suitable additional up-to-the-minute prefix (discussed in Cresswell 2015, and in Last 2018). This current (re)encounter between geography and the humanities, however, emerges in relation to new stimuli and manifests particular interests and approaches in the context of a 'convergence of material social change and cognitive adjustments' (Dear 2011: 310). While such social and intellectual change may justify distinguishing the GeoHumanities from earlier engagements with the humanities or from current Cultural Geographies, this reflects neither schism nor opposition but rather extension, complement and deepening of directions, topics and practices of geographical research. This, then, is not a reincarnation of cultural geography but rather a novel intervention into the academic lexicon. Four approaches to research have tended to characterise the subfield and indeed have served to mark the divergence in focus from other modes of cultural research in geography. These approaches relate to interdisciplinarity; creativity; embodied, sensory and affective attunements; and digital technologies.

First, the GeoHumanities are positioned importantly as having an interdisciplinary orientation. The new label captures a zeitgeist for conversations beyond conventional disciplinary concerns and practices (Richardson 2011). Such conversations include those within disciplines, such as across physical and human geography, or quantitative and qualitative epistemologies; those across academic disciplines, most notably between geography and the arts and humanities; and those between the academy and other fields of practice. There is, in much of the work under this banner, a commitment, whether expressly stated or not, to the connections between knowledge and action, to understanding practice or to engaging with and in activism (e.g. Skinner and Masuda 2013).

Such engagement with nonacademic practitioners, and particularly with creative artists, has become a second characteristic of work seen as GeoHumanities. This (re)turn to the creative captures the activities of geographers who interpret artworks through a spatial frame (Magrane 2015; Saunders 2010); who collaborate with creative artists at different stages in the research process (Askins and Pain 2011; Atkinson and Robson 2012); who curate creative outputs and exhibitions; or who are themselves creative artists (de Leeuw and Hawkins 2017; Eshun and Madge 2016; Johnston and Pratt 2010; Madge 2014). Moreover, there are calls for creative geography to embrace a broad definition of creativity to include skilled crafts from knitting through bothying to taxidermy (Hawkins 2018; Hunt 2018; Patchett 2016). A creative geography also underwrites the growing enthusiasm for exploration and experimentation in how and what we produce as research outputs, including in and beyond the conventions of our written outputs (DeLyser and Hawkins 2014; Jacobsen and Larsen 2014; Lorimer 2018).

Much work in the GeoHumanities is also inspired by increased theoretical and methodological attention within geography to emotion (Anderson and Smith 2001;

Davidson et al. 2007; Harrison 2015), embodiment (Longhurst 2001), affective atmospheres (Anderson 2009) and the senses. The sensory turn in geography, in particular, (Pocock 1993; Parr 2010) has precipitated a shift from occularcentric understandings of space and place to bodies as 'instruments … through which we view the world' (Parr 2010: 1) and has been accompanied by a surge in participatory and mobile methods (the sound walk, smell walk, etc., Büscher et al. 2011) and, more recently, the development of multisensory mapping (McLean 2017). It is, perhaps, this attention to bodies, atmospheres and corporeal attunement, used in conjunction with more traditional humanities scholarship and digital technologies, that particularly informs the emergence of the hybrid GeoHumanities scholarship.

Interdisciplinarity and creativity also combine in informing the third approach, the mobilisation of recent and emerging digital technologies to enable a whole suite of new practices of research. These new practices include how we encounter and research our cultural worlds, arguably reconfiguring our cultural practices and spawning a family of new forms of digital humanities (see Berry and Fagerjord 2017). Bringing spatial considerations into the rapidly growing field of digital humanities was central to geographers' intimations of a new subfield of GeoHumanities, as is particularly evident in the collection by Dear et al. (2011). The use of digital technologies has overlap with exploratory creative practices, not least in the spaces afforded for varied forms of communication and engagement. The ever-growing availability of digital technologies and the associated digital sources of data allow experimental analyses in new and exciting ways, such as through distributed, participatory modes of spatial mapping (Dunn 2007; Pain 2004; Perkins 2014) or through creative, playful and performative approaches (Osborne et al. 2019).

There is within the development of the GeoHumanities something of a split between work inspired by data and the digital humanities and that which wrestles with issues of creativity, practice and performance. In the former we often see links between the physical sciences and the humanities, with the use of geocoded data and an emphasis upon the creative potential of geographic information systems (GIS) (see, e.g. in Dear et al. (2011) by Trevor Harris and colleagues or Stephen Young). The latter focuses upon a more arts and cultural approach, often calling upon site-specific performance and emphasising a sense-based approach to spatial understandings (see, e.g. chapters in Dear et al. (2011) by artists Ursula Biemann or Philip Govedare and the work of the Royal Holloway Centre for the GeoHumanities at the University of London). This variety is exciting as it allows the GeoHumanities to collate a wide range of spatially focused research which, in its diversity, is perfectly placed to tackle the multifarious issues of contemporary society and house the current push to 'blur and breach boundaries' (Forsyth 2011: 251) through interdisciplinary scholarship. Nonetheless, there is arguably more attention given to the potential of engagement with the creative or digital arts than to what it means to engage the wider conceptualisation of the humanities. While there is no space to rehearse the extensive debates on the nature and value of the humanities ongoing in the contemporary academy, we want to keep in sight the broad project of the humanities as concerned with human experience and meaning-making in all its diversity across times and settings and in tandem with other organic and inorganic entities and how, as humans, we variously process and change, document and express our own condition. The humanities bring a particular attention to aspects of human experience that have gained greater traction within geography and other social sciences in recent decades. These include, for example, explorations of emotions, embodiment, temporality and distributed agency. The humanities also offer modes of critically interrogating our key concepts and of reflecting on different genres of writing our academic outputs. While work in the modes of this widely conceived humanities is abundant in contemporary geographically informed work, it could be more explicitly brought within the purview of the 'gap-stone' of GeoHumanities.

Medicine, Health and the Humanities

Geographers are not alone in exploring a revitalised relationship with the arts and the humanities at this time of 'the maelstrom of changing material and mental worlds' (Dear 2011: 312). The positivism of conventional science is associated with enormous gains in the field of medical science, and yet, of all the sciences, medicine and medical practice in its broadest sense has arguably the most uncomfortable relationship with this epistemological base. Medical practitioners are daily confronted with the uncertainties of their science, the messy unruly interrelationship of bodies and environments and the associated emotions of their objects of study. This is perhaps most evident in relation to mental health practice, and it is no accident that the pioneers of critical medical sub-fields within the social sciences often came from a psychiatry background. At the same time, medical knowledge has its roots as much in the careful documenting of specific case histories as it has in more recent technologies of survey and experimental design.

The field of the medical humanities, now expanded to Medical and Health Humanities, had its roots in this malaise by health professionals with their own knowledge base. Their own engagement with the creative arts prompted an interest in the potential of relationships between such engagement and patient-centred practices including empathy (Macnaughton 2009) and patient narratives (Charon 2006). This initial medicine-led engagement with the humanities, as serving medicine's core purpose and as instrumental in producing health professionals better attuned to their patients' needs, has expanded dramatically in the last two decades beyond the narrow confines of the clinic and the consultation. The contemporary Medical and Health Humanities includes research led from and contributing primarily to the questions and concerns in the humanities, research examining the political dimensions to health and medicine and research grounded in patient-led experiences. The intervention into the field to advance a 'Critical Medical Humanities' (Viney et al. 2015; Whitehead et al. 2016) calls for understanding the relationships possible between medicine or health and the humanities as entangled (after Barad 2007), neither hierarchical nor oppositional, but fluid in its boundary crossing to enable productive possibilities (Whitehead and Woods 2016). A Critical Medical Humanities, then, concerns:

> …how medical and health-related knowledge, care, intervention, education and research are extensively, complexly and unevenly distributed throughout social life, deeply and irrevocable entangled in the vital, corporeal and physiological commitments of biomedical research. (Viney 2016: 114)

Much of a contemporary medical and health humanities manifests similar intersections with the intellectual and social change logged by Dear et al. (2011) for the GeoHumanities. The various turns in social theory to digital technologies, visual cultures and the material world that have come to the fore in informing geographical thinking are evident in shifts within the humanities and the medical humanities from culture, mind and language onto nature, bodies and things (Whitehead and Woods 2016). The profound influences of disability studies, feminist and queer approaches through the humanities and the social sciences are evident in the orientations of the Medical and Health Humanities as well as prompts for work specifically in these areas (see, e.g. Burke 2016; Evans and Cooper 2016; Foster and Funke 2018). The Medical and Health Humanities emerged from and remains rooted in the academic and medical traditions of the Global North, and what-is-more, predominantly in an anglophone tradition. The new Millennium, however, has seen an expansion in non-anglophone Europe and in parts of Asia and Africa. A recent first collection of essays on the Medical and Health Humanities across Africa provide a sensitive attunement to the challenges of interdisciplinary work on what the editors term 'the edgelands', the hierarchies of power across institutions, languages, researchers and associated inequalities in economic resources and digital access (Tsampiras et al. 2018).

The specific intersection of geography with the Medical and Health Humanities has a relatively limited history, with just two collections to date explicitly claiming the relationship (Atkinson et al. 2015; de Leeuw et al. 2018). Nonetheless, there is a rich history of research at the interface of the geographies of health and cultural geographies (Gesler and Kearns 2001) including the substantial body of work on therapeutic landscapes and spaces (Conradson 2005; Gesler 1992; Williams 2007). Other cultural encounters inform health-related explorations through emotional geographies (Davidson et al. 2007), historical geographies (McGeachan et al. 2012; Moon et al. 2016; Philo and Pickstone 2009), literary geographies (Curtis and Tonnellier 2005; Gesler 2000), sensory geographies (Bell 2016; Foley 2017), non-representational theory (Andrews 2018; Hall and Wilton 2016) and the creative arts both as researcher (Atkinson and Scott 2015; Bingley 2012; Parr 2006, 2007) and as practising creative artist (de Leeuw 2012, 2019). There is an important interface between the geographies of health and a wider conceptualisation of the humanities that informs interrogations of key concepts such as care (Atkinson 2016; Milligan and Wiles 2010), identity (Thien and Del Casino Jr. 2012) or vulnerability (Coyle and Atkinson 2019) and intersections with related fields such as neuroscience (Fitzgerald and Callard 2015) or biosciences and animal studies (Greenhough 2010). Finally, geographies of health have a long tradition of privileging first-person accounts of experience in a diverse range of times, places and settings. It is, then, perhaps surprising as the label of the GeoHumanities has gained currency across the discipline of geography, that researchers in the geographies of health have underplayed their vibrant engagements with the humanities, ignoring these emerging geographies in recent overviews of the field (see, e.g. Brown et al. 2018).

GeoHumanities and Health

This collection of essays seeks to stage an explicit intervention in further foregrounding the richness of work entangling medicine and health with the concerns of geography and of the humanities. In this, our collection builds on, downwards and outwards, the previous excursion at these intersections (Atkinson et al. 2015; de Leeuw et al. 2018). The earlier collection brings geographical thinking to the Medical Humanities, through the field's dedicated *Journal of Medical Humanities*, in order to better explore the 'worlding' of health, healing and wellbeing (Atkinson et al. 2015: 4). The direction of conceptual travel is perhaps opposite in the 2018 collection, located in the new journal *GeoHumanities*, and offered as a range of creative and narrative possibilities that reframe understandings of what constitutes health, disease, care, experience and the places and spaces in which these unfold (de Leeuw et al. 2018). This current third collection continues to celebrate the common exploratory and experimental spirit evident in these predecessors and the wider GeoHumanities. We honour those characteristics that may be viewed as the heart of a GeoHumanities project, however fluid and permeable we hope that to be, and particularly those of interdisciplinarity, creativity, digital potentials and attention to the sensory world. We intend for this collection not only to showcase how engaging medicine and health through and around the gap-stone stile of GeoHumanities enriches our understandings and practices, but also to extend these dialogues through embracing both a broad conception of the humanities and a critical approach in our intellectual and practical pursuits.

We have explicitly sought to bring together thinking, practice and writings across different disciplines, with voices represented from history (Tsampiras, Walsh), from geography (Bell, Duff, Hoyez et al., McGeachan and Parr, Foley), from sociology (Bradby), from ethnology (Hansson), from environmental humanities (Overy) and from interdisciplinary teams of academics (Davies and colleagues) and hybrid research-practitioners and nonacademic practitioners (Andrews and Wilson, Herron and colleagues, Tan and Atkinson), including creative work from those who are more commonly known for their research, such as the two poems from poet-geographer Sarah de Leeuw, reflecting on her father's illness, and selected from her recent collection *Outside, America* (De Leeuw 2019).

We have sought contributions across the subtly varied inflections of historical and intellectual trajectories of different countries. In this, we pursued authors working in contexts that have recognisably embraced a reinvigorated dialogue across medicine, health and the humanities and who are attentive to spatial considerations.

There has been explicit and intentional dialogue by both Canadian and British geographers of health with the Medical and Health Humanities (Atkinson et al. 2015; de Leeuw et al. 2018; Whitehead and Woods 2016), and several of the contributions to this collection, including our own, come out of these contexts (Andrews and Wilson, Bell, Davies et al., de Leeuw, Herron et al., McGeachan and Parr). Medical and health humanities are also flourishing in Ireland (Foley, Walsh), Australia and New Zealand (Duff) and in the Nordic countries. This is particularly the case in Sweden, attracting humanities facing social scientists (Bradby) as well as humanities scholars and including those in 'cultural sciences' (*kulturvetenskaper*) which is a distinctive category in Swedish academia that captures interdisciplinary research between culture and science, including social and technological studies (Hansson). The Medical and Health Humanities have less visibility in the global south, but teaching modules and research programmes in medical humanities are increasingly being developed including in China and South Asia (Tan from Singapore) and Sub-Saharan Africa (Overy, Tsampiras from South Africa). Tan, Overy and Tsampiras underscore the desire of the GeoHumanities to attend to diversity and contextual specificity, in these cases through explorations of elder care in Singapore, environmental emergency in various settings and policy debates and formulations in South Africa. In some cases, our authors are writing in a second language (Hansson, from Swedish; Hoyez et al. from French) or from a context of multiple official languages where English was historically imposed (Andrews and Wilson, Herron and colleagues, and de Leeuw in Canada; Foley, Walsh in Ireland; Tan in Singapore; Tsampiras in South Africa). While the language of the collection is English, we recognise, celebrate and allow for how different historical linguistic engagements extend and diversify its expression. These varied pathways through national, disciplinary and linguistic filters bring approaches to issues of health from different angles, through fresh questions, perspectives, sources of data, embedding scholarship and styles of writing.

The range of substantive topics encompasses more recognisably medical and health concerns such as ageing (de Leeuw; Herron and colleagues; Tan and Atkinson), living with asthma (Hansson), HIV/AIDS policy (Tsampiras) or obstetrical procedures (Walsh), more recognisably geographical concerns of health such as sensing nature (Bell; de Leeuw), emergencies of environmental health (Overy), cartographies (Foley) or care beyond the health sector (Duff) through to emerging themes such as the histories of place (Andrews and Wilson; de Leeuw), embodied migration (Hoyez et al.), the post-human in relation to animal research (Davies et al.), the nature of evidence (Bradby) and humanities-led practices of pedagogy (McGeachan and Parr). Our contributors draw on diverse sources of information and practice. Many use first-person accounts as a primary empirical source for capturing experience and embodiment, often together with observations and mobile methods (Bell; Herron et al.; Tan and Atkinson) and other modes of visual documentation (Andrews and Wilson; Duff). Engagements with new digital technologies inform sources of data (Bradby works with literature that grew from an initial blog series; Davies et al. access film online) and contemporary but more intimate cartographic approaches (Foley). Several authors draw on and integrate reflections on their own stories, practices and experiences, often in tandem with a range of other sources (Foley; Hansson; Hoyez et al.; McGeachan and Parr; Overy). Historians conventionally draw on secondary sources and archives, here in the form of the documentation of policy debates, public statements and media coverage (Tsampiras; Walsh), while social scientists draw in material from literature, film and online posts (Bradby, Davies et al.) as well as from creative practices of their own and their collaborators (De Leeuw; Herron et al.; Hoyez et al.; Overy; Tan and Atkinson). Our contributors engage different styles of structuring and presenting their material and thoughts. The conventions of social science or of humanities scholarship are evident, respectively, in the presentation of first-person accounts (Andrews and Wilson; Bell; Herron et al.) and archives (Tsampiras). These, however, are often combined and infused with creative movement, visual or literary material variously used as illustration (Duff; Tan and Atkinson), process (Herron et al.; Hoyez et al.; Tan and Atkinson), method (McGeachan and Parr) and data (Foley; Hoyez et al.). Our social scientists using literary and cinematic sources as data weave these into the social science conventions, generating a writing style more akin to the humanities essay (Bradby; Davies et al.). Other authors offer their contributions in an explicit storytelling form (Walsh), albeit interwoven with a social science genre (Andrews and Wilson's oral histories or Hansson's Autoethnography), or through the creative media of writing and photography (De Leeuw; Foley; Overy). Five contributions, three essays and two poems, engage intentionally experimental structures and styles. Foley explores the new intimate cartographies of health through five short 'vignettes', a form of brief case study, that both illustrate and generate his reflections. McGeachan and Parr present their pedagogic practices, experiences and student engagements through a range of materials organised as a kind of acrostic of the field study site, Gartnavel. And Overy, as a researcher-activist in environmental humanities, provides a photographic essay combined with reflection and activist commentary. De Leeuw's elegant poetry draws into its lines and coheres into the person of her father the diverse considerations across other essays, including memories, passions, bodily and imaginary senses, doings, movements and the vibrancy of life and everyday living.

There are, of course, a myriad of ways to structure and organise a collection of this kind. We have chosen to organise the essays into what can be seen as an epistemological grouping, by bodies, voices and practices. In doing so we acknowledge that each chapter can lay claim to more than one of these labels and indeed, essays have been shifted around during the process of compiling the collection. Nonetheless, these substantive headings enable a pooling of concepts, approaches and styles. The following discussion develops the potential that each of these groups has to offer.

Bodies

While almost all of the contributions in this collection concern bodies as sites of health and ill-health, the first five essays focus their scholarly attention directly onto bodies. These essays coalesce through the ways in which they account for the material, sensory and more than physical capacities of bodies. The valorisation of the materiality of bodies, their sensory and emotional capacities and their primacy in experience and world-making is an underpinning intellectual influence for the renewed engagements between geography and the humanities and the humanities and medicine and health seen over recent decades. This section is fronted by de Leeuw's poem 'How to Convey that Dying Really is Lonely'. The poem, written in a staccato form of couplets, underpins the sense of disjuncture across and from the parts of her father's life. The situated memories, habits and passions from his past resonate with the essay of Andrews and Wilson, and the bodily and sensory engagements with that of Bell, even while contrasting her work in exploring an alienation of being out of the place or home that was once enjoyed. This alienation has some echo with Hansson's places that can become hostile in relation to bodily changes. Herron and her colleagues and Hansson both attend to social relationships as variously beneficial and challenging, and these are similarly foregrounded in de Leeuw's poem where, with close family far away, her father, like the turtle in the poem, is dependent on the kindness and care of strangers.

Embodied experience is a common theme through all the contributions. This is explored explicitly through the expressions of sensory and emotional encounters with the specificities of given times and places. Bell's intervention challenges the privileging both of sight within sensory geographies and of the picturesque, aesthetic or sublime in relation to experience of nature by bringing a focus on the 'more-than-visual' in experiences of wellbeing. A broadened attention to the multiplicity of ways of 'being, sensing or moving through the world' enables a 'zest' for curiosity, connection and movement in nurturing wellbeing. While Bell documents sensory and embodied experiences in the moment, Andrews and Wilson confront how to document sensory experience in the past. Although historians debate whether experience can ever be read from an archival record, oral histories gathered with those still alive can engage past times and places with particular health and wellbeing traces. Using photographs as prompt and record of the historical bombing of an English seaside town during World War II, Andrews and Wilson reanimate the past through sensory recollections not to document what happened but what it felt like to live through this and how this resonates in the present as, following Maddrell, an 'absence-presence' (2013: 502) haunting from the past. Current and recollected sensory embodied experiences of particular places combine in Hansson's combination of Autoethnography and other first-person accounts of living with asthma and allergies. The body and especially the lungs are typically understood as having their own agency in the life of the individual, shaping and responding to the physical and social world around it; Hansson argues that from the perspective of the embodied self, it is the physical and social world as well that has agency inscribing the body with memories and embodied knowledge of places best avoided and social situations demanding trade-offs between health and fitting-in. These critical places where, for Hansson and others, body, environment and health enact an asthmatic episode reveal practices for hiding the reactive body to 'pass' socially and that are silenced within broader debates in health. As such Hansson's essay aligns with the interests of Hoyez et al., albeit in the very different context of migration, with a politics of disadvantage, exclusion and social culpability. Hoyez et al. ask and explore how they might move beyond their conventional methodological practices of *describing* the human condition through words to *showing* the human condition through non-verbal and visual approaches. To this end, migrant families had the opportunity to collaborate with a photographer and picture their story. This emphasis on the human body generated an exhibition with which the authors of this chapter engage in terms of what can be learned about the complexities of shifting social landscapes, social inequalities and their management by nation states. This engagement with the creative arts is also taken up by Herron and colleagues who explore the importance of embodied movement and non-verbal bodily communication through attention to sensory experiences of dance as a means of creating social inclusion for those living with dementia. The dance intervention as a social encounter draws out the notion of a relational citizenship by recognising corporeality as primal source of experience and expression. Through this focus on bodies and well-being, Herron et al. produce a novel conceptualisation of care and of social inclusion which sits between research from geography on the embodied ways of knowing, communicating and experiencing society and research in dementia care which is weighted towards access to services and support.

Voices

The humanities and the social sciences have both conventionally attended to first-person tellings and to the variations in accounts across different constituencies, but medicine and health have privileged these only more recently and arguably only in relatively limited ways. Voice in contemporary medicine takes the form of user-led research and user feedback. This mobilisation reflects two tacit assumptions. First, a well-established disjunction between professional and lay positions within health care is only slightly reworked. While inclusions of user voice into decision-making and feedback do help rework power relations, an understanding of the voices of the professionals as similarly subjective and contextual is less evident. Second, experience is often treated as straightforwardly accessed through simple first-person accounts, with little interrogation of the situated nature of accounts in terms of personal and contextual histories, setting and social embeddedness. Voice then is absolutely crucial to an emerging GeoHumanities and Health both as method in accessing experiential knowledge that is nuanced and situated, and as politics in defining which voices are to be heard and how they are interpreted.

Essays in this section continue to extend our conventional understandings through staging encounters between different voices in different sites. A key theme emerging from these encounters is the need to embrace a notion of indeterminacy. The uptake of narrative into medicine has arguably been the most successful intervention from the humanities to date. Bradby problematises the limited understandings and uses of narrative as a 'stripped-down, documentary form' to complement the medical evidence base, challenging us to reflect on which narratives are viewed as acceptable and trustworthy and therefore to be included in decisions surrounding policy and practice. The focus on voice is, therefore, particularly strong in Bradby's work as she engages an autobiographical end-of-life account to show how an attention to story, as the more unreliable cousin to narrative, can allow for the 'contradictory, fantastic metaphorical elements of experience' that are not easily subsumed within the epistemological orderliness of biomedicine. As a sociologist, Bradby's attention to stories neatly bridges between geographies of health and existing work in the GeoHumanities which focuses upon tales and 'small stories' (Lorimer 2003). In a similar vein, but in vastly different arena, Davies and colleagues work with the medium of film to explore the ways in which meaning is translated between patient, public and animal bodies. As such, they utilise artistic expression to interrogate the interface of science and society and, in doing so, show how the GeoHumanities areas uniquely placed to bring attention to practice and acceptance of indeterminacy in our understandings of the production of knowledge in medicine and health. Voice here is used to deal with indeterminacy, but rather than making space to extend the understandings of any one body, this research tracks voice between bodies, highlighting and embracing the messiness of communicating between species. This rendering of disciplinary and subject borders and voices as more fluid, and more open, enables ways to be increasingly accepting of the not yet known or not quite understood.

Recognition of the importance of indeterminacy in the uncharted spaces of interdisciplinary encounters together with a critical interrogation of the geographies of borders and blame also characterise the historical scholarship of Tsampiras in relation to HIV/AIDS policies in South Africa. Her historical sensibility focuses on revealing the roots of dominant voices and ideas and the trajectories through which these become accepted norms within complex and interrelated systems of power and inequity and impact on people's everyday lives and deaths. In doing so, she embraces an attitude of respectful sharing across multiple disciplinary domains, including the physical and socio-political geographies of landscape, the archaeology of archives, the contours of the body and the psychosocial power of emotion. This complicating of established categories and their implication in practices of inequality continues in Duff's explorations of homelessness on the streets of Melbourne. Duff challenges as neither necessary nor beneficial the separation of attention to welfare interventions from attention to everyday struggles of health-in-the-making. In giving renewed attention to the latter of these, Duff turns to the social and affective experiences and accounts that can enliven attention to the ecologies, contexts and dwelling practices in situations of particular disadvantage. This breaking down of borders and boundaries is methodological, conceptual and substantive in these chapters, with voice effecting border crossings and openings in diverse ways.

A historical sensibility features in a number of our essays (e.g. Andrews and Wilson; Bradby; McGeachan and Parr; Tsampiras) and the analytical practice of the historian can attend to the specificities of language, tone and exposure as topics related to medicine and health emerge and take hold within a popular imagination. A number of essays implicitly hold different positions in relation to a concept of truth. A focus on the particularities of sensory and emotional experiences treats truth as less happenings and timelines than as how these felt (Andrews and Wilson; Bell; Hansson) and infused by the complex, uncertain unknowability inherent to experience (Bradby). Walsh, in contrast, interrogates the potential problematics associated with media-based regimes of truth. Her account of the creation of truth, the steadfast belief in false truths and the danger of moving into a post-truth era focuses upon shocking revelations of malpractice across several different public sectors in modern Ireland. The presentation and distortion of truth are sensitively set against lived experience of those whose self-identity had become inseparably linked to survivorship and activism. As an orthodoxy of scholarly contestation of the very notion of truth risks underwriting social media's facilitation of fake news, a renewed questioning of how we relate to constructions of truth and the misrepresentations of history becomes urgent and timely.

Practices

Our third category of practice reflects the commitment of engagements between the humanities and medicine and health to confront the significant challenges in supporting and teaching issues of health, care particularly with regard to addressing the inequalities of highly differentiated health experiences. Both the GeoHumanities and the Medical and Health Humanities have confronted these kinds of challenges through direct or collaborative relations with various forms of practice and particularly those of the creative arts. Despite long-standing debates surrounding the relationship between theory and praxis, academics often underestimate our own practices in terms of intervention, pedagogy or activism. The techniques of careful analysis of policy documents, the reflections on module content and activities or the processes through which research itself is developed and disseminated can all stage forms of intervention into the range of contemporary challenges for health and health inequalities.

A number of our chapters continue this emerging tradition, not all located in the practice section. We start this section with a second poem from poet-geographer Sarah de Leeuw, which, as well as exemplifying a creative practice, echoes the focus on 'doing' that has characterised much of the engagement of GeoHumanities with creativity (Hawkins 2011). Her poem reverberates with the practices of everyday life, from the vibrant feminine urban colours in the big city to the chilly masculine early morning patterns in a small town, and the sharpness of new spring shoots of life that contrast her father's declining energy and painful effort in his everyday movement. As well as the creative writing by de Leeuw, the collection includes several collaborations between researchers and creative practitioners (Andrews and Wilson; Herron et al.; Hoyez et al.) and hybrid researcher-practitioners (Overy; Tan and Atkinson). The contributions by hybrid researcher-practitioners in this section each elaborate a different transdisciplinary practice. Overy, a hybrid photographer-environmental humanities scholar and activist, cites Barthes (2000: 38) on photography as, 'subversive not when it frightens, repels or even stigmatises, but when it is pensive, when it thinks'. Overy positions this alongside a contemporary photographic enthusiasm for landscapes without human life as being 'reclaimed' by nature. He enjoins the implied 'unravelling' of modernity with Barthes' pensive practice in relation to nuclear emergencies, health threats and establishing risk and safe zones. Overy thus echoes earlier themes in the collection relating to indeterminacy, borders and critical places, engaging them here through both photography and activism. While Overy's essay leans towards the practitioner and activist, Tan, a visual artist and researcher working in collaboration with a geographer (Atkinson), leans more towards his research persona in reflecting specifically on his practice as a 'caring artist' in a nursing home. Tan and Atkinson tease out, through dialogue with existing literatures, the different modalities of caring practice evident within the nursing home that resonates with themes in the collection relating to embodied experiences, multiplicity of voices and health-in-the-making. In this, explorations of the use of visual documentation (Overy), processes of involvement in photography (Hoyez et al.) or participation in other creative activities (Herron et al.; Hoyez et al.; Tan and Atkinson) may be seen not only as forms of creative practice but of creative activism.

The challenge of practising a transdisciplinary GeoHumanities pedagogy across the academic borders of medicine and the humanities is referenced by Tsampiras in her essay and is something she has written about more fully elsewhere (Tsampiras 2018). Here, McGeachan and Parr operationalise their aim for 'hopeful pedagogies' that can shorten the time for theoretical development to translate into pedagogic practice through a fieldwork week related to mental health geographies and detail their experiences in their contribution. Through a creative focus on site and practice, they explore and illustrate what a GeoHumanities facing pedagogy might look like. Although working within one discipline, that of geography, their students interact with multiple sources of information, including embodied and mobile experiences in the site of a former mental health institution and archival and visual materials including first-person accounts, video and formal medical records. In this they intentionally experiment with ways of encountering and knowing health as entangled with place, history and norms of medical and social practice.

Our collection closes where arguably the GeoHumanities began, with an encounter with cartography, creative engagements and spatial literacies. Foley introduces five cartographic 'vignettes' of health in action that enable embodied, sensory and emotional expressions both through participation in the projects or, in some case, through resisting participation in these more intimate mappings. Foley's essay touches on the diverse themes throughout the collection from the materiality of bodies, through voice and power, to creativity and activism ultimately highlighting that, 'maps will always be needed' and benefit of the GeoHumanities is their potential to 'blur… the traditional and exploratory in cartographic terms'.

Conclusion

The GeoHumanities offer a conceptually poignant space for experimentation, and this collection itself takes up that offer, as intimated in Alice Oswald's poem, to 'be almost abstracted' (Oswald 1996, 32) in contemplating the opportunities for productive entanglements in the 'gap-stone stile' between the humanities, geography, medicine and health. These multiple encounters across disciplines, histories, methods and topics serve to problematise often taken-for-granted research agenda in terms of key questions, sources of information and forms of argumentation. Privileging an expanded attention to bodies and voices, creativity and practices, words and visuals brings to the foreground more complex geographies of health through intimate accounts, indeterminate trajectories, destabilised and reformulated identities, expressive and activist mediations. The virtue of lingering in this gap inheres not only in enriching our comprehension of human life and its myriad entanglements but in pursuing a practice, whether creative, pedagogic or activist, that is more securely grounded in all our messy and incoherent complexities.

Bringing together this breadth of encounter between the GeoHumanities and Health demands flex in the epistemological and ontological engagements at play and makes necessary a space for future dialogues from which to weave research at the interplay of these diverse disciplinary, historicised and linguistic modes of research and practice. This collection is, therefore, as *Envisioning Landscapes* was once described, intended to be a 'base map for further conversations' (Forsyth 2011: 252). We wish to stimulate a continued line of scholarship which attends to the flourishing intersection between geography, medicine and health and the humanities – a gap in which innovations in theory, politics, data and practice can explore and expand the potential for a transdisciplinary intellectual culture.

Acknowledgements Sarah Atkinson is supported through Wellcome Trust Development Grant, WT 209513.

References

Anderson, B. (2009). Affective atmospheres. *Emotion, Space and Society, 2*, 77–81.

Anderson, K., & Smith, S. (2001). Editorial: Emotional geographies. *Transactions of the Institute of British Geographers, 26*, 7–10.

Andrews, G. (2018). *Non-representational theory and health: The health in life in space-time revealing*. London: Routledge.

Askins, K., & Pain, R. (2011). Contact zones: Participation, materiality, and the messiness of interaction. *Environment and Planning D Society and Space, 29*, 803–821.
Atkinson, S. (2016). Care, kidneys and clones: The distance of space, time and imagination. In A. Whitehead et al. (Eds.), *The Edinburgh companion to the critical medical humanities* (pp. 611–626). Edinburgh: Edinburgh University Press.
Atkinson, S., & Robson, M. (2012). Arts and health as a practice of liminality: Managing the spaces of transformation for social and emotional wellbeing with primary school children. *Health & Place, 18*, 1348–1355.
Atkinson, S., & Scott, K. (2015). Stable and destabilised states of subjective wellbeing: Dance and movement as catalysts of transitions. *Social & Cultural Geography, 16*, 75–94.
Atkinson, S., Foley, R., & Parr, H. E. (2015). Spatial perspectives and the medical humanities. *Journal of Medical Humanities, 36*(1), 71–81, special issue.
Barad, K. (2007). *Meeting the universe halfway: Quantum physics and the entanglement of matter and meaning*. Durham: Duke University Press.
Barthes, R. (2000). *Camera lucida*. London: Vintage.
Bell, S. L. (2016). The role of fluctuating soundscapes in shaping the emotional geographies of individuals living with Ménière's disease. *Social & Cultural Geography, 18*, 831–850.
Berry, D. M., & Fagerjord, A. (2017). *Digital humanities*. Cambridge: Polity Press.
Bingley, A. (2012). Exploring experiences of illness and recovery through tactile art. In M. Paterson & M. Dodge (Eds.), *Touching space, placing touch* (pp. 71–88). London: Routledge.
Brown, T., Andrews, G. J., Cummins, S., Greenhough, B., Lewis, D., & Power, A. (2018). *Health geographies: A critical introduction*. Chichester: Wiley Blackwell.
Bunkše, E. V. (1981). Humboldt and an aesthetic tradition in geography. *The Geographical Review, 71*, 127–146.
Bunkše, E. V. (1990). Saint-Exupéry's geography lesson: Art and science in the creation and cultivation of landscape values. *Annals of the Association of American Geographers, 80*, 96–108.
Burke, L. (2016). On (not) caring: Tracing the meanings of care in the imaginative literature of 'alzheimer's epidemic'. In A. Whitehead et al. (Eds.), *The Edinburgh companion to the critical medical humanities* (pp. 596–610). Edinburgh: Edinburgh University Press.
Büscher, M., Urry, J., & Witchger, K. (Eds.). (2011). *Mobile methods*. Abingdon: Routledge.
Buttimer, A. (1976). Grasping the dynamism of the lifeworld. *Annals of the Association of American Geographers, 66*, 277–292.
Buttimer, A. (1990). Geography, humanism and global concern. *Annals of the Association of American Geographers, 80*, 1–33.
Charon, R. (2006). *Narrative medicine: Honoring the stories of illness*. New York: Oxford University Press.
Conradson, D. (2005). Landscape, care and the relational self: Therapeutic encounters in rural England. *Health & Place, 11*, 337–348.
Coyle, L. A., & Atkinson, S. (2019). Vulnerability as practice in diagnosing multiple conditions. *Medical Humanities*. https://doi.org/10.1136/medhum-2017-011433.
Cresswell, T. (2015). Space, place, and the triumph of the humanities. *GeoHumanities, 1*, 4–9.
Cresswell, T., & Dixon, D. P. (2015). Imagining and practicing the geohumanities: Past, present and future. *GeoHumanities, 1*, 1–3.
Curtis, S., & Tonnellier, F. (2005). Medicine, landscapes, symbols: "The country doctor" by Honoré de Balzac. *Health & Place, 11*, 313–321.
Daniels, S., DeLyser, D., Entrikin, J. N., & Richardson, D. (Eds.). (2011). *Envisioning landscapes, making worlds: Geography and the humanities*. London: Routledge.
Davidson, J., Smith, M., & Bondi, L. (Eds.). (2007). *Emotional geographies*. Farnham: Ashgate.
De Leeuw, S. (2012). *Geographies of a lover*. Edmonton: NeWest Press.
De Leeuw, S. (2019). *Outside America*. Gibsons: Nightwood Editions.
De Leeuw, S., & Hawkins, H. (2017). Critical geographies and geography's creative re/turn: Poetics and practices for new disciplinary spaces. *Gender, Place & Culture, 24*, 303–324.

De Leeuw, S., Donovan, C., Schafenacker, N., Kearns, R., Neuwelt, P., Merrill Squier, C., et al. (2018). Geographies of medical and health humanities: A cross-disciplinary conversation. *GeoHumanities, 4*, 285–334.
Dear, M. (2011). Afterword. In M. Dear, J. Ketchum, S. Luria, & D. Richardson (Eds.), *Geohumanities: Art, history, text at the edge of place* (pp. 309–314). New York: Routledge.
Dear, M., Ketchum, J., Luria, S., & Richardson, D. (Eds.). (2011). *GeoHumanities: Art, history, text at the edge of place*. New York: Routledge.
DeLyser, D., & Hawkins, H. (2014). Introduction: Writing creatively – Process, practice, and product. *Cultural Geographies, 21*, 131–134.
Dunn, C. E. (2007). Participatory GIS – A people's GIS. *Progress in Human Geography, 31*, 616–637.
Entrikin, J. N. (1976). Contemporary humanism in geography. *Annals of the Association of American Geographers, 66*, 615–632.
Eshun, G., & Madge, C. (2016). Poetic world-writing in a pluriversal world: A provocation to the creative (re)turn in geography. *Social & Cultural Geography, 17*, 778–785.
Evans, B., & Cooper, C. (2016). Reframing fatness: Critiquing 'obesity'. In A. Whitehead et al. (Eds.), *The Edinburgh companion to the critical medical humanities* (pp. 225–241). Edinburgh: Edinburgh University Press.
Fitzgerald, D., & Callard, F. (2015). Social science and neuroscience beyond interdisciplinarity: Experimental entanglements. *Theory, Culture & Society, 32*, 3–32.
Foley, R. (2017). Swimming as an accretive practice in healthy blue space. *Emotion, Space & Society, 22*, 43–51.
Forsyth, I. (2011). Review of Dear et al. GeoHumanities: Art, history, text at the edge of place. *Scottish Geographical Journal, 127*, 251–253.
Foster, S. L., & Funke, J. (Eds.). (2018). Feminist encounters with the medical humanities. *Feminist Encounters: A Journal of Critical Studies in Culture and Politics, 2*(2), 14, Special issue.
Gesler, W. (1992). Therapeutic landscapes: Medical issues in light of the new cultural geography. *Social Science and Medicine, 34*, 735–746.
Gesler, W. (2000). Hans Castorp's journey-to-knowledge of disease and health in Thomas Mann's the Magic Mountain. *Health & Place, 6*, 125–134.
Gesler, W., & Kearns, R. (2001). *Culture/place/health*. London: Routledge.
Greenhough, B. (2010). Citizenship, care and companionship: Approaching geographies of health and bioscience. *Progress in Human Geography, 35*, 153–171.
Hall, E., & Wilton, R. (2016). Towards a relational geography of disability. *Progress in Human Geography, 41*, 727–744.
Harrison, P. (2015). After affirmation, or, being a loser: On vitalism, sacrifice, and cinders. *GeoHumanities, 1*, 285–306.
Hawkins, H. (2011). Dialogues and doings: Sketching the relationships between geography and art. *Geography Compass, 5*, 464–478.
Hawkins, H. (2018). Geography's creative (re)turn: Toward a critical framework. *Progress in Human Geography*. https://doi.org/10.1177/0309132518804341.
Hunt, R. (2018). On sawing a loaf: Living simply and skilfully in hut and bothy. *Cultural Geographies, 25*, 71–89.
Jacobsen, M., & Larsen, S. C. (2014). Ethnographic fiction for writing and research in cultural geography. *Journal of Cultural Geography, 31*, 179–193.
Johnston, C., & Pratt, G. (2010). Nanay (mother): A testimonial play. *Cultural Geographies, 17*, 123–133.
Last, A. (2018). An experiment in teaching GeoHumanities. *Mutable Matter*. https://mutablematter.wordpress.com/2018/06/03/an-experiment-in-teaching-geohumanities/. Accessed 21 Mar 2019.
Livingstone, D. N. (1992). *The geographical tradition: Episodes in the history of a contested enterprise*. Oxford: Blackwell.
Longhurst, R. (2001). *Bodies: Exploring fluid boundaries*. London: Routledge.

Lorimer, H. (2003). Telling small stories: Spaces of knowledge and the practice of geography. *Transactions of the Institute of British Geographers, 28*, 197–217.

Lorimer, H. (2018). Dear departed: Writing the lifeworlds of place. *Transactions of the Institute of British Geographers*. https://doi.org/10.1111/tran.12278.

Macnaughton, J. (2009). The dangerous practice of empathy. *The Lancet, 373*(9679), 1940–1941.

Maddrell, A. (2013). Living with the deceased: Absence, presence and absence-presence. *Cultural Geographies, 20*, 501–522.

Madge, C. (2014). On the creative (re)turn to geography: Poetry, politics and passion. *Area, 46*, 178–185.

Magrane, E. (2015). Situating geopoetics. *GeoHumanities, 1*, 86–102.

Marston, S., & de Leeuw, S. (Eds.). (2013). Creativity and geography: Toward a politicized intervention. *The Geographical Review, 103*, iii–xxvi.

McGeachan, C., Forsyth, I., & Hasty, W. (2012). Certain subjects ? Working with biography and life-writing in historical geography. *Historical Geography, 40*, 169–185.

McLean, K. (2017). Smellmap: Amsterdam—Olfactory art and smell visualization. *Leonardo, 50*, 92–93.

Milligan, C., & Wiles, J. (2010). Landscapes of care. *Progress in Human Geography, 34*, 736–754.

Moon, G., Kearns, R., & Joseph, A. (2016). *The afterlives of the psychiatric asylum: Recycling concepts, sites, and memories*. London: Routledge.

Osborne, T., Warner, E., Jones, P. I., & Resch, B. (2019). Performing social media: Artistic approaches to analyzing big data. *GeoHumanities*. https://doi.org/10.1080/2373566X.2018.1543552.

Oswald, A. (1996). *The thing in the gap-stone stile*. London: Faber and Faber.

Pain, R. (2004). Social geography: Participatory research. *Progress in Human Geography, 28*, 652–663.

Parr, H. (2006). Mental health, the arts, and belongings. *Transactions of the Institute of British Geographers, 31*, 150–166.

Parr, H. (2007). Collaborative film-making as process, method and text in mental health research. *Cultural Geographies, 14*, 114–138.

Parr, J. (2010). *Sensing changes: Technologies, environments, and the everyday, 1953–2003*. Vancouver: University of British Columbia Press.

Patchett, M. M. (2016). The taxidermist's apprentice: Stitching together the past and present of a craft practice. *Cultural Geographies, 23*, 401–419.

Perkins, C. (2014). Plotting practices and politics: (im)mutable narratives in OpenStreetMap. *Transactions of the Institute of British Geographers, 39*, 304–317.

Philo, C., & Pickstone, J. (2009). Unpromising configurations: Towards local historical geographies of psychiatry. *Health & Place, 15*, 649–656.

Pocock, D. C. D. (1981a). Place and the novelist. *Transactions of the Institute of British Geographers, 6*, 337–347.

Pocock, D. C. D. (1981b). Sight and knowledge. *Transactions of the Institute of British Geographers, 6*, 385–393.

Pocock, D. C. D. (1988). Geography and literature. *Progress in Human Geography, 12*, 87–102.

Pocock, D. (1993). The senses in focus. *Area, 25*, 11–16.

Porteous, D. (1985). Literature and human geography. *Area, 17*, 117–122.

Richardson, D. (2011). Introducing the GeoHumanities. In M. Dear, J. Ketchum, S. Luria, & D. Richardson (Eds.), *GeoHumanities: Art, history, text at the edge of place* (pp. 3–4). New York: Routledge.

Saunders, A. (2010). Literary geographies: Reforging the connections. *Progress in Human Geography, 34*, 436–452.

Skinner, E., & Masuda, J. R. (2013). Right to a healthy city? Examining the relationship between urban space and health inequity by aboriginal youth artist-activists in Winnipeg. *Social Science and Medicine, 91*, 210–218.

Thien, D., & Del Casino, V. J., Jr. (2012). (Un)healthy men, masculinities and the geographies of health. *Annals of the Association of American Geographers, 102*, 1146–1156.
Tsampiras, C. (2018). Walking up hills, through history and in-between disciplines: MHH and health sciences education at the tip of Africa. *Medical Humanities, 44*, 270–280.
Tsampiras, C., Mkhwanazi, N., & Hume, V. (Eds.). (2018). Medical and health humanities in Africa: Inclusion, access, and social justice. *Medical Humanities, 44* (4), Special issue.
Tuan, Y.-F. (1974). *Topophilia*. Englewood Cliffs: Prentice-Hall.
Viney, W. (2016). Getting the measure of twins. In A. Whitehead et al. (Eds.), *The Edinburgh companion to the critical medical humanities* (pp. 104–119). Edinburgh: Edinburgh University Press.
Viney, W., Callard, F., & Woods, A. (2015). The critical medical humanities: Embracing entanglement, taking risks. *Medical Humanities, 41*, 2–7.
Whitehead, A., & Woods, A. (2016). Introduction. In A. Whitehead (Ed.), *The Edinburgh companion to the critical medical humanities* (pp. 1–31). Edinburgh: Edinburgh University Press.
Whitehead, A., Woods, A., Atkinson, S., Macnaughton, J., & Richards, J. (Eds.). (2016). *The Edinburgh companion to the critical medical humanities*. Edinburgh: Edinburgh University Press.
Williams, A. (Ed.). (2007). *Therapeutic landscapes*. Farnham: Ashgate.
Withers, C. W. J. (2001). *Geography, science and national identity: Scotland since 1520*. Cambridge: Cambridge University Press.

Part I
Bodies

How to Convey That Dying Really Is Lonely

Sarah de Leeuw

It matters how, for instance, your daughters
do not visit, only calling once – maybe twice –

a month from far away, from a city with a river
you refuse to call a river, flanked with paved

sidewalks, not shores or banks, a gassy water
highway, nothing wild or untamed left, only

reintroduced salmon, opposite of those you
miss fishing in northern streams, a dog at

your side, now you watch him every morning
after you've taken a flight of pills knowing

he will outlive you, knowing the snow outside
may be the last snow you'll see, you can't see

the cells as they metastasize to pelvis just
like you can't see the jays, the chickadees or

the flickers, but you hear them like you dream
of cancer being white, an invisible racket

of noise you have to fall asleep with, knowing
it's all inside you. You wake up to the news

of an Olive Ridley sea turtle carried north
on ocean currents from Pacific coast Mexico,

life cooling away, found by crab fishermen
near Alaska, a slack-beaked face dappled jade

green, skin like light on a tropical sea,
how a woman named him Frank, wrapped

him in a blanket to warm him, how that didn't
work so she drove his body around town

in the hatchback of a rusting Honda Civic,
a sample specimen, a dead animal far from home.

From: de Leeuw, S. (2019). *Outside America* (pp. 22–23). Gibsons, BC: Nightwood Editions.

Chapter 2
Sensing Health and Wellbeing Through Oral Histories: The 'Tip and Run' Air Attacks on a British Coastal Town 1939–1944

Gavin J. Andrews and Viv Wilson

Abstract This chapter explores how the immediacy and feeling of past times and places, and how they are important to health and wellbeing, might be conveyed through the humanities tradition of oral history. It reports on 17 oral history interviews conducted between 2005 and 2012 with senior residents of the town of Teignmouth, Devon, UK, who experienced the World War II coastal homefront first-hand, and in particular the numerous 'tip and run' air attacks that took place. The chapter focuses on how respondents recollect events as sensory encounters. It argues that these encounters are important experiences on their own right, involving their own level and form of knowing time and place, but also in terms of how they interplay with meanings consciously attributed to them. Whilst encounters can re-emerge like 'hauntings' from the past in the process of respondents recalling and conveying, they might equally act like pre-emptive affective facts, helping researchers feel, empathise with and ultimately reanimate what it felt like to be there.

Keywords Sensory encounters · Affect · Memories · War · United Kingdom

Introduction: The Sensory Turn

> Sensory studies involves a cultural approach to the study of the senses and a sensory approach to the study of culture. It challenges the monopoly that the discipline of psychology has long exercised over the study of the senses and sense perception by foregrounding the sociality of sensation… scholars from across the humanities and social sciences have, over the past few decades, successively turned their attention on the sensorium. (Howes 2013:1)

G. J. Andrews (✉)
McMaster University, Department of Health, Aging and Society, Hamilton, ON, Canada
e-mail: andrews@mcmaster.ca

V. Wilson
Bridge View, Teignmouth, Devon, UK

S. Atkinson, R. Hunt (eds.), *GeoHumanities and Health*, Global Perspectives on Health Geography, https://doi.org/10.1007/978-3-030-21406-7_2

The social sciences and humanities have been gripped in recent years by a 'sensory turn'. It is a turn that recognises that human knowledge is not only something consciously learned, but also something formed processually through the body: it's internal/natural/biological properties and specifically through it doing things with other bodies and objects (Andrews 2018). It is a turn that realises how human senses pick up and register the material world, driving human responses to, co-evolution with, and knowledge of it. Moreover, it realises how human senses become embedded in ways of life and critical to the way society works (Andrews 2018). Hence, there has been substantive attention in scholarship to sound/auditory experiences and cultures (Bull and Back 1993), gustatory/taste experiences and cultures (Korsmeyer 2005), olfactory/smell experiences and cultures (Drobnick 2006), sight/visual experiences and cultures (Edwards and Bhaumik 2008) and tactile/touch experiences and cultures (Classen 2005). Notably the sensory turn has been a truly interdisciplinary endeavour as evidenced, for example, by diverse contributions – pertinent to the current chapter – in the fields of health studies (e.g. Allen-Collinson and Hockey 2015; Fullagar 2017; Jones 2012; Sparkes 2017; Sparkes et al. 2018; Bissell 2009, 2010; Brown 2017; Gorman 2017; Middleton 2010; Solomon 2011) and modern history (e.g. Crang and Tolia-Kelly 2010; Dixon and Straughan 2010; Jones 2011; Lahiri 2011; Lauwrens 2012; Matless 2005; Rycroft 2007; Smith 2007), both of which have been contributed to substantially by geographical, sociological and anthropological imaginations (Howes 2013; Low 2012).

This essay makes a modest contribution to this ever-expanding sensory literature. Adding to emerging scholarship on sensory methodologies (Mason and Davies 2009; Pink 2015), via a case study of life on the small town homefront during World War II, it shows how the humanities tradition oral history might recall sensory engagements with past times and places that have particular health and wellbeing impacts and consequences. Moreover, how oral history might help 'reanimate' these times, places and impacts in novel ways. We start, however, by first reviewing the development of oral history methodologies to the current stage at which sensory concerns and approaches have newly emerged.

The Oral History Tradition

As both a qualitative methodology and research approach, the humanities tradition of oral history is making an increasingly significant contribution to the academic study of modern times (see Perks and Thomson 2006; Thomson 2008; Yow 2014). In simple terms, oral history involves talking to people – very often older people – in order to add to the historical record, ranging from the recollection of prominent or rare events, to those that make everyday life. Indeed, through interpersonal conversational practices, oral history brings to life the past whilst at the same time preserving it for future reference.

As Thomson (2007) describes, as a tradition oral history can be traced back over six decades. Its development can be characterised by consecutive paradigm shifts, each leaving a lasting mark on scholarship and culminating on the approaches employed today by scholars. The first of these involved a general emergence in the 1950s and 1960s of memory as a source of 'peoples' history'. The increasing popularity of eyewitness accounts at the time marked a notable challenge to archival research which had dominated since the nineteenth century at the birth of history as a humanities discipline (Thomson 2007). The second shift, occurring from the 1970s onwards, was probably most significant in terms of the intellectual development of oral history, and involved approaching memory and subjectivity more critically (as Thomson suggests, it being largely a response to archival historians' concerns over the reliability of oral accounts). The legacy of this second shift is twofold. On one level, whilst recognising the parameters and limitations of oral accounts – being, for example, partial, selective, myth-laden, biased by situation and the unconscious (Norquay 1999) – the expectation of complete objectivity and transparent statements of 'truth' in oral history waned. Instead a consensus opinion emerged that oral accounts can be celebrated as important cultural representations and constructions with value as such (Gluck 1999). On another level, the need for highly reflective method, analysis and writing in oral history was recognised, for example, in the contexts of interview techniques (Morris 2002), the construction of respondents' narratives (Errante 2000; Norquay 1999), the blending of respondents' narratives and academic narration (Good 2000; Jones 2004) and associated issues of authority and meaning (Frisch 1990; Portelli 1991; Thomson 2003).

The third shift, occurring from the late 1980s onwards, has involved changing roles for oral historians and changing ways in which they go about their research, all realised in four trends (see Thomson 2007): (i) Many oral historians have become active in public life, often acting as advocates for their respondent groups and supporting particular social and political agendas. (ii) The digital revolution has created new ways of finding, recording, storing and analysing oral accounts. (iii) Social scientists have become increasingly involved in oral history research, facilitated somewhat by the 'humanities turn' across many disciplines. Whilst many of these scholars have been anthropologists and sociologists (Dunaway and Baum 1984), a number of human geographers have become involved in oral history research, telling the histories of peoples and places at various scales (see Andrews et al. 2006; Andrews 2017; Cameron 2001; Butler 2007; George and Stratford 2005; Jenkins and Ward 2001; Jones and Fowler 2007; Kwan 2008; Lorimer 2003; Matless et al. 2007; Miles and Crush 1993; Ní Laoire 2007; Riley and Harvey 2007a, b). (iv) Important in the current context, emerging theoretical debates have focused on an array of performance and relational issues in the research encounter and process, including the role of the self, intersubjectivity and power (Abrams 2010). One outcome of these debates has been increasing attention in studies on how the past is known through the senses (Butler 2007; Damousi and Hamilton 2017; Hamilton 2011, 2017). It is this proposition that the current chapter 'tests' in a particular empirical context.

The Setting and Study

Even modest, ordinary places can have notable moments and histories, and that is certainly the case for Teignmouth, a small seaside resort and port with a population of 15,000 located on the South Devon, UK, coastline (see Fig. 2.1). At the time of writing in 2018, Teignmouth was the subject of a film *The Mercy* about the failing mental health and ultimate death of Donald Crowhurst, a novice sailor who famously cheated during a 1968 round-the-world yacht race (Tomalin and Hall 2017). As Andrews and Kearns (2005) suggest, before that health emerged strongly in Teignmouth in key periods through military invasions and incursions, the infusion of infectious disease, the growth of the therapeutic resort, periods of wartime and war effort, retirement migration and growth and decline in care services.

As Andrews (2017) notes, during the World War II, Teignmouth was subjected to sudden and often severe air raids. Small fighter aircraft attacked in high-speed low-level sorties. This was typical of what became known as the 'tip and run' air campaign conducted by the Luftwaffe at the time: a strategy which aimed to destroy specific targets (such as railways, shipping facilities, coastal defences, energy supplies and manufacturing) in UK south coastal towns (including also Hastings, Torquay, Bournemouth and Eastbourne) (Goss 2010). In practice however, a lack of

Fig. 2.1 Teignmouth, Devon, UK

accurate intelligence and target finding technologies on aircraft meant that most tip and run raids affected civilian life and housing; this though serving a secondary objective - to decrease morale in local populations and exact a degree of revenge for RAF attacks on Germany (Goss 2010). Indeed, between the evening of 7 July 1940 and the afternoon of 29 May 1944, Teignmouth experienced 22 air raids. Over 1000 incendiaries and 79 high explosives were dropped on the town killing 79 people and injuring 151. Of the town's 3000 houses, over 200 were either completely destroyed or damaged beyond repair, whilst less than 800 were left unscathed (Andrews et al. 2006; Andrews 2017).

In terms of data collection, 17 interviews were completed with people who had lived in Teignmouth during and since World War II, 12 being conducted in 2005, 2 in 2007 and 3 in 2012. Most of the respondents were recruited by a local historian – the second author of this chapter – who had come to know them, or know of them, during her longtime work in the town. With the exception of two who were in residential care, all respondents were living independently in their own homes, which also served as the venues for interviews. In terms of their ages during wartime, the average age at the commencement of World War II was 15 and at the end was 21. However, a considerable variation in the age meant that some were children throughout the war ($n = 5$), some grew from teenagers into adults ($n = 6$) and others were adults throughout ($n = 6$). With the exception of one respondent all were female; this reflecting life expectancy and the domestic role of women who did not fight in the military forces. The majority of interviews took approximately 1 hour to complete, but the exact length depended on how much the particular respondent wished to say and the length of time that they were able to dedicate to the process.

On the whole, conversational practice was free-flowing and unhindered, and most respondents required only the most basic of prompts to elaborate on their practices and experiences. As suggested earlier, issues surrounding truthfulness and reliability are well- acknowledged and well-debated in oral history research (so will not be repeated here). Needless to say however, attention was paid, where possible, to pick up and account for such things as exaggerations, false memories, selective recall and even trivialising. This involved, for example, assessing respondents' general communication traits, comparing various respondents' stories with each other and with official records. Respondents were never asked to comment specifically on, and recall, sensory happenings and feelings in their interviews, this being something that happened quite organically during conversations, and something that was noticed by one of the authors during analysis undertaken for two earlier publications (see Andrews et al. 2006; Andrews 2017). Indeed, the realization that oral histories are often quite sensory led to the interviews being revisited and re-analysed in early 2018 with their sensory content – and the current chapter – specifically in mind. Basic thematic analysis was conducted (i.e. familiarization, drawing out initial codes, creating firm codes and subcodes), which established the particular findings subsections that follow.

Findings

Sensing Restrictions and Change

Change was something mentioned by respondents in all of the interviews: change being multifaceted and something almost always sensed whether or not opinionated upon. One change, for example, came in the form of spatial restrictions resulting from the citing of anti-aircraft guns and anti-invasion measures in the town. Here two respondents convey their memories of hard barriers. Whilst they certainly possessed opinions on their lives being restricted and on the thinking/justifications of officialdom enforcing this, the size, shape and form of barriers were at the fore of their comments:

> I was annoyed at the barbed wire stretched across the prom and metal barriers on the beach but I did accept them. They looked out of place compared to the sea.

> Concrete tank traps were installed across the top of the beach near the ferry point. They got in the way. Coasters had big twenty-five feet long kite balloons attached to them which would blow about in the wind. One was struck by lightning and I saw it burn up.

Other changes concerned new consumer regulations and rules of conduct. Respondents, for example, talked about related sounds, sights and tastes:

> Blackout rules at night meant that it was always very dark with no street lights and no cars about. If anyone else approached through the dark they would say 'Good evening' but it was as if they were only talking to a cigarette or a walking set of eye whites. But I always carried a torch with a square of brown paper to hold over the bulb just in case the keys were dropped or any other dire necessity arose.

> There was rationing so you were only allowed small amounts of food per person. But you could make something really nice and tasty if you were inventive and put your mind to it. Like stews, pies – everything thrown in and padded with potato.

These kinds of restrictions, whether manifest in obvious material objects, or more subtle rules and codes, are described in the homefront literature. Rather than being concerns of academic studies, it has often been left to popular history texts to convey their sensory consequences and experiences, often through the occasional illustrative inclusion of oral history narratives (e.g. Gardiner 2004).

Another form of change was social: more and different people entering the town particularly foreign armed forces, initially for defence and eventually in preparation for D-day and the Allied invasion of Europe. Although respondents offerred their past opinions on this, more often than not their comments involved spectacle and a certain affective energy reanimated in the present through the passionate way the stories were told. Two respondents stated:

> We had Indian soldiers and Americans. Some in camps outside the town and others in huts on hotel lawns. They did drills and things which were ordered but they walked around Teignmouth in the day and evening, in their spare time. They went in the pubs which were heaving.

> Busy busy. Americans were everywhere, evacuees, pack mules, army trucks, the Navy doing sea trials. There was a huge influx of people filling in.

Sensing in this context is about the contribution of interpersonal affective energy, which ultimately helps lead to increased group identity, unity and collective spirit and action: a common concern of behavioural psychology literature but typically not of historical literature (see Reicher 2001).

Sensing Air Raids

The most impactful and intense sensory events to occur during the war, and those with direct and serious health consequences were the air raids (see Figs. 2.2 and 2.3). These were recalled by all respondents. 'Big' sounds, sights and bodily feelings were common conveyances. Three respondents commented:

> In a raid I [a teacher] was thrown around by the power of the bombs. The children were screaming in fear for their lives at the cannon shell and machine gun fire. Eight of them had diarrhoea.

> There was a daylight raid and I was working behind the counter at the post office. You could hear the sound of machine gun fire, crack crack. Two soldiers burst through the swing doors and landed on the floor. It was funny....

> I witnessed an air raid from camp field on the cliffs. The plane went straight past us at our level as we stood watching on raised ground. I was a bit nervous but curiosity overcame fear. The speed he was going, he would not have seen us though.

Fig. 2.2 Bomb damage

Fig. 2.3 Bomb damage and war effort

Such sensory experiences of war and conflict have often been focused upon in the historical literature, although most typically in the context of battlefronts and conflicts directly between armed forces (see Smith 2015). The obvious lesson of the current study hence is the realization that they are just as relevant and impactful at the interface of military action and civilian life.

Destruction of material objects and human bodies was a common theme, respondents often recounting events with a sense of fear re-emerging in their voices:

> I witnessed two events when victims were injured and taken away. Once I saw a local woman that I knew have her leg blown right off. On another occasion, a girl was walking up the lane and was blown by a bomb blast into the bakery back door. She died later.

> The roof of our house was blown off and open so you could see the sky. I have never forgotten looking upwards in fear and seeing the sky….

Restriction and claustrophobia were common themes, resulting from the places frequented for safety reasons during air raids. Of course, claustrophobia was referred to not in terms of being an anxiety 'disorder' (e.g. Rachman and Taylor 1993) but in popular nonclinical terms. The difference from an anxiety disorder definition being that respondents' experiences were not brought on by 'irrational' thinking – the surrounding danger and risk being very real. Two stated:

> A brick-built shelter stood on the village green between the bowling green and the clock. It was about thirty by ten feet and could hold about fifty people. Very tight. At home we had a Morrison shelter. Everyone in the house crushed into it the night Shaldon was bombed. I just hoped it would soon be over. I desperately wanted to get out and go back to bed. There was a sensation of the ceiling moving up and down after the bomb dropped.

> During a raid I had my first experience of fainting. I was one of five people crammed into the cupboard under the stairs where the coal was kept. I had a sense of claustrophobia.

There was often a sense of foreboding and doom: darkness and silence being common themes in interviews. As the psychology literature tells us, base feelings such as these are common in how survivors experience and remember traumatic events (McNally 2005). Four commented:

> Night raids were the worst, more frightening in the dark, and loss of sleep added to the feeling of confusion. We did not have a shelter and gathered downstairs in the dark until All Clear.

> Raids always shook me up and I felt greatly upset because people were buried. It was my worst fear to be buried alive under a building.

> I often saw planes out over the sea and was struck by how silent they were. The pilots meant business. If they could kill you they would be happy.

> During raids we all sat quiet and listened to it all. We didn't talk, we just worried. After the all-clear I would think, OK- it's all over.

Sensing Mitigation and Aftermath

Mitigation was a common theme that emerged during interviews. Although mitigation, by its very nature, involves conscious strategic thinking and implementation, its sensory aspects also came to the fore. Locating, understanding and reacting during raids were discussed often, whilst the importance of shelter/cover, home and open space emerged as key spatial attributes. Recalling sensory experiences, three respondents stated:

> When a bomb had obviously been dropped, I would try to work out its location from the bang and how close it was... I would join the men in the street, watching with the searchlights. I recall seeing dozens and dozens of planes overhead on their way to bomb Exeter or Plymouth. But I also learned to distinguish an enemy plane through its distinctive hum.

> If we heard a plane, we sheltered in the hedge. But if only the siren went, we returned home immediately. My mother always sang songs during the raids. Usually children's songs and she often rocked my younger brother in her arms. After the raids, all the neighbours would run around to one another asking if they were alright.

> I knew to duck under tables or furniture if a raid was happening right then. The best place to be was on the outskirts of the town so you could run to the fields where it was quiet and safe.

Aftermath, both immediate and longer term, was a strongly emerging theme in interviews. Emotions were worked through cognitively but also through sensory experiences in a very basic way. Three respondents commented:

> My outstanding memory is after the school raid. Huge wooden cupboards full of equipment had fallen over and lots of damage. I will never forget the faces of the mothers racing to find out if their child was dead or alive.

> I remember rubble after raids, the smell of smoke. Wood everywhere – lots of damage everywhere. Terrible.

> After a raid I was alright until I saw my Mum, then I burst into tears. But it was because blood was running down her leg from a piece of shrapnel. She made light of it but a nurse came to our home to remove the shrapnel and dress the wound. I watched the blood and the process in detail and still remember it in detail now.

Importantly, although many respondents talked about the emotional consequences of what they went through and how they coped, two mentioned serious long-term impacts on their health and wellbeing. One commented (see also Andrews 2017:742):

> After a raid involving the deaths of people in their own homes very close to mine, I heard a baby cry under the rubble. It was the catalyst to my decision that there was no point in going on with life. The world seemed to have gone mad. I walked, fully clothed into the river and was found unconscious by two fishermen. I gave a promise to my mother that I would never do it again. Not long afterwards, I met and married my first husband but was widowed within three weeks. Less than a year later my brother was killed during a raid on Teignmouth and once again I was devastated. My family all moved out into the countryside to get away from the raids...

> At the end of the war, I married a soldier and went up country to live, but after a number of years, I returned home without him. I had suffered two nervous breakdowns and saw a psychiatrist who asked me to draw or write whatever came into my head. I drew a slice through an orange and filled the segments with words such as 'fear' and 'anger', the emotions that ruled my life for some time.

These experiences, and lives lived in the wake of them, occurred in a time where post-traumatic stress disorder (PTSD) was yet to become formally recognised and treated (Jones and Wessely 2005; Apfel and Simon 1996). Hence, it is quite possible that some of the respondents suffered from what would now be recognised as PTSD throughout their adult lives.

Sensing Normality

Air raids were powerful physical, sensory and emotional experiences, yet life on the homefront was a 24-hour, 7-day a week experience for respondents for 6 long years. Hence, at the time it became their new normal and everyday. A common theme was 'out and about' and the sensory experiences involved. Two respondents explained:

> We enjoyed the town and countryside as much as we could, and life went on as normal most of the time. Home was so important as a refuge however. Once the school bell rang, you went home and Mum was always there. The bell was the signal to rush home.
>
> Sometimes we walked up to Guppy's pig farm at Higher Holcombe and were allowed to pick up windfall apples. We filled a steel basin with blackberries and the farmer cut us a cauli or cabbage or swede to take home. Mum could make a tasty stew out of a cardboard box. She would send me to Star Supply Stores to order 6d worth of bacon bones and told me to tell them, "don't skin all the meat off them!" She soaked lentils and split peas overnight then added carrots and onions in a big cast iron pot.

Indeed, such places could be understood as everyday spaces of wellbeing (e.g. Andrews et al. 2014) or as 'therapeutic landscapes' with particular active sensory qualities (e.g. Lea 2008).

Duty as normal/everyday was another theme, often raised by those who undertook official roles associated with the war effort. Two respondents explained:

> Once a week I worked on the phones at the report centre at Bitton House. They gave me a tin hat and a tiny pocket torch and I followed instructions according to what colour alert prevailed. Red, and I phoned the fire brigade.
>
> I did the Fire Watching at night in the school even after it was damaged because no one told me not to bother. It was eerie and probably not safe. A soldier came in once and asked me what I was up to. On New Year's Eve I returned early to do my stint and the team took a wind-up gramophone and played Silent Night as the New Year dawned.

However, even normal life could be suddenly suspended by events. A respondent recalled:

> Nothing deterred us boys from investigating bomb sites and searching for souvenirs. We went into fields near Labrador bay to dig out bits of grenades, sticky bombs and mortars. There were mines in the cliffs and a soldier got blown up. Bits of uniform hung around in the trees. It was gruesome.

Another theme that emerged under the theme of normality is how past events are re-sensed in the present during everyday encounters, both purposeful and accidental. Three respondents commented on this phenomenon:

> I hated putting the gas mask over my face. After a long time when nothing much happened, I did not always carry it with me. I still hate the smell of rubber.
>
> I loved the rubbery smell from my Mickey Mouse gas mask and recently enjoyed evoking old memories by sniffing one belonging to a friend. It was kept in the cupboard under the stairs and I never had to use it except for the gas mask drill at school. It was very hot to wear.
>
> Even now, when I go out on the Den, I can see the cook house in the pavilion. I remember the Christmas party the soldiers invited us to in 1943.

The sensory triggering of past events is a common theme in psychology and gerontology literatures; of concern being how it can be mitigated and best dealt if connected to mental health problems (e.g. Hilton 1997). For most respondents in the current study however, it was a relatively unproblematic part of their personal recollection and even nostalgia.

Conclusions

This study has indicated how oral histories methods might garner sensory recollections of past times and places, reanimating them in the present. Respondents in the study conveyed, without specific guiding questions or prompting, a wide range of sensory encounters which formed part of their time on, and telling of, the British coastal homefront. As respondents told their stories, these came through, helping them convey what had happened to them. These are important experiences and levels of knowing in their own right but often connect to more complex beliefs, emotions and meanings attached to the specific time and place (as evidenced by opinions often being included in recollections of physical encounters). For us the researchers – and we hope eventual readers of this chapter – these recollections help understand on a physical level what it felt like to be present. Thinking about this theoretically, as McHugh (2009) argues, the past is not dead simply because it is not currently happening. Instead the past can exist in a virtual dimension, its affects and other feelings reactivated by storytelling; listeners using their own reference points to create their own visual images and versions. Indeed, for McHugh, past events and experiences can re-emerge, affectively resonating in the present as 'hauntings', although not ever being legitimised by an outcome in the present (Amin 2004; Malpas 2012).

Thinking about what all this means for future scholarship, the emphasis on senses and sensations in contemporary academia is part of a reformed engagement with the body, not so much concerned with how the body is constructed (an interest of biology and other health sciences), or represented and known (an interest of humanistic and social constructionist social science), but instead concerned with the 'being' and potential of bodies through material and other body relations: senses being critical to how the body co-evolves with things (Andrews 2018). With such a posthumanist intellectual project, the conventional wisdom has been that inquiry has to be somewhat 'anti-biographical' (Thrift 2008). We argue however, particularly given the current research findings, that people can, and frequently do, tell stories that are not only about themselves in a 'deep and meaningful' self-conscious sense, but also about the sensory and affective happenings they were exposed to. This is likely because narrative/stories are always framed at least in part by earlier performances and experiences that are sensed, intersubjective, relational and often shared. Indeed, when recalling the past, most people tend to convey at least some of what actually happened and what it felt like, in addition to what they thought of it, or why it had happened: this merely being part of storytelling. Of course, in the current study, restrictions, rapid changes and bombing raids are powerful physical occurrences, so in some ways it is not surprising that sensory experiences came to the fore in interviews. But for the aforementioned theoretical reasons, arguably oral history might possess broad potential as a contributor to sensory studies in the geohumanities.

If this intellectual project is to be pursued, scholars will need to pay close attention to a number of issues. First, they will need to consider how oral history methods might learn from dedicated sensory methodological innovations, such as sensory interviewing and sensory ethnography, deployed – unlike the current study – where researchers are purposefully digging for sensory insights (Mason and Davies 2009; Pink 2015). Second, they will need to think about – and about the relative merits of – how sensory oral histories can both tell stories *of* health and wellbeing in the past, but equally how there is health and wellbeing *in* the process of telling the past (the latter drawing close to reminiscence as a therapeutic practice (see Andrews 2017; Skinner et al. 2015)). Third and finally, on a practical level is the issue of how to pool resources and expertise to undertake sensory oral histories and maximise their quality and contribution. In the current study, the combination of a geographer and a local historian provided compatible skills, although challenges are always encountered when crossing disciplinary divides. Nevertheless, whatever the next priority or debate might be, how past health and place might be told and reanimated by oral histories deserves serious recognition.

Acknowledgement Figure 2.1, the view of Teignmouth, was provided by local photographer Paul Blythe. Many thanks Paul. Figures 2.2 and 2.3 are from The Viv Wilson Archive (second author).

References

Abrams, L. (2010). *Oral history theory*. London: Routledge.

Allen-Collinson, J., & Hockey, J. (2015). From a certain point of view: Sensory phenomenological envisionings of running space and place. *Journal of Contemporary Ethnography, 44*, 63–83.

Amin, A. (2004). Regions unbound: Towards a new politics of place. *Geografiska Annaler: Series B, Human Geography, 86*, 33–44.

Andrews, G. J. (2017). The relational making of people and place: The case of the Teignmouth World War II homefront. *Ageing & Society, 37*, 725–752.

Andrews, G. J. (2018). *Non-representational theory & health: The health in life in space-time revealing*. London: Routledge.

Andrews, G. J., & Kearns, R. A. (2005). Everyday health histories and the making of place: The case of an English coastal town. *Social Science & Medicine, 60*, 2697–2713.

Andrews, G. J., Kearns, R. A., Kontos, P., & Wilson, V. (2006). 'Their finest hour': Older people, oral histories, and the historical geography of social life. *Social & Cultural Geography, 7*, 153–177.

Andrews, G. J., Chen, S., & Myers, S. (2014). The 'taking place' of health and wellbeing: Towards non-representational theory. *Social Science & Medicine, 108*, 210–222.

Apfel, R. J., & Simon, B. (1996). *Minefields in their hearts: The mental health of children in war and communal violence*. New Haven: Yale University Press.

Bissell, D. (2009). Obdurate pains, transient intensities: Affect and the chronically pained body. *Environment and Planning A, 41*, 911–928.

Bissell, D. (2010). Placing affective relations: Uncertain geographies of pain. In B. Anderson & P. Harrison (Eds.), *Taking-place: Non-representational theories and geography* (pp. 79–98). London: Routledge.

Brown, K. M. (2017). The haptic pleasures of ground-feel: The role of textured terrain in motivating regular exercise. *Health & Place, 46*, 307–314.

Bull, M., & Back, L. (Eds.). (1993). *The auditory culture reader*. Oxford: Berg.

Butler, T. (2007). Memoryscape: How audio walks can deepen our sense of place by integrating art, oral history and cultural geography. *Geography Compass, 1*, 360–372.
Cameron, L. (2001). Oral history in the Freud archives: Incidents, ethics, and relations. *Historical Geography, 29*, 38–44.
Classen, C. (2005). *The book of touch*. Oxford: Berg.
Crang, M., & Tolia-Kelly, D. P. (2010). Nation, race, and affect: Senses and sensibilities at national heritage sites. *Environment and Planning A, 42*, 2315–2331.
Damousi, J., & Hamilton, P. (2017). *A cultural history of sound, memory, and the senses*. New York: Routledge.
Dixon, D. P., & Straughan, E. R. (2010). Geographies of touch/touched by geography. *Geography Compass, 4*, 449–459.
Drobnick, J. (2006). *The smell culture reader*. Oxford: Berg.
Dunaway, D. K., & Baum, W. K. (Eds.). (1984). *Oral history: An interdisciplinary anthology*. Nashville: The American Association for State and Local History and Maryland: Rowman & Littlefield.
Edwards, E., & Bhaumik, K. (2008). *Visual sense: A cultural reader*. Oxford: Berg.
Errante, A. (2000). But sometimes you're not part of the story: Oral histories and ways of remembering and telling. *Educational Researcher, 29*, 16–27.
Frisch, M. (1990). *A shared authority: Essays on the craft and meaning of oral and public history*. Albany: State University of New York Press.
Fullagar, S. (2017). Post-qualitative inquiry and the new materialist turn: Implications for sport, health and physical culture research. *Qualitative Research in Sport, Exercise and Health, 9*, 247–257.
Gardiner, J. (2004). *Wartime: Britain 1939–1945*. London: Headline.
George, K., & Stratford, E. (2005). Oral history and human geography. In I. Hay (Ed.), *Qualitative research methods in human geography* (pp. 189–202). Oxford: Oxford University Press.
Gluck, B. (1999). Reflections on oral history in the new Millennium: Roundtable comments. *Oral History Review, 26*(2), 1–27.
Good, F. (2000). Voice, ear and text: Words and meaning. *Oral History Association of Australia Journal, 22*, 102.
Gorman, R. (2017). Smelling therapeutic landscapes: Embodied encounters within spaces of care farming. *Health & Place, 47*, 22–28.
Goss, C. (2010). *Luftwaffe fighter-bombers over Britain: The tip and run campaign 1942–43*. Mechanicsburg: Stackpole Books.
Hamilton, P. (2011). *The Proust effect: Oral history and the senses*. Oxford: Oxford University Press.
Hamilton, P. (2017). Intimate strangers: Multisensorial memories of working in the home. In J. Damousi & P. Hamilton (Eds.), *A cultural history of sound, memory, and the senses* (pp. 194–211). London: Routledge.
Hilton, C. (1997). Media triggers of post-traumatic stress disorder 50 years after the Second World War. *International Journal of Geriatric Psychiatry, 12*, 862–867.
Howes, D. (2013). The expanding field of sensory studies. https://www.Sensorystudies.Org/Sensorial-Investigations/The-Expanding-Field-Of-Sensory-Studies/. Accessed 21 Mar 2019.
Jenkins, A., & Ward, A. (2001). Moving with the times: An oral history of a geography department. *Journal of Geography in Higher Education, 25*, 191–208.
Jones, R. (2004). Blended voices: Crafting a narrative from oral history interviews. *Oral History Review, 31*, 23–42.
Jones, O. (2011). Geography, memory and non-representational geographies. *Geography Compass, 5*, 875–885.
Jones, P. (2012). Sensory indiscipline and affect: A study of commuter cycling. *Social & Cultural Geography, 13*, 645–658.
Jones, R., & Fowler, C. (2007). National élites, national masses: Oral history and the (re)production of the Welsh nation. *Social & Cultural Geography, 8*, 417–432.

Jones, E., & Wessely, S. (2005). *Shell shock to PTSD: Military psychiatry from 1900 to the Gulf War*. London: Psychology Press.
Korsmeyer, C. (2005). *The taste culture reader: Experiencing food and drink*. Oxford: Berg.
Kwan, M. P. (2008). From oral histories to visual narratives: Re-presenting the post-September 11 experiences of the muslim women in the USA. *Social & Cultural Geography, 9*, 653–669.
Lahiri, S. (2011). Remembering the city: Translocality and the senses. *Social & Cultural Geography, 12*, 855–869.
Lauwrens, J. (2012). Welcome to the revolution: The sensory turn and art history. *Journal of Art Historiography, 7*, 1. https://arthistoriography.files.wordpress.com/2012/12/lauwrens.pdf. Accessed 21 Mar 2019.
Lea, J. (2008). Retreating to nature: Rethinking 'therapeutic landscapes'. *Area, 40*, 90–98.
Lorimer, H. (2003). Telling small stories: Spaces of knowledge and the practice of geography. *Transactions of the Institute of British Geographers, 28*, 197–217.
Low, K. E. (2012). The social life of the senses: Charting directions. *Sociology Compass, 6*, 271–282.
Malpas, J. (2012). Putting space in place: Philosophical topography and relational geography. *Environment and Planning D, 30*, 226–242.
Mason, J., & Davies, K. (2009). Coming to our senses? a critical approach to sensory methodology. *Qualitative Research, 9*, 587–603.
Matless, D. (2005). Sonic geography in a nature region. *Social & Cultural Geography, 6*, 745–766.
Matless, D., Oldfield, J., & Swain, A. (2007). Encountering soviet geography: Oral histories of british geographical studies of the USSR and Eastern Europe 1945–1991. *Social & Cultural Geography, 8*, 353–372.
McHugh, K. E. (2009). Movement, memory, landscape: An excursion in non-representational thought. *GeoJournal, 74*, 209–218.
McNally, R. J. (2005). *Remembering trauma*. Cambridge: Harvard University Press.
Middleton, J. (2010). Sense and the city: Exploring the embodied geographies of urban walking. *Social & Cultural Geography, 11*, 575–596.
Miles, M., & Crush, J. (1993). Personal narratives as interactive texts: Collecting and interpreting migrant life-histories. *The Professional Geographer, 45*, 84–94.
Morris, G. (2002). Asking personal questions. *Oral History Review, 29*, 57–60.
Ní Laoire, C. (2007). To name or not to name: Reflections on the use of anonymity in an oral archive of migrant life narratives. *Social & Cultural Geography, 8*, 373–390.
Norquay, N. (1999). Identity and forgetting. *Oral History Review, 26*, 1–22.
Perks, R., & Thomson, A. (2006). *The oral history reader*. London: Routledge.
Pink, S. (2015). *Doing sensory ethnography*. London: Sage.
Portelli, A. (1991). *The death of Luigi Rrastulli and other stories: Form and meaning in oral history*. Albany: State University of New York Press.
Rachman, S., & Taylor, S. (1993). Analyses of claustrophobia. *Journal of Anxiety Disorders, 7*, 281–291.
Reicher, S. D. (2001). The psychology of crowd dynamics. In M. A. Hogg & S. Tindale (Eds.), *Blackwell handbook of social psychology: Group processes* (pp. 182–208). Malden: Blackwell.
Riley, M., & Harvey, D. (2007a). Talking geography: On oral history and the practice of geography. *Social & Cultural Geography, 8*, 345–351.
Riley, M., & Harvey, D. (2007b). Oral histories, farm practice and uncovering meaning in the countryside. *Social & Cultural Geography, 8*, 391–415.
Rycroft, S. (2007). Towards an historical geography of nonrepresentation: Making the countercultural subject in the 1960s. *Social & Cultural Geography, 8*, 615–633.
Skinner, M. W., Cloutier, D., & Andrews, G. J. (2015). Geographies of ageing: Progress and possibilities after two decades of change. *Progress in Human Geography, 39*, 776–799.
Smith, M. M. (2007). Producing sense, consuming sense, making sense: Perils and prospects for sensory history. *Journal of Social History, 40*, 841–858.
Smith, M. M. (2015). *The smell of battle, the taste of siege: A sensory history of the Civil War*. New York: Oxford University Press.

Solomon, H. (2011). Affective journeys: The emotional structuring of medical tourism in India. *Anthropology & Medicine, 18*, 105–118.

Sparkes, A. C. (Ed.). (2017). *Seeking the senses in physical culture: Sensuous scholarship in action*. London: Routledge.

Sparkes, A. C., Brighton, J., & Inckle, K. (2018). 'It's a part of me': An ethnographic exploration of becoming a disabled sporting cyborg following spinal cord injury. *Qualitative Research in Sport, Exercise and Health, 10*, 151–166.

Thomson, A. (2003). Sharing authority: Oral history and the collaborative process: Introduction. *Oral History Review, 30*, 23–26.

Thomson, A. (2007). Four paradigm transformations in oral history. *Oral History Review, 34*, 49–70.

Thomson, A. (2008). Oral history and community history in Britain: Personal and critical reflections on twenty-five years of continuity and change. *Oral History, 36*, 95–104.

Thrift, N. (2008). *Non-representational theory: Space, politics, affect*. London: Routledge.

Tomalin, N., & Hall, R. (2017). *The strange last voyage of Donald Crowhurst*. London: Quercus.

Yow, V. R. (2014). *Recording oral history: A guide for the humanities and social sciences*. Lanham: Rowman & Littlefield.

Chapter 3
Bodies at the Crossroads Between Immigration and Health

Anne-Cécile Hoyez, Clélia Gasquet-Blanchard, and François Lepage

Abstract There is a growing concern for the intensification of health problems among immigrants who have recently arrived in European countries in relation to the deleterious impacts of the poor living conditions in which they find themselves and the constraints on accessing healthcare services. Taking a biopolitical approach in which immigrants' bodies are not passive vectors and receptors of illness, we question why the immigrant body figures in our social worlds as the sign of a difference. In order to redraw immigrant pathways of healthcare and healing in the light of their daily contexts and realities, we privilege a multidisciplinary approach, connecting social geography with humanities disciplines and mixing different research materials arising from ethnographic research work and photography. We first illustrate how immigrant women re-appropriate their own bodies and analyse their place in the world as socially constructed objects within a social space in which carers, social workers, women themselves and their relatives and families are shaped and emerge together. We then expand our reflections on the subject of immigration as social geographers by adding a new path through photography to our "conventional" qualitative methodological tools in order to explore emotions and body concerns. By going beyond the argumentative role of research in the human sciences, and by going beyond scientific postures seeking to "speak for", we have tried to bring something other than scientific proof or critical testimony about pregnancy and maternity in migration.

Keywords Immigrants · Pregnancy · Bodies · Photography · France

---

A.-C. Hoyez (✉)
Université de Rennes, Unité Mixte de Recherche – Espaces et Sociétés (UMR-ESO), Rennes, France
e-mail: anne-cecile.hoyez@univ-rennes2.fr

C. Gasquet-Blanchard
École des Hautes Études en Santé Publique (EHESP), Saint-Denis, France
e-mail: clelia.gasquet@ehesp.fr

F. Lepage
Photographer and photojournalist, Paris, France

S. Atkinson, R. Hunt (eds.), *GeoHumanities and Health*, Global Perspectives on Health Geography, https://doi.org/10.1007/978-3-030-21406-7_3

As an introduction to our reflection, in the field of geo-humanities (Dear et al. 2011; Atkinson et al 2015a, 2015b), on healthcare access for immigrants in Europe, we would like to relate an anecdote which happened, in the field, to one author of this chapter. This anecdote tends to indicate that there are times when research on healthcare and the body can impact the researcher in a way that possibly prompts him/her to approach his/her object of study with more humanity. The researcher in question had made an appointment a few days before to conduct an interview with the director of a support centre for immigrants. As she arrived at the centre on the agreed date and time, the researcher, who was 8 months pregnant at the time, was greeted as follows:

> Oh…but it's you! my colleague told me "there's a pregnant woman here who wants to see you". I got worried!!! I thought for a minute that it was another pregnant woman from Eastern Europe who'd come here to give birth! [she laughs]. (Director of a support centre for immigrants, Rennes, June 2008)

At this time when the researcher first met the director in question, she had just begun her fieldwork. The research involved immersing herself in social welfare centres with a view to gaining a better understanding of the conditions of access to healthcare among immigrants in France and focusing attention both on the immigrants themselves and on the working conditions of professionals and volunteers operating in various support centres. That day, the director's reaction to the researcher's (pregnant) body is indicative of stereotypical representations of all kinds applied cumulatively by professionals to immigrants and which would not normally be revealed in the context of an interview. In this particular place – the support centre for immigrants – a pregnant, white woman could only be an Eastern European immigrant; the researcher embodies the figure of female immigration in France. Clearly, beyond the stereotypes and ethical questions raised by this situation (which it is not our place to judge here), this fieldwork prompted an in-depth reflection on how we should conduct our research: should we concentrate primarily on the professionals' "deficiencies", which are representative of the difficulties they experience on a daily basis in managing the humanitarian emergencies and difficult situations that hinder their work and make it more challenging? Should we bring to light and analyse the identity labels applied to immigrants in those support and healthcare centres? It was above all the researcher's physical appearance which caused the director to assume she was an East European migrant. This leads us straight to the question of the link between immigration and health in Europe today: what causes carers and social workers to "fear" the arrival of immigrants, especially pregnant immigrant women? Is it the fear of an uncontrollable "wave of immigration"? Is it the fear that this would place an additional financial burden on public services? Is it the anxiety about the growing number of very complex situations in which individuals face a combination of challenges such as precarity, exile, poverty, uncertainty concerning their legal status and the prospect of a growing family? Is it a combination of all three factors? At the same time, professionals in the field express significant concerns and conflicts related to the impact of restrictive public policies, to the increasingly tight control of public spending and to the management of populations (including immigrants).

In Europe, contexts of immigration tend to be analysed in terms of "crisis", managed in a temporality of urgency in different parts of the territory. At the same time, the public sector itself, including health and social welfare, is in "crisis" as a result of major economic restructuring marked by severe budgetary restrictions and sector liberalization. But are we dealing with "immigration crisis" or an "immigrant reception crisis"? Who or what is actually posing a threat to the healthcare and social welfare systems? Is it the arrival of immigrants or the neo-liberal economic policies and programmes? Moreover, we know from research on migration that the routes used by immigrants to reach Europe are increasingly dangerous; a significant number of NGOs report an increasing number of deaths at Europe's borders as well as an intensification of health problems among recently arrived immigrants to European countries. In addition, several researchers have highlighted the harmful consequences for health and well-being of the precarious living conditions of newcomers once in Europe (Thomas 2016). However, as Didier Fassin (2001) points out, scholars should bear in mind the fact that, in relation to these alarming finding, an immigrant's body is not merely a passive vector or recipient of disease. In this regard, part of our work as researchers is to examine (1) what is involved when dealing with the body, disease and suffering and (2) how, in our social worlds, immigrants' bodies serve as a marker of difference.

In this chapter, we will explore the complexity of the situations which, we, as health geographers specialized in migration issues, are confronted with by gradually shifting from a social geography to a geo-humanities approach. In our research, we aimed at stepping away from "describing" with words towards showing the way human conditions are in places. In this way, we wanted to connect our knowledge as social scientists with humanities disciplines. We chose to develop this through arts, for their potential in creativity and engagement, and especially with photography. Using an approach based on the human body, what can we learn about social inequalities and their management by nation states? What do we know about the healthcare pathways of immigrants and about their access to healthcare in their everyday lives (Hoyez and Thomas 2016)? What method would make it possible to examine and analyse these phenomena while taking greater account of the human dimension in the processing of scientific information? This chapter is the result of a research project which investigated access to healthcare for immigrants in France, in which we took into account the situations of both ill people and non-ill people whose precarious living conditions impact their health or whose life events require that they receive medical and social attention, as in the case of pregnancy, to which we will pay particular attention here. In a context where migratory pathways are stretching (both in time and space), and where the number of female immigrants is growing, we explore how women manage to empower themselves, to overcome the identity labels applied to them throughout their journey and which, for the most part, impact them negatively. Within their life histories and immigration pathways, it is their bodies that are at stake. In the first step of this chapter, we illustrate how immigrant women re-appropriate their own bodies and analyse their place, as socially constructed objects, in constituting the social space in which carers, social workers, women themselves and their relatives and families are shaped and emerge together. In a second step, we expand our reflections on the subject of immigration as social geographers by adding a new path through photography to our "conventional" qualitative methodological tools in order to explore emotions and body concerns.

Immigrant Women's Bodies in a Context of Restrictive Policies

The pregnant women we met in the course of our research have very different histories: some fled a country at war, both alone or with their male partner or family, and may have either joined relatives in France or had no contacts in this country, France not necessarily having been their intended destination country initially. Other women moved to join a spouse who had already come to work in France, while others came to pursue higher education. Whatever the socio-economic context in which they found themselves, their individual migration is undertaken as part of a collective destiny shared with parents or even children left behind or close ones scattered in different countries. All mention the impact on their everyday experiences and affects of the spatial dimension of this situation of, on the one hand, separation from part or all of their families and/or, on the other, of their moving closer to other family members, the coming together of siblings who had previously been living far from one another or the development of strong friendships in the host country. These experiences of distance and mobility mark each stage of their existence.

Migration journeys can be long, costly, dangerous and punctuated by a varying number of short stays in cities or other places that serve as more or less chosen stages. During these journey stages, they may have been subjected to sexual violence through rape, kidnapping, domestic slavery, excision, insults or discriminatory attitudes, and these are experiences unlikely to emerge and difficult to identify in an interview situation. This raises the question of how we, as researchers and likely the umpteenth people asking for testimonies, intrude into the most troubled realms of these immigrants. Moreover, once women arrive in France, they face other gender-related problems in terms of the highly fragmented labour market which offers few job opportunities for them; even those who are hired are always, initially at least, given underqualified jobs (Schmoll 2011). However, amidst the variety of female migratory trajectories, one specific figure stands out as being both sometimes particularly exposed and sometimes particularly protected compared with others that of pregnant women or women with children. The stages and events that punctuate the trajectories of pregnant women can be analysed by focusing our attention on their body and its transformations as well as on their emotions.

Those women are in an ambivalent position. They are more likely to be exposed to gender-based violence or to processes of differentiation or social downgrading, but they are also the object of specific policies, such as those for family protection and child protection, whereby they can receive some welfare benefits to which men are not entitled. Illegal immigrant women – these "precarious subjects" who are not recognized by the system (Butler 2004) – are more likely to fall victim to differentiation or discrimination processes, or even to be excluded from certain areas of the social world, especially in public spaces and in professional worlds (Fournand 2009). From an intersectional point of view, they are, more than other women, subjected to domination through racial/ethnic, gender and social stereotyping. However, pregnancy strongly impacts the conditions in which these women live. Pregnancy gives these women access to common rights, which will influence their living conditions while they are pregnant and throughout their child's infancy (Virole 2016). Women themselves must apply for those services, but nonetheless, the support they can then access is designed to empower them and will help them to be recognized by healthcare professionals and institutions as subjects with rights in ways not the case before pregnancy. Through these support programmes, pregnant women benefit from regular pre- and postnatal care services, which many of them had never experienced in their country of origin. It is our responsibility, as researchers, to examine the reasons and processes through which pregnancy and motherhood help women be recognized as subjects with rights, and to emphasize that women's pregnant bodies and infants can help them to assert themselves in their lives as women.

The migratory experiences of these immigrant women, as they related them to us, gave rise to many discussions about opportunities to change their circumstances. Their stories are marked by various shifts and major changes in their situation that occurred over a relatively short period of time when they became pregnant. Such is the case of Milly who, now that she was pregnant, knew that she would be given priority on the waiting list for housing and would have stable accommodation during the last two trimesters of her pregnancy and the first months of her child's life:

> For me it's okay now, with the baby, I'm going to be entitled to housing when I get out of here. But what will happen to Ahmed? He is all alone, no child, no housing. It is for him that we must fight now.

She is aware of the fact that her personal, intimate, family event is, in fact, a social event that marks the end of years of residential precariousness, during which she, as a single woman, had never had priority on the housing waiting list.

The bodies of these immigrant pregnant women become a place, a geographic level at which to examine and report on individual experiences and on how corporealities are taken into account in institutions. The body is then placed at the core of social, institutional and political issues through control of the migrant body. Obviously, the way in which pregnant immigrant women are treated and cared for is tinged with moral overtones. Indeed, as noted in our introduction to this chapter, pregnant immigrant women can be labelled with derogatory stereotypes. There is a belief, as displayed in the opening vignette, that some of them come to France to give birth in order to take advantage of the supposedly more supportive social welfare system, but nonetheless, whatever happens, most will be helped and supported and, if they are not, public controversy will ensue. Moreover, the women use a number of techniques that contradict the common notion that, because of their conditions as immigrants living in precarity, they lack any capacity for action (Fournand 2008). For pregnancy follow-ups, ultrasound readings, giving an opinion on the possibility of an epidural, planning to breastfeed or not, feeding a very young child, carrying children, etc., these women have proved very resourceful. The arrival of a child is mostly experienced as a blessing, and the descriptions we were given of the mother-child relationship are based on a large variety of emotions and sensitivities, which pleads for a sensitive and humane approach to inequalities and discrimination.

Bodies Combining Disadvantages: Pregnant, Undocumented and Poor Women

Several studies have been conducted about social and health inequalities, particularly those that develop during pregnancy for women living in precarity. Precarity is not the only prism through which the trajectories of pregnant immigrant women are examined, but it is an important part of the realities encountered (see Box 3.1). Addressing precarity experienced by these women is not just a matter of describing their economic and social living conditions or of assessing their access to healthcare rights and to stable housing. However, studying these two events (pregnancy and migration) when they occur within the same temporal framework requires that we situate our remarks in this context, so as to better analyse the effects of precarity on pregnancies and their outcomes, to better identify what is universal in the situation of precarity during pregnancy and what turns out to be different for immigrant women.

Box 3.1 Field Data

The NGO Médecins du Monde (2016, 2017) points out in its recent reports that the precarious administrative situation of women puts them at greater risk: 18.1% are asylum seekers and 52% are undocumented. The pregnant women we met in MdM healthcare centres and their partner clinics have very limited or no access to healthcare coverage (68%) and consequently to prenatal care (over 40% had no access to prenatal care in 2016 and 58.4% in 2017, and 45.5% of pregnant women do not receive regular prenatal care). Their living conditions have particularly harmful effects on their health: the majority of them indicate that they are living with a third party (55.6%), and it is not known what this type of living arrangement involves, 8.1% are homeless, and 48.4% consider that they live in temporary and unstable accommodation. Finally, women who have received care at an MdM clinic report that they go out as little as possible for fear of being arrested. Thus, the fear of being arrested by the police acts as a further significant obstacle to their access to healthcare in specialized centres.

In the medical field, precarity, for all populations combined, is known to be associated with an increase in pregnancy and neonatal complications and with an increase in the length of hospital stays (Gayral-Taminh 2005). There is a correlation between precarity and substandard or non-existent prenatal care and therefore the increase in perinatal complications (Lejeune 2008). As seen in Box 3.1, women's situation as immigrants can further contribute to the difficulties and vulnerability associated with precarity in general. Several studies conducted in northern countries show that immigrant populations originating from Asian, North African and sub-Saharan African have a higher risk of neonatal mortality and of prematurity compared with non-immigrant populations and immigrants of other origins (Gagnon et al. 2009; Gissler et al. 2009; Malin and Gissler 2009; Racapé 2010). In addition, the birth of a child causes significant social, economic and psychological changes that place the mother in an even more vulnerable position, especially if it occurs in a situation of migration (Battaglini 2002). This is reinforced in the context of violence that currently exists in many of the regions immigrants have to cross in order to get to their destination (Médecins du Monde 2016, 2017; Report of the Samusocial Observatory 2018).

The negative effect of precarity aggravates the illnesses that can occur during pregnancy in terms of care through the possibilities for effective and efficient use of the healthcare system. In France, the National Commission on Child Birth and Health (CNNSE, 2014) highlights that women living in precarity use the healthcare system less and later. Several reasons explain this: (i) a late initiation of prenatal care involving late diagnosis of pregnancy, fewer prenatal consultations, fewer ultrasounds and so forth due to the daily difficulties encountered by immigrant women living in precarity, difficulties which often prevent women from being aware of or paying attention to the early symptoms of pregnancy; (ii) irregular attendance at prenatal care through the difficulties for women in attending all the scheduled appointments or to have additional examinations performed, particularly if this entails attending a different unit other than the clinic they normally attend, due to their poor knowledge of an area in which they have only recently arrived; and (iii) frequent use of hospital emergency services, given these are the only places immigrant women can identify easily.

These difficulties are among the general factors of disadvantage associated with characteristics of the socio-spatial contexts in which women operate: poor knowledge of the area and of the location of the healthcare facilities, geographical isolation, socio-economic context of the area of residence, use of specialized support services, preference for carers who share the same geographic origins or the same language and can advise in favour or against different types of consultation, support provided through personal networks of friends or family who may or may not help them familiarize themselves with healthcare facilities, and, finally, all the characteristics of the person including age at the time of pregnancy, education, income, single parent or not and physical and psychological health (Battaglini 2002:2). These factors are combined with those that affect immigrant women more specifically including social (language, support and presence of the partner, family and friends), financial (the type of financial resources available to the woman) and administrative (the type of health insurance, type of residence document). The situation of vulnerability in which women find themselves – mentioned in other studies (Médecins du Monde 2015) – has proved recurrent, although not systematic, among the women we met.

The women we met experienced, after their pregnancy was confirmed, different changes in their social lives. Some of them mentioned going through marital and family breakups, which often caused them to leave their place of residence. At the same time, others reported developing a closer relationship with their spouse. The social network is the first line of protection against the immediate effects of a marital breakdown, but where networks are geographically dispersed, the women must navigate their pregnancy within a fragmented social world. Even when surrounded by relatives, these women can sometimes be exposed to one or more risks due to their status as an isolated immigrant dependant on a defined group of family and acquaintances. For example, the pregnant women who receive medical care from the Paris Samu Social (emergency social services) often report a high level of social of isolation exactly at a time when moral support is essential and which can further hinder their access to healthcare. These risks are combined with the mobility limitations experienced by pregnant women and mothers of small children (Fournand 2009). This isolation can be observed through a distance/proximity dialectic. Geographic distance from one's home country can foster proximity with one's compatriots in the host country, a proximity which would not have developed in other circumstances, and which can contribute to the development of new forms of social relationships, or even of new functions in the family, as illustrated by the roles taken on by some fathers, roles they would not have had in their country of origin.

Having moved away from their usual social network, these women find themselves separated from those who would have been able to give them the psychological and practical support they needed as pregnant women, such as a mother, sisters, spouse and friends. They then surround themselves with, and find resources and support among, their new neighbours, compatriots and caregivers, which enables them to recreate a safe environment for the duration of their pregnancy.

Risk factors also include the difficulties of communication between pregnant immigrant women and carers. These difficulties are not only related to the patient's level of proficiency in the language but also to the contexts of communication with the medical and social care teams. The information exchanged between the various team members must include not only information about the patient's healthcare history but also about other important dimensions of her life, such as physical and psychological health, social condition, food and housing. These communication difficulties can lead to delays in the provision of care, or to inappropriate treatments, and to a poor understanding or ignorance of the healthcare system. Among pregnant immigrant women, we have found that certain barriers prevent them from receiving optimal healthcare, which can be reinforced by personal and family-related vulnerabilities that may lead to dysfunctions in the natural birth process and even in the establishment of a bond between parent and child (Davoudian 2007). Finally, the difficulties linked to the negative stereotypes and representations of pregnant immigrant women applied by health professionals must be taken into account. Negative representations of patients with unstable residence permits can lead to the denial of treatment by carers. More generally, the medical profession has been known to engage in processes of ethnicization or racialization of women, which can lead to inadequate and differential care and, thereby, increase the risks of complications (Sauvegrain et al. 2017).

Geo-Humanities: Building a Bridge Between Geographies and Art Photography to Replace Women's Bodies in Sensitive Contexts

Geographers have for a long time developed approaches for observing and analysing the interrelations between spatial configurations and social phenomena. Several methodologies are often combined and used to account for those interactions, including approaches borrowed from ethnography and sociology. In our qualitative research, we conducted a large number of in-depth interviews, in situ observations and collections of visuals, including photographs. Yet, inexperienced in this art form and reduced to shooting photographs in a high-speed continuous mode with our pocket digital devices, we collected a few visual documents that could be used in publications or that could be given back to participants. However, in 2018, we had the opportunity to collaborate with François Lepage, a photographer and author on immigration issues, and this was just the push we needed to move towards a geo-humanities approach. In parallel with our field observations, which are fairly typical in the field of social geography, we have developed an interest in artistic approaches and the conversations, experiences and insights they can produce. For a year, while nearly 200 immigrants lived in a squat, which was eventually legalized (Hoyez et al. 2019), we conducted surveys and observations among the inhabitants of this place to fully understand the sensitive and everyday aspects of life in a squat and in precarious accommodation. Meanwhile, François Lepage, who for several years has turned his lens towards migration phenomena, set up a photo studio and invited the squat residents to come and have their photos taken whenever they wanted (see Box 3.2). We worked in parallel with François, interviewing and photographing women. By paying close attention, together, to the faces of the women, men, children and families who lived in this place, we gradually adopted a humanities approach to our research through artistic engagements. These were not the snaps that we captured with our pocket devices; these images were more, said more and produced very different encounters. These "art" photographs brought something extra to the research. They helped to go beyond the approaches commonly used in the social sciences to examine social and spatial inequalities in healthcare, discriminatory processes affecting immigrant women and their partners and children or "French-style" pregnancy care pathways. These photographs bring us face to face with women, men, mothers and fathers of children who pose together proudly. Inevitably, they challenge the descriptive categorizations that we, as researchers, often apply. Could we still justifiably call the people we met "immigrants" in a generic way? No, we could not, because when we look at these photographs, we see, first and foremost, people, inhabitants, whose dwelling occupancy pattern reminds us of the foundations of the Right to the City of H. Lefebvre (1968). The photography and the exchanges between the academic and artistic worlds have placed the human being at the heart of our analysis. This approach repositions emotions and affects in the foreground of research. In going beyond scientific protocols and reintegrating emotions, the photographic process gives pride of place to the reflexivity of researchers and to its impacts on the human sensitivity with which we conduct research work. The more significant change, however, was that, beyond these frameworks established between researchers and the artist, those people who were photographed intervened as actors in a collaborative process. François Lepage wished not only to "take photos" but also to "give pictures" to the inhabitants. The shots were in fact intended primarily to be distributed to those photographed, with the intention to share something concretely. This posture has distorted the conventional relationship between the photographer and the photographed, shifting it from a relationship of "observer/

observed", in which the act of photographing formally captures a particular moment, to an interactional relationship in which the protagonists are participants without having to answer a question or an order defined a priori. The photographer no longer stands as a professional of the image, but takes on the role of an artist who thwarts the materialist universe of photographs. In the same way, the charitable organizations working with those living in this place indicated to us that the photo studio was part of the "cultural" coordination organized for the inhabitants in that it was not a project to photograph everyone, but rather a proposal or an opportunity to have a photograph. These, then, are pictures of inhabitants, not just the photos of the photographer. This distinction is important because it emphasizes the ambition to make the inhabitants active in their presentation of themselves. They had a total freedom of pose and free choice of the day when they would show up as the studio was set up permanently. Here, photography is not only a social act or a way of meeting in one moment. It is, instead, a process of mutual trust and sharing, across the short, medium and long term, because the photography participates in the identity reconstruction of people, both for themselves and for those who come to see them through the associated photography exhibitions. The process re-humanizes identities that the media too often distance and even render invisible.

Box 3.2 François Lepage Presents His Photographic Approach

In the summer of 2016, I set up a photographic studio in the squat known as "Les Jardins de la Poterie" in Rennes. An ephemeral dwelling in which 170 exiled people lived for a year, half of them children.

At first, people pass by, a little surprised. "What are you doing"? – "Is it compulsory"? They leave and then come back. Alone, or with their families, they walk around the studio, dressed in their Sunday best and prepared. They assume their own poses.

I lose control of the photo shoot session. I go with the flow.

From the very first pictures, I am disturbed. The shot doesn't correspond to the image I had of them. Perhaps I had imagined them "unkempt and threatening" just as they would have been depicted on an anti-foreigner Nazi propaganda poster, or else I've been conditioned to see them that way. Under the flashes, their joy and light shine through. They remind me of Leonard Cohen: "There is a crack, a crack in everything, that's how the light gets in".

This photographic project has led to many encounters: with people from all corners of a planet fragmented by walls and barbed wire, from Mongolia to Congo, from Albania to Chechnya; with associations that give direction to human beings who have lost their bearings (at UTUD, A roof is a right; MRAP, and many others); and with social science researchers from the University of Rennes 2 and the Institut d'Études Politiques, who are working on "ephemeral housing".

Fig. 3.1 Young Afghan girl

The people who have been photographed have agreed to use the rendering for public exhibition purposes. All the people agreed, with the exception of one girl who especially wished not to be recognized by her high school classmates. She nevertheless accepted after having discussed with François Lepage who proposed to "veil" the photograph (Fig. 3.1).

The total freedom given on the shooting allowed us to probe the desires of use of photography by the inhabitants. In the short term, desires were very varied: the use of a new photo for a Facebook profile (Fig. 3.2), the wish to send a photo back to family who have remained in their country of origin and the display of a nice family photo as an everyday visible object (Figs. 3.3, 3.4, and 3.5). In the medium term, the permission to broadcast the photograph for public display was also supported. In the view of the inhabitants, it was necessary that their photographs circulate in order to testify to their conditions of existence as immigrants and, thereby, to be able to express themselves publicly on this subject. Most of the people took part in the events organized around the exhibitions through exhibition opening events, forums and debates, along with the charity organizations that accompany them on a daily basis. Thus, making their own photographs available at the exhibitions served as a medium for talking to observers, that is the photographer, the researchers and the wider audience, and connecting with them. Consequently, the act of presenting oneself physically, of going beyond the posture of the subject of photography and embodying immigration phenomena during various public events, brought a lot of meaning and spontaneity. Beyond the words of conversation, debates or exchanges were the moments of sharing food, songs and tales that emerged. For example, the first vernissage gave rise to some interesting scenes in terms of reversing social

Fig. 3.2 Young Afghan boy

Fig. 3.3 Mother and her son

Fig. 3.4 A family from Mongolia

roles in different places: the hall of the university library was filled with groups of children and bathed in the smell of donuts and fried chicken as the immigrants offered meals to the exhibition visitors for almost 2 hours. We heard a cheerful Congolese woman say to one of the vice presidents of the university: “Do you like my donuts? It’s good! Eat, you’re so skinny! And take, take all, you will give to your children”. This possibility of going beyond social roles was also appreciated by the academic staff, who were delighted to find themselves shaken up in their everyday functions by people who they otherwise might not have met.

During and following the exhibitions, of which there had been four at the time of writing in early 2019, the confrontation and the exchange with the public’s gaze allow us to go further in our work and our knowledge of the everyday experience of the inhabitants of the Jardins de la Poterie. The series of events around the exhibition in different public places gave us the opportunity to take the time to focus on details that might have eluded us. By taking part in public debates, we could note, for example, that, even as people posed during the photography shoots, they would hold or touch each other, in search of contact, and were all held to each other during the shooting and that the children did not come exclusively to be photographed with

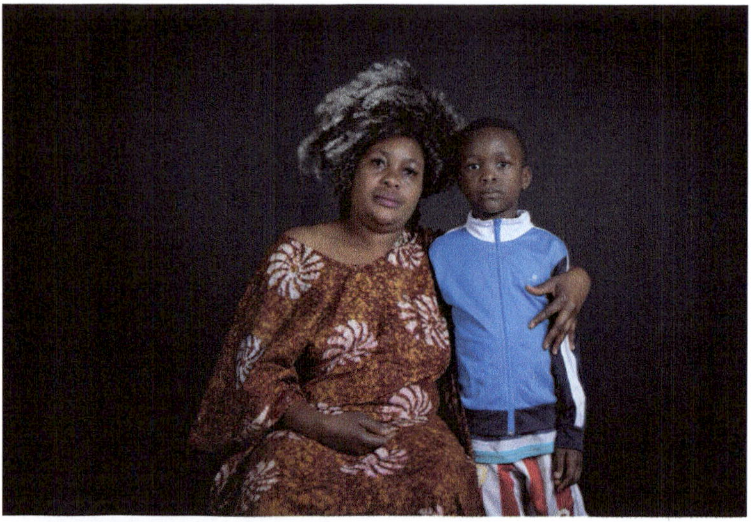

Fig. 3.5 Mother and her son

Fig. 3.6 Two young friends

family members but with friends (Fig. 3.6). This enjoins us to study more precisely the body-sensitive relationships, interpersonal relationships and/or collective feelings of belonging.

For the inhabitants, the photographic approach was greeted positively, mainly because it did not impose a frame that obliged the residents to "give" something a priori and because it did not answer an institutional order put in advance. It is important, however, that this issue is treated ethically by photographers and researchers. Indeed, if the approach was initially disinterested financially and free from any institutional injunction, the "fundraising" which allowed the circulation of the exhibition had to respect the political and critical positions of the inhabitants, the photographer and the researchers. Our financial support has been found in public university institutions that claim to carry out research on the hospitality, reception and integration of all people within the territory, in this instance, France, and by charitable organizations supporting immigrants and homeless people. It is not conceivable, in the long term, that this research is conducted, for example, through European Union institutional funds, a body considered as abusive by immigrants not only in the European territory but also in their countries or regions of origin. Research, creation and critical intent cannot be dissociated in this type of research in the geo-humanities.

Conclusion

In this chapter, we have traced the passage from a "conventional" research in social geography to a research methodology enriched through the contributions of an artistic approach. By going beyond the argumentative role of research in the human sciences and by going beyond scientific postures seeking to "speak for", we have tried to bring something other than scientific proof or critical testimony about pregnancy and maternity in migration. We have also sought to trigger patterns of space and places hidden by academic traditions and introduce them into the artistic posture.

This collaborative work opens new perspectives for research, but also for the art world, which could develop more "geographically aware" approaches (Creswell, Dixon 2017). Indeed, the "spatial shift" does not only concern the social sciences but also the artistic world, insofar as we share common interests, particularly in the influence of places on the lives of individuals, their mobility, the relation of individuals and groups to their territories and the multi-scalar approach. Through the example of pregnancy, we show how the woman's body appears as the first envelope of the developing human being. This has enabled us to explore the relations between the accessibility of spaces and the female experience of pregnancy, understood as a unique experience occurring in a limited temporality, that of pregnancy. This temporality, nevertheless, makes it possible to re-examine the bodily experience (change in the centre of gravity) in physical and psychological terms (fulfilling pregnancy, cumbersome pregnancy) and its relationship with the spatial experience. We have used here a sensitive approach to a bodily modification that alters one's relation to space as the modified body changes at the same time as the woman's direct relation to space, to places in terms of the places of well-being which are therefore regularly visited or the converse of anxiogenic places that are therefore avoided and to territories through the policies for the provision of prenatal healthcare assistance to women and particularly to immigrant women. This sensitive approach helps us to re-examine and analyse our data via the women's accounts while at the same time examining the investigative relationship and the researchers' own involvement.

References

Atkinson, S., Evans, B., Woods, A., & Kearns, R. (2015a). The 'medical' and 'health' in a critical medical humanities. *Journal of Medical Humanities, 36*, 71–81.

Atkinson, S., Foley, R., & Parr, H. (2015b). Introduction: Spatial perspectives and medical humanities. *Journal of Medical Humanities, 36*, 1–4.

Battaglini, A. (2002). « Quand migration et maternité se croisent: perspectives des intervenantes et des mères immigrantes », Service social, n°1, pp. 35–69.

Butler, J. (2004). *Precarious life: The powers of mourning and violence*. Verso.

CNNSE. (2014). Commission nationale de la naissance et de la santé de l'enfant, Organisation de la prise en charge et de l'accompagnement des femmes en situation de précarité ou de vulnerabilité, 64 p.

Cresswell, T., & Dixon, D. P. (2017). GeoHumanities. In D. Richardson, N. Castree, M. F. Goodchild, A. Kobayashi, W. Liu, & R. A. Marston (Eds.), *International encyclopedia of geography: People, the earth, environment and technology*. https://doi.org/10.1002/9781118786352.wbieg1169.

Davoudian, C. (2007). Maternité d'exil. *La clinique lacanienne, 12*(1), 69–73. https://doi.org/10.3917/cla.012.0069.

Dear, M., Ketchum, J., Luria, S., & Richardson, D. (Eds.). (2011). *GeoHumanities: Art, history, text at the edge of place*. New York: Routledge.

Fassin, D. (2001). Quand le corps fait loi. la raison humanitaire dans les procédures de régularisation des étrangers. *Sciences Sociales et Santé, 19*, 5–34.

Fournand, A. (2008). Le corps des femmes, enjeu géopolitique. *Géographie et Cultures, 68*, 63–80. http://gc.revues.org/907. Accessed 21 Mar 2019.

Fournand, A. (2009). La femme enceinte, la jeune mère et son bébé dans l'espace public. *Géographie et Cultures, 70*, 79–98. http://gc.revues.org/2320. Accessed 21 Mar 2019.

Gagnon, A. J., Zimbeck, M., & Zeitlin, J. (2009). Migration to western industrialised countries and perinatal health: A systematic review. *Social Science & Medicine, 69*, 934–946.

Gayral-Taminh, M. (2005). Caractéristiques socio-démographiques et risques périnatals des mères en situation de précarité. *Journal de Gynécologie Obstétrique et Biologie de la Reproduction, 34*, 23–32.

Gissler, M., Alexander, S., Macfarlane, A., Small, R., Stray-Pedersen, B., Zeitlin, J., et al. (2009). Stillbirths and infant deaths among migrants in industrialized countries. *Acta Obstetricia et Gynecologica Scandinavica, 88*, 134–148.

Hoyez, A.-C., & Thomas, F. (2016). Socio-spatial dimensions of healthcare for newly arrived migrants. In F. Thomas (Ed.), *Handbook of migration and health* (pp. 158–172). Cheltenham: Edward Elgar.

Hoyez A.-C., Gasquet-Blanchard C., Bergeon C. (2019). Health system and immigrants – a focus on urban France. In G. De Verteuil (dir) (Ed.), *Handbook of global urban health*. Taylor and Francis.

Lefebvre, H. (1968). *Le droit à la ville*. Paris: Anthropos, Ed. Du Seuil, Coll. Points.

Lejeune, C. (2008). Précarité et prématurité. *Journal de Pédiatrie et de Puériculture, 21*(8), 344–348.

Malin, M., & Gissler, M. (2009). Maternal care and birth outcomes among ethnic minority women in Finland. *BMC Public Health, 9*, 84.

Médecins du Monde. (2015). New report on access to healthcare for the most vulnerable in a europe in social crisis. Focus on pregnant women and children. *Médecins du Monde*. https://mdmeuroblog.files.wordpress.com/2014/08/mdm-access-to-healthcare-europe-2014-def.pdf. Accessed 21 Mar 2019.

Médecins du Monde. (2016). Synthèse du rapport de l'observatoire de l'accès aux droits et aux soins de la mission France. 11 pages.

Médecins du Monde. (2017). Synthèse du rapport de l'observatoire de l'accès aux droits et aux soins de la mission France. 11 pages.

Racapé, J. (2010). Inégalités sociales de santé périnatale des mères migrantes en Belgique. Revue de santé publique et d'épidémiologie, 64–2.

Report of the Samusocial observatory. (2018). L'accès aux soins des « migrants » en Île-de-France Une enquête auprès des centres d'hébergement d'urgence migrants en Île-de-France, au printemps 2017. (Isabelle Siffert, Aude Cordone, Mégane Réginal et Erwan Le Méner) Observatoire du Samusocial de Paris, 125 p.

Schmoll, C. (2011). Le corps des autres dans l'Europe des frontières. *Outis. Revue de Philosophie (Post-)Européenne, 1*, 121–128.

Sauvegrain, P., Azria, E., Chiesa-Dubruille, C., & Deneux-Tharaux, C. (2017). Exploring the hypothesis of differential care for African immigrant and native women in France with hypertensive disorders during pregnancy: A qualitative study. *BJOG : An International Journal of Obstetrics and Gynaecology, 124*(12), 1858–1865.

Thomas, F. (Ed.). (2016). *Handbook of migration and health*. Cheltenham: Edward Elgar.

Virole, L. (2016). Devenir mère, devenir sujet? parcours de femmes enceintes sans-papiers en France. *Genre, Sexualité & Société, 16*. http://journals.openedition.org/gss/3862. Accessed 21 Mar 2019.

Chapter 4
Beyond Therapy: Exploring the Potential of Sharing Dance to Improve Social Inclusion for People Living with Dementia

Rachel Herron, Mark Skinner, Pia Kontos, Verena Menec, and Rachel Bar

Abstract A growing body of research suggests that the arts (e.g. music, dance, painting) can improve the lives of people living with dementia. Much of this research has focused on measuring the impact of arts-based programs on "behaviour", cognition, and emotional states. Given the positive health outcomes that have been demonstrated, the instrumental use of the arts to generate social and behavioural changes is increasingly adopted as a non-pharmacological approach to managing and/or improving behaviour, cognition, and emotional states in dementia care. Little attention has been given to other benefits of engaging with the arts such as empowerment, embodied self-expression, and sociability. In this essay, we examine the potential of an innovative dance program developed by Canada's National Ballet School and Baycrest Health Sciences, as a holistic approach to enriching the lives of people with dementia and their carers. Drawing on a multi-method pilot study

R. Herron (✉)
Assistant Professor, Department of Geography, Brandon University, Brandon, MB, Canada
e-mail: HerronR@ubrandon.ca

M. Skinner
Professor, Trent School of the Environment, Trent University, Peterborough, NH, Canada
e-mail: markskinner@utrent.ca

P. Kontos
Professor, Toronto Rehabilitation Institute – University Health Network, Toronto, ON, Canada
e-mail: pia.kontos@uhn.ca

V. Menec
Professor, Department of Community Health Sciences, University of Manitoba, Winnipeg, MB, Canada
e-mail: verena.menec@umanitoba.ca

R. Bar
PhD Candidate, Ryerson University, Canada and Canada's National Ballet School, Manager, Health and Research Initiatives, Toronto, ON, Canada
e-mail: rbar@ubs-enb.ca

S. Atkinson, R. Hunt (eds.), *GeoHumanities and Health*, Global Perspectives on Health Geography, https://doi.org/10.1007/978-3-030-21406-7_4

employing observations, diaries, focus groups and interviews, we explore how dance can foster a relational environment that fully supports embodied selfhood.

Keywords Arts · Ageing · Empowerment · Sociability · Canada

Introduction

Dementia is a progressive neurodegenerative condition that affects multiple cognitive domains including memory, language, attention, and behaviour. It is estimated that 46.8 million people are living with some form of dementia worldwide (Alzheimer's Disease International 2015). As the population ages, the number of people living with dementia is increasing and so too is the need for programs and environments that enable people living with dementia to express themselves, relate to others, and live well. Research on arts-based health programs (such as music, dance, and painting) have demonstrated their value in terms of improving physical functioning, mood, balance, and cognition among people living with dementia (Abreu and Hartley 2013; Blankevoort et al. 2010; Coubard et al. 2011); however, only a very small number of studies have begun to explore the transformative potential of the arts in terms of challenging dominant ideas about dementia and supporting individual's diverse capacities to participate (Dupuis et al. 2016; Kontos et al. 2017). This area of inquiry is important given the lack of opportunities for meaningful engagement and participation that people living with dementia often face across a range of settings.

Negative stereotypes that assume people living with dementia are incapable of learning and developing, and making meaningful contributions to their own lives and the lives of others, foster social exclusion and marginalization (Kontos et al. 2018). Many people living with dementia internalize such stigmatizing views and withdraw from social activities, fearing the embarrassment of making a mistake (Fortune and McKeown 2016). These experiences of exclusion are often exacerbated by broader structural barriers to participation such as a lack of public transportation and other infrastructure supporting the mobility, safety, and inclusion of people living with dementia (Clarke and Bailey 2016). Arts and leisure activities may offer individuals an opportunity to resist stigmatizing views of dementia (Innes et al. 2016), and if carefully facilitated, they may also address broader structural barriers to participation and inclusion.

Dance programs offer people living with dementia opportunities to interact meaningfully through embodied movement and non-verbal forms of communication (Beard 2011; Kontos and Grigorovich 2018a). The expressive capabilities of people living with dementia are frequently undermined by discourses that focus on the cognitive decline associated with dementia without considering the body's powers of expression. Research has demonstrated that people living with dementia can express their memories and emotions, and relate to others through song and dance

(Kontos and Grigorovich 2018b). In addition, depending on the mode of dance, participation can provide opportunities for physical touch and connection, and as such, dance is a powerful corporeal and relational means of engaging with the world and with others. Given this, dance is an important medium through which to explore social inclusion as a place-based, embodied, and relational process.

In this chapter, we draw together literature on social inclusion and embodiment to inform an exploration of the potential of dance to support life enrichment for people living with dementia. We draw on findings from pilot studies of Sharing Dance Seniors, a joint venture between Canada's National Ballet School (NBS) and Baycrest Health Science that is underway in Peterborough, Ontario, and Brandon, Manitoba. The purpose of the pilots was to investigate how, and in what ways, dance can promote social inclusion for older people, including those living with dementia. In doing so, we join other scholars who argue that when the arts are restricted to their instrumental application as a therapeutic tool to improve "behaviors", scholars and practitioners fail to realize the full potential of the arts to enrich the lives of people living with dementia and to promote social inclusion (Kontos and Grigorovich 2018a, b). These pilot studies highlight the value of a geohumanities approach—integrating geographical and gerontological perspectives on social inclusion with the arts and philosophy of embodied selfhood—to address exclusion and enhance quality of life for people living with dementia.

Conceptualizing Social Inclusion and the Body

There is a large body of social science, humanities, and geographic literature that has explored the complex and dynamic nature of social inclusion for people living with mental health problems (Parr 2008), disabilities (Hall and Wilton 2015), and older adults (Van Regenmortel et al. 2016; Warburton et al. 2013). We have drawn together these disparate literatures to develop a conceptualization of social inclusion that is attentive to the role of structural and environmental factors as well as specific social encounters and sites of inclusion. For the purposes of this chapter, we provide a general definition and explore the key factors influencing social inclusion. We then discuss the need to examine more seriously embodied actions and feelings to understand social inclusion among people living with dementia.

There are many varied definitions of social inclusion. In general, the concept refers to the ability to meaningfully participate in relationships and activities within the context of a particular environment (Wilton et al. 2018). Social inclusion is a process and an outcome that emphasizes eliminating barriers such as lack of resources, rights, goods, and services required to participate (Levitas et al. 2007). In their work with older adults, Walsh et al. (2012) suggest that social inclusion is best understood by looking at the dynamic linkages between social relations, access to services, financial resources, transportation and mobility, safety and security, macro-economic forces, place and community characteristics, individual capacities, and life-course trajectories. They provide a multilayered, multidimensional approach to inclusion that emphasizes the relationship between individual and structural factors in producing inclusion. While their framework is successful in illustrating how people and organizations can shape programs, practices, policies, and attitudes to more fully include those who experience exclusion, it is not attentive to the embodied and experiential nature of inclusion.

Hall and Wilton (2015) argue that social inclusion is an experience and feeling of belonging and that presence and participation alongside others does not necessarily produce social inclusion. Taking this further, Wilton and Evans (2016) demonstrate that experiences and feelings of social inclusion can be understood by examining social encounters and the material surroundings, relational norms, and types of activities that shape these encounters in ways that challenge or promote inclusion. They argue that people's encounters, or moments of connection, provide an important starting point for exploring the processes of social inclusion in particular places and times. Small events, such as a quick conversation, and the circumstances surrounding the event can help us to understand how social inclusion is constituted or undermined. Their work is similarly attentive to the multidimensional, multi-scaler, and relational nature of social inclusion; however, like Walsh et al. (2012), they do not incorporate the body and embodiment in their conceptualization of social inclusion.

Social inclusion is felt in the body and is often related to bodily movements and practices in social encounters (e.g. facial expressions, hand gestures, and bodily expressions of class). Although what people think and say to others is a part of the process of negotiating social inclusion, social encounters also involve a myriad of actions that may be considered pre-reflective or affectual (see Andrews 2018). For example, the touch of another person may produce a sensation in the body and as the body is affected the individual may modify their action, and in doing so, affect the embodied movements of another body (Andrews and Grenier 2015). As bodies move through the world, many of their actions are pre-reflective—we do not think about everything we do before we act. As such, attention to the body is critical to understanding social inclusion as a relational process that unfolds between bodies; this is particularly the case for people living with dementia given that pre-reflective intentionality is a primary agential source of their interactive and communicative practices (Kontos 2012).

Critical gerontologists have argued that people living with dementia retain "pre-reflexive sources of perception… despite even severe cognitive impairment" (Kontos and Grigorovich 2018b, p. 40) and that a focus on the embodied self-expression of people living with dementia is critical to understanding their desires, agency, and meaningful engagement with the world. Scholars working in this area have critiqued the mind/body dualism that emerged in the Enlightenment period which regards the body and emotions as subordinate to the mind and reason, and suggest that cognition is the defining feature of personhood (Kontos and Martin 2013; Millet 2011). Kontos and Martin (2013) argue that there is more to being a person than thinking: agency, intentionality, and expression are sustained by, and persist through, the body and its movements. Kontos (2004) refers to this pre-reflective agency as embodied selfhood, which "highlights our intrinsic corporeality of being-in-the world, a positioning and directionality that is primordial and socio-cultural, sources of self-expression which are themselves always embedded in and intertwined with a shared world" (Kontos and Grigorovich 2018b, p. 164). Kontos has done extensive theoretical and empirical research that foregrounds pre-reflective ways of being-in-the-world when examining selfhood (for a review see Kontos and Martin 2013). Other examples have been described by Herron and Wrathall (2018) in their exploration of care relationships. They explored how bodily signs of distress and pain from wincing to striking out are acknowledged as ways of communicating needs and desires in care relationships. This research collectively demonstrates that embodied selfhood is fundamental to all human interaction and communication, and is foundational to everyday existence, particularly for individuals living with dementia. However, there remains a need for more research on strategies for mobilizing sociocultural structures and economic resources to nurture and support embodied selfhood in dementia care settings and in the community more broadly. Dementia care research has tended to focus on access to services and community support, narrowly defined, failing to recognize the role of the arts in supporting the rights and inclusion of people living with dementia in the world around them.

Kontos et al. (2017) argue that the arts have the potential to support embodied selfhood and to advance a relational citizenship approach to empowering and engaging people living with dementia. Relational citizenship is a new model that recognizes that corporeality is a fundamental source of self-expression, interdependence, and reciprocal engagement, and highlights the ethical imperative to fully support such embodiment through institutional policies, structures, and practices. As Kontos and Grigorovich (2018a, b) have argued, the arts can powerfully support relational citizenship by nurturing embodied capacities, senses, and interconnectedness. Their work on musicality with people living with dementia demonstrates how musicality "emanates from our corporeal depths"; recognizing corporeality and relationality as fundamental to self-expression brings a new and critical dimension to understanding the importance of music in the context of dementia and problematizes the restriction of music to its instrumental use as a therapy to improve behaviour or cognitive functioning in dementia (Kontos and Grigorovich 2018b, p. 46). Drawing on this work, we suggest that dance has similar potential and that recognizing the body's expressive capacitates is critical to understanding and supporting the social encounters that come to constitute social inclusion. In this chapter, we look at the potential of an NBS-Baycrest Sharing Dance Seniors program to provide a setting that supports embodied selfhood and social inclusion.

Insights from the Baycrest NBS Sharing Dance Seniors Programs

Sharing Dance Seniors is a suite of dance programs currently being developed as a joint venture between Baycrest Health Sciences and Canada's National Ballet School (NBS). The programs are designed to make dance accessible to seniors with a range of physical and cognitive abilities, including people with dementia. The Sharing Dance Seniors programs follow a protocol that outlines a series of dances (e.g. warm-up, dances of upper and lower extremities, narrative dances, mirroring and improvisation, and singing), each with specific goals and rationale. More importantly, although Sharing Dance Seniors may prove to be therapeutic for physical, psychological, or social challenges, the developers of the programs do not consider them to be a "dance therapy". The programs have been developed with a focus on dance objectives rather and therapeutic objectives and are taught by professional dance teachers rather than occupational therapists or other healthcare professionals. The guiding premise of these programs is that dance is an inherently and universally beneficial activity for humans across the lifespan. Thus, when high-quality dance programming is made accessible to people who may experience barriers to accessing traditional opportunities to dance, they too will experience the physical, social, emotional, and cultural benefits of dancing. This premise is made explicit in the guiding principles that inform all of NBS' Sharing Dance programs (see Table 4.1).

The Baycrest NBS Sharing Dance Seniors programs are still evolving as research and evaluation continue to inform their development. At this time, the programs are offered at two levels. Level one is described as accessible to people with significant cognitive and/or physical impairment, and was developed specifically with the intention of making quality dance programming accessible to people living with dementia. The program was developed by working with residents of Baycrest's

Table 4.1 Canada's National Ballet School's guiding principles: programming for adults and seniors

Principles	Adults and seniors
Dance is for everyone	Age should not diminish access to high-quality, inclusive, and accessible dance programming
Dance is fun	When made accessible, dance activities are fun and motivate participants to move
Dance is creative	Participants apply the elements of dance and the creative process to generate movement and express themselves creatively. This provides participants with opportunity for agency over their own movement and the opportunity to create and lead
Dance is healthy	Inherent in dance are physical, cognitive, and psychological benefits that can contribute to health and well-being across the lifespan
Physical and emotional safety	Participants feel safe exploring, taking risks, and trying new things
Positive relationships	Participants build positive, meaningful relationships with program leaders, peers, and caregivers
Targeted training	Dance teacher and facilitator training supports and develops the capacities to share high-quality dance programming with as many seniors as possible
Dance sector engagement	Dance teachers embrace the principles of program delivery in community settings and act as partners, mentors, and advocates
Sustainability	Programs are practical and sustainable. Sustainability is supported through collaborative program design and the use of digital technologies
Evaluation and research	Programs employ an evidence informed, iterative approach to content and design. Research partners provide an understanding of program impact on stakeholders (participants, facilitators, organizations, etc.)

Note: These principles have been reproduced with the permission of Canada's National Ballet School

long-term care facility in Toronto, Ontario. The class protocol is danced in 45-minute sessions. Early on in the development of this program (working with residents in long-term care), it was decided that the program should be danced entirely from a seated position to reduce the risk of falling. Although many people with dementia are able to dance from a standing position, some people living with dementia may have difficulties determining if dancing from a standing position is a safe choice. In contrast, level two of Sharing Dance Seniors was designed for seniors with minor physical and/or cognitive limitations. In these classes, both seated and standing options are presented to participants to decide for themselves what feels most appropriate.

Along with developing the Sharing Dance Seniors programs, NBS and Baycrest began to consider more inclusive delivery models of the programs that could maximize dissemination and grant more people access to the program. They began exploring video-streaming classes recorded weekly on-site at NBS in Toronto, Ontario, and streamed to community sites where seniors meet to dance the program together. In each community where the program is offered, an on-site facilitator oversees safety, the setup of the room and the stream, and helps foster an encouraging environment. After each class, the facilitator is sent an online feedback form. Feedback from the classes is then incorporated into upcoming classes so that the program is responsive to the needs and desires of the participants (including participants' musical preferences, names, and other information gathered by the facilitator).

In 2016, NBS partnered with researchers at Trent University and Brandon University, as well as community partners, with an interest in hosting the program using the video-streaming format to evaluate the experiences of participants, the effectiveness of the program, and the challenges of expanding the program. After receiving ethics approval from the university ethics review boards, the community-based research team held a series of meetings to discuss the delivery of the program in two study regions: Peterborough, Ontario, and Brandon, Manitoba. The first pilot of the program began in April 2017, with Trent University, NBS, and Community Care Peterborough working collaboratively to test streaming group sessions of Sharing Dance Seniors (level 2) over 8 weeks in one site. Drawing on data collected in this first phase, the program was expanded and subsequently delivered twice to five sites in the region during September of 2017 and March of 2018. Data from the Peterborough pilot informed the first phase of Sharing Dance Seniors (level 1) piloted in Brandon, Manitoba. In January 2018, Brandon University, NBS, and the Alzheimer's Society worked collaboratively to test the first video-streamed delivery of Sharing Dance Seniors to a group of community dwelling people living with dementia. The program took place at the Alzheimer's Society office in Brandon, Manitoba. The 8-week program ran from January 2018 to March 2018 and was scheduled to coincide with the Alzheimer Society's existing support group for people living with dementia. Sharing Dance Seniors ran for 45 minutes and was immediately followed by a snack break and their usual support group. For the purposes of this chapter, we focus on data collected in phase one of the Brandon pilot study.

The two pilot studies are unified by an identical method of data collection and program evaluation, including observational methods, participant diaries, focus groups, and interviews to understand the experiences of people living with dementia participating in the dance program. Researcher observations included weekly descriptive notes about the physical setting, participants' actions and interactions, the sequence of activities and responses to the dance instructors on screen, and communication between the on-site facilitator and the participants. In the Brandon pilot, diaries were completed with participants immediately after the dance session, with the support of a research assistant, and included participant descriptions of what happened in the dance session each week as well as participant evaluations of their favourite elements of the program, any challenges they encountered, and how they felt during the program that week. After the program, focus groups were conducted with participants to explore how they experienced the program as a group. Pre- and post-program interviews were also conducted with each individual participant who completed the diaries to gather contextual information and explore their reflections regarding their experiences with the program (for more information about the research protocol and ethics, see Skinner et al. 2018). Taken together, this combination of data sources allowed the research team to explore the relational and embodied experiences of participants in the dance program as they relate to promoting social inclusion.

Overall, six participants agreed to be involved in weekly observations of the dance program (two females and four males) as well as a follow-up focus group. Three people living with dementia and their carers also participated in pre- and post-program interviews as well as weekly diaries. In the section that follows, we focus primarily on the narratives of three participants in the program. We begin by introducing the participants and their carers (using pseudonyms to protect their anonymity) and their social encounters prior to the program. In the subsequent paragraphs, we highlight embodied expressions and encounters we observed and discuss their role in supporting social inclusion.

Understanding Feelings and Embodied Encounters

Alice is 65 years old and she lives with her husband, Ben, in a home they recently purchased in the small city of Brandon. She obtained a university degree and worked in the public sector for the majority of her working life. Alice and her husband have a total of five children from previous marriages, but they do not see them often. She has some close friends in Brandon as well as some who live outside the city who keep in touch with her through phone calls. She frequently joked with the research assistants during her encounters with them. Although she was very sociable during dance sessions, her husband explained that she sometimes experiences "anxiety and gets more mixed up around too many people and people she don't know". She feels comfortable going for coffee with old friends and going to the Alzheimer's Society support group. When asked about being lonely, her husband said he does not consider either of them lonely right now. He added, "I think it was harder at the beginning". Getting a diagnosis and accessing support from the Alzheimer's Society has helped to alleviate some of the social isolation and exclusion initially associated with the condition for Alice. As Alice explained:

> When I found out that I had dementia, I didn't want to go see the doctor. So I enjoy getting out with other people that are in the same group. I know everybody is different and has different issues, but it's nice to be with people who sort of get it… I originally thought, oh I don't need to go. This is silly; I can do this at home. But now I am always asking when I get to go to the dance sessions.

Over the course of the program, Alice documented changes in the group and her experience of the group in her diary. In her second week, she observed "Some of the people have loosened up;" their bodies and interactions with one another were more relaxed. She noted that it is good to see everyone participating, especially a new member that had just joined the group. She remarked, "The new guy seems to really fit in" and "Our group sort of sticks together in a way". Although it was difficult for her to elaborate on her feelings, she shared a sense of cohesiveness with the group. In the same entry, she also commented that "It feels really good being here and doing something rather than just being plopped at home watching TV". Her comments highlight how the dance program and the Alzheimer's Society are experienced as a good place associated with engaging activities rather than one-way consumption of a program. Even though Alice was sitting in front of a television taking in the program, her attention was on the people in the room with her. She continued to comment about the "new fella" and she hoped he would continue participating. When asked about other participants and how she got along with them, she acknowledged some conflicting desires and opinions among the group but suggested that "Everybody has their own opinions but music seems to be a go-to thing for everybody". Her sense was that the music facilitated relationality in spite of differences among participants. She noted that sometimes people get "distracted", but the music seemed to bring people in unison with one another. In the post-program interview, she also linked her sense of how nice everyone was and how enjoyable the program was to a set of shared problems related to the disease and perhaps even a shared acceptance of one another. She said, "everybody else is, they're sort of the same, we've all got the same kind of issues. So it's not like... so people kind of knew what problems you have…".

Alice also explained that the in-class facilitator played an important role in facilitating positive social interaction. She said, "She finds a way to get everyone involved". During the focus group, another participant, Helen, shared a similar sentiment saying, "she [facilitator] made us feel important". When prompted during the post-program interview, Alice explained "how she works with people. She's so good with that and just you know she'd definitely be checking to see if you're doing okay and if you're looking and that." Her comment underscores the importance of embodied interaction such as eye contact in making participants feel they belong.

Charlie is 70 years old and he lives with his 68-year-old wife Delores. He went to a community college and worked in the public sector. He has a daughter and grandchildren who he sees regularly. He used to volunteer at the hospital regularly, but has not done so for several years. He explained that he has difficulty walking and keeping his balance. When asked if he feels isolated, he says "In a way". Delores explains that he goes to an adult day program and the Alzheimer's Society support group, and that he has an activity worker come into the home once a week. He adds "my days are gone" and when he misses the day program he says "I don't miss it. I just want to get back to my norm, so I can do things I want to be able to cut the lawn this summer. I want to be able to do all the things I usually did". Although Charlie and his wife keep busy, he longs for more purposeful activity.

At first, Charlie's response to Sharing Dance Seniors was more ambivalent than Alice's and some of the other participants', in part because of some of his physical limitations. He initially found participation difficult because he was not able to do all the moves. Through weekly feedback to the instructor in Toronto and the on-site facilitator, the program was modified, and Charlie's response became more favourable. Part way into the program, he said "Well it's good exercise. I think it's good to have the music with it because if you have some rhythm with you it helps a lot". He explained that the rhythm helped him to move in spite of physical limitations, and it also affirmed his sense of self and family identity; "I was brought up in a musical family", he said in his second and fourth diary entries. He also confirmed that the instructor's comments helped him: "The instructor was right; I shouldn't push myself too hard to do certain movements". In contrast to Alice, he seemed to connect more with the on-screen instructor. During our observations, we noted that he watched her intently and when asked about the instructor in the interview, he said, "That was good. She was from Toronto eh?" During the sessions, the on-screen instructor would talk about events in Toronto and ask about events in Brandon. Charlie was always quick to chime in with his own remarks. It seemed important to him that he was connecting to someone somewhere else. In the focus group, he commented, "oh yeah, she knew who I was". He felt that the instructor could see him as he related to her. Most importantly, his comments suggest that connecting across settings through video streaming is not only possible, but it may also give people with dementia an opportunity to feel engaged with the world beyond their immediate environment.

Ed is 82 and lives with his wife Fiona who is 74. They each obtained university degrees and worked as professionals. They have two daughters living in the area that they see on holidays and some friends in the community; however, they explained that their social life has changed significantly with retirement and the onset of Ed's dementia. Fiona said "we feel a wee bit trapped here and used to get away in the winter more than we do now". She adds that Ed used to go to breakfast with former colleagues and he has stopped going. They also stopped going to church and volunteering. In his first diary entry, Ed responded to questions about what he enjoyed about the program saying, "Well it sure beats watching anything on television" and in his second entry he said "Sitting around and doing nothing makes me feel like I'm back in grade one". In the first few weeks of the program, he was concerned about his lower back and he shared a sense that he "could have done it [the dance program] better". While he noted that he needed to remind himself that the program was not competitive, his initial comments suggest some self-consciousness about his body and performance. As the program progressed, he no longer made these comments. In his observations about the group, he noted that "the humour is always kind" and "Conversation flows quite easily in the group. It's a friendly group". He also referred to the environment as "non-threatening". Not unlike Alice, he expressed a sense of being part of the group: "It's a group of people that know me and recognize me". He linked his sense of belonging to the social environment and comfort he felt rather than to a shared experience of the disease. He commented that he liked that everyone has an opportunity to "do what they do" and he enjoyed watching them and trying to pick up what they were doing. Importantly, the Sharing Dance program was structured to include mimicking a partner's actions; this allowed Ed, who was fairly quiet, to relate to others through embodied movement and mimicry.

The diary entries and pre- and post-program interviews with participants contribute to scholars' understandings of the factors influencing feelings of inclusion within specific program settings (Hall and Wilton 2015). One of the factors influencing the success of the program was participants' shared awareness of the dementia experience as well as a more general sense of knowing one another. These factors may have been strengthened by the program; however, they are also a product of the ongoing Alzheimer's Society support group. Specific to the dance program, the on-site facilitator played an important role in encouraging active participation through her movements while the on-screen instructor sought to connect with participants remotely through sharing knowledge of place. While previous studies (e.g. Walsh et al. 2012) have highlighted the importance of local place history and knowledge in social inclusion, this program suggests that local knowledge can be mobilized from a distance to provide place-based support. Building on the work of Kontos and Grigorovich (2018b), participants shared a sense that music supported positive social encounters; although they did not necessarily expand upon why this was the case in their diaries and interviews, the focus group and our observations provide further insights about the moments of encounter and laughter that music and movement together helped foster.

The dance program supported structured and spontaneous encounters where participants related to one another. For example, acting was a part of the program design and an important part of how participants related with one another. During the focus group, the participants spoke about the western narrative dance and many of the focus group participants recreated the dance by joining together and making motions as if they were pushing a saloon door open and peering out under their hands. They made eye contact with others around the focus group table and laughed and smiled. In contrast to exercise programs, narrative dances provided a structure for participants to engage creatively with one another. Participants also connected more spontaneously through their actions. For example, in one of the dance sessions, two participants were observed accidentally touching arms during a movement; this resulted in some laughs and playful taps back and forth between participants. Participants' proximity to one another and the playful nature of the program helped to support these moments of connection.

Encounters within the dance program were not exclusively positive and playful. Furthermore, it was sometimes difficult to interpret ambiguous actions that may begin as playful in nature and take on different meanings to different parties over time. A particular example of this sort of interaction occurred midway through the program when George tied a scarf (used as a prop in a partner dance) around Charlie's neck. Charlie responded, "he's just jealous he doesn't look as good" [laughs]. George playfully whipped Charlie in the face with his scarf. The instructor was unsure of George's intent and how Charlie was feeling, but nonetheless hesitated to intervene. The following week George continued teasing Charlie about his shoes and referred to tying the scarf around his neck the week before. There was some laughter but the AS staff member interjected and tried to refocus George when he made reference to the interaction between the two men the previous week. This encounter highlights the dynamic nature of inclusion as a process that is constantly being negotiated between individuals in a particular context. In this context, the program staff played an important role in shaping the norms of social interaction (and boundaries) within the program environment and mediating what was interpreted as social and physical harm. Incidents such as these highlight the constant work of trying to sustain an inclusive environment, the importance of paying attention to how bodies move in relation to others in these environments, and the challenges of interpreting and understanding these movements.

Discussion and Concluding Comments

In this chapter, we have drawn on preliminary data from a pilot study of Sharing Dance Seniors to examine the potential of dance to improve social inclusion and embodied selfhood for people living with dementia. Although other scholars note that social inclusion is difficult to achieve (Wilton et al. 2018), a focus on embodied movement can help us to better understand social inclusion as a place-based, relational, and embodied process. By examining embodied encounters alongside participant reflections, we traced how moments of relating to others in a program environment can produce feelings of inclusion over time. Drawing on the stories and actions of Alice, Charlie, and Ed, and their carers, we illustrate how dance can foster a relational environment that supports embodied selfhood through acting, touch, eye contact, and mimicry. These different forms of self-expression enabled participants to relate in meaningful ways within the program environment.

Consistent with Hall and Wilton's (2015) work, our preliminary findings illustrate the value of separate spaces in fostering inclusion among the group as well as the potential of such spaces and programs to enable people living with dementia to engage in different ways with the broader community. For example, in a post-program interview, an Alzheimer's Society staff member noted that one couple started going to community dances as a result of the program. Programs such as these can foster feelings of inclusion related to activities that translate beyond the program environment. They may offer people living with dementia opportunities to express themselves and to relate to others in the community.

Although studies have supported the instrumental use of arts-based programs to generate social and behavioural changes, we argue that programs such as Sharing Dance Seniors can provide a holistic approach to enriching the lives of people living with dementia. Rather than view dance as a tool for managing symptoms and behaviors, scholars and practitioners ought to view opportunities to express oneself, relate to others, and engage creatively with the world as fundamental to citizenship and inclusion. Ultimately, such programs can challenge stigmatizing assumptions that people with dementia are incapable of learning, developing, and being creative and support the development of more inclusive spaces and attitudes (see also Tan and Atkinson, this volume). Although we highlight the potential of the arts in this regard, our future work will also explore the limits of this program and specific community sites in supporting social inclusion for people living with dementia.

Finally, our study illustrates the value of integrating perspectives from geography, gerontology, and the arts to address social injustices and advance citizenship and quality of life for people living with dementia. Critical gerontologists and geographers are increasingly interested in how people relate to others and express themselves through their actions and not just what they say and think (Andrews 2018; Andrews and Grenier 2015). Such actions and expression are critical to understanding processes of inclusion and exclusion as they unfold in time and space. Ultimately, addressing exclusion and the spatialization of difference requires that geographers engage with the arts and humanities to explore creative means to foster social encounters in ways that respect embodied selfhood and empower people living with dementia.

References

Abreu, M., & Hartley, G. (2013). The effects of salsa dance on balance, gait, and fall risk in a sedentary patient with alzheimer's dementia, multiple comorbidities, and recurrent falls. *Journal of Geriatric Physical Therapy, 36*, 100–108.

Alzheimer's Disease International. (2015). *The global impact of dementia: An analysis of prevalence, incidence, cost and trends*. London: Alzheimer Disease International.

Andrews, G. J. (2018). *Non-representational theory and health: The health in life in space-time revealing*. London: Routledge.

Andrews, G. J., & Grenier, J. M. (2015). Ageing movement as space-time: Introducing non-representational theory to the geography of ageing. *Progress in Geography, 34*, 1512–1534.

Beard, R. L. (2011). Art therapies and dementia care: A systematic review. *Dementia, 11*, 1–24.

Blankevoort, C. G., Van Heuvelen, M. J., Boersma, F., Luning, H., De Jong, J., & Scherder, E. J. (2010). Review of effects of physical activity on strength, balance, mobility and ADL performance in elderly subjects with dementia. *Dementia and Geriatric Cognitive Disorders, 30*, 392–402.

Clarke, C., & Bailey, C. (2016). Narrative citizenship resilience and inclusion with dementia: On the inside or on the outside of physical and social places. *Dementia, 15*, 434–452.

Coubard, O. A., Duretz, S., Lefebvre, V., Lapalus, P., & Ferrufino, L. (2011). Practice of contemporary dance improves cognitive flexibility in aging. *Frontiers in Aging Neuroscience, 3*, 13. https://doi.org/10.3389/Fnagi.2011.00013.

Dupuis, S. L., Kontos, P., Mitchell, G., Jonas-Simpson, C., & Gray, J. (2016). Re-claiming citizenship through the arts. *Dementia, 15*, 358–380.
Fortune, D., & Mckeown, J. (2016). Sharing the journey: Exploring a social leisure program for persons with dementia and their spouses. *Leisure Sciences, 38*, 373–387.
Hall, E., & Wilton, R. (2015). Alternative spaces of "work" and inclusion for disabled people. In C. Grover & L. Piggott (Eds.), *Disabled people, work and welfare* (pp. 315–333). Bristol: Policy Press.
Herron, R. V., & Wrathall, M. A. (2018). Putting responsive behaviours in place: Examining how formal and informal carers understand the actions of people with dementia. *Social Science & Medicine, 204*, 9–15.
Innes, A., Page, S. J., & Cutler, C. (2016). Barriers to leisure participation for people with dementia and their carers: An exploratory analysis of carer and people with dementia's experiences. *Dementia, 15*, 1643–1665.
Kontos, P. (2004). Ethnographic reflections on selfhood, embodiment and alzheimer's disease. *Ageing & Society, 24*, 829–849.
Kontos, P. (2012). Rethinking sociability in long-term care: An embodied dimension of selfhood. *Dementia, 11*, 329–346.
Kontos, P., & Grigorovich, A. (2018a). Dancing with dementia: Citizenship, embodiment, and everyday life in long-term care. In S. Katz (Ed.), *Ageing and everyday life: Materialities and embodiments* (pp. 163–179). Bristol: Policy Press.
Kontos, P., & Grigorovich, A. (2018b). Rethinking musicality in dementia as embodied and relational. *Journal of Aging Studies, 45*, 39–48.
Kontos, P., & Martin, W. (2013). Embodiment and dementia: Exploring critical narratives of selfhood, surveillance, and dementia care. *Dementia, 12*, 288–302.
Kontos, P., Miller, K. L., & And Kontos, A. P. (2017). Relational citizenship: Supporting embodied selfhood and relationality in dementia care. *Sociology of Health & Illness, 39*(2), 182–198.
Kontos, P., Grigorovich, A., Dupuis, S., Jonas-Simpson, C., Mitchell, G. J., & Gray, J. (2018). Raising the curtain on stigma associated with dementia: Fostering a new cultural imaginary for a more inclusive society. *Critical Public Health*. https://doi.org/10.1080/09581596.2018.1508822
Levitas, R., Pantazis, C., Fahmy, E., Gordon, D., Llyod, E., & Patsios, D. (2007). *The multidimensional analysis of social exclusion*. London: Department for Communities and Local Government and Social Exclusion Task Force.
Millet, S. (2011). Self and embodiment: A bio-phenomenological approach to dementia. *Dementia, 10*, 509–522.
Parr, H. (2008). Mental health and social space: Towards inclusionary geographies? MA: Blackwell.
Skinner, M.W., Herron, R.V., Bar, R. J., Kontos, P., & Menec, V. (2018). Improving social inclusion for people with dementia and carers through sharing dance: a qualitative sequential continuum of care pilot study protocol. *BMJ Open*. https://doi.org/10.1136/bmjopen-2018-026912.
Van Regenmortel, S., De Donder, L., Dury, S., Smetcoren, A. S., De Witte, N., & Verté, D. (2016). Social exclusion in later life: A systematic review of the literature. *Journal of Population Ageing, 9*, 315–344.
Walsh, K., O'Shea, E., & Scharf, T. (2012). *Social exclusion and ageing in diverse rural communities*. Galway: Irish Centre for Social Gerontology.
Warburton, J., Ng, S. H., & Shardlow, S. M. (2013). Social inclusion in an ageing world: Introduction to the special issue. *Ageing & Society, 33*, 1–15.
Wilton, R., & Evans, J. (2016). Social enterprises as spaces of encounter for mental health consumers. *Area, 48*, 236–243.
Wilton, R., Fudge Schormans, A., & Marquis, N. (2018). Shopping, social inclusion and the urban geographies of people with intellectual disability. *Social & Cultural Geography, 19*, 230–252.

Chapter 5
Critical Places and Emerging Health Matters: Body, Risk and Spatial Obstacles

Kristofer Hansson

Abstract The essay presents and develops the concept of "critical places" and how it can be used when studying the everyday experience of living with long-term sickness and/or disability. The concept analyses the duality of both physical risk and social benefit and how they can collide in one specific place and create a bodily situation where the individual needs to act. The concept of "critical places" explores the phenomenological thought about *doing* and happening in specific situations, and, as such, the concept can also be seen as an ethnographic method. As a more theoretical concept, "critical places" can be used for a hermeneutic analysis of risk-taking, hiding from stigma, identity formation, power relations in a specific place and so on. I have used the concept in a couple of Swedish text concerning disability, and in these texts the concept has been elaborated with theories from geographies (Alftberg et al. Inledning: ljudmiljöer, kulturella praktiker och hörselnedsättning. [Introduction: sound environments, cultural practise and hearing impairment]. In: Ljud tar plats: Funktionshinderperspektiv på ljudmiljöer [Sound takes place: disability perspective on audio environments] 11. Lund: Department of Arts and Cultural Sciences, Lund University, 2016). The concept has also been used by Meghan Cridland ("May contain traces of": an ethnographic study of eating communities and the gluten free diet. Lund: Department of Arts and Cultural Sciences, Lund University, 2017) in her ethnographic study of eating communities and by Niclas Hagen (A molecular body in a digital society: from practical biosociality to online biosociality. In: The atomized body – the cultural life of stem cells, genes and neurons. Lund: Nordic Academic Press, 2012) in his study about people living with Huntington's disease. In this article, I will also develop the concept with my new research project about disability and accessibility, a 3-year project that will start in 2018. The article will introduce the concept of "critical places" to an international arena.

Keywords Asthma · Breathing · Ethnography · Navigating risk · Sweden

K. Hansson (✉)
Lund University, Department of Arts and Cultural Sciences, Lund, Sweden
e-mail: kristofer.hansson@kultur.lu.se

S. Atkinson, R. Hunt (eds.), *GeoHumanities and Health*, Global Perspectives on Health Geography, https://doi.org/10.1007/978-3-030-21406-7_5

Introduction

Health is not easy to study. Most ethnographers have an experience of this; it is as if health is invisible, and so there is nothing to observe. Therefore, most ethnographers try to find those moments in which the unexpected, the sudden or the different occurs. In such moments, health all of a sudden springs to life and emanates as being something that not only can be understood but also experienced. In other words, health is transformed into something that has a voice and can be heard. This chapter defines these moments of time and place as *critical places*[1] and offers this as a concept that can be used both as a methodological perspective for fieldworking ethnographers and as a theoretical tool for those in the medical and geo-humanities to comprehend health.

The aim of the chapter is to connect a medical humanities perspective on health to critical places. The starting point is to raise my own voice and to clarify my experiences of health in relation to the long-term illness of asthma. I will therefore start this chapter with an autobiographical, self-reflexive narrative that aims to more precisely pinpoint what kind of health experiences can be envisaged through this methodological perspective. The self-reflexive narrative is seen as a methodological opportunity to explore and study those health practices that need a voice. Next, I present examples from my research with young people living with asthma and allergies in which critical places emerged as playing an important role. Based on these autobiographical and ethnographic descriptions, the chapter develops the concept of critical places and how it can be used when studying the everyday experiences and practices of health when living with long-term illness. I outline three central themes in order to understand critical places: body, risk and spatial obstacles. The themes are presented after the self-reflexive background with which the chapter begins.

A Self-Reflexive Narrative

There is a recognised challenge for the ethnographic researcher in observing and studying people's innermost experiences and decisions as, for example, in relation to health. Such experiences do not always result in any obviously related actions or verbal expressions; they are not even always conscious for the people we study. But a way forward is to start using a self-reflexive perspective by going to and drawing on our own experiences as part of the ethnographic process. The self-reflexive perspective has a long tradition in the humanities associated with its positioning as a

[1] I have used the concept in a couple of Swedish text concerning illness, disability and health, and in these texts the concept has been elaborated with theories from culture geography and ethnology (Hansson 2007a; Alftberg et al. 2016). The concept has also been used by my colleagues at Lund University, namely, Cridland (2017) in her ethnographic study of eating communities and Hagen (2012) in his study about people living with Huntington's disease.

political project (Wendell 1996).[2] In a humanities-facing health research, it affords a method to put words and give action to the experiences of life with long-term illness.

The self-reflexive approach that is used in this chapter is inspired by Husserl's statement: to the "things themselves!" (Husserl 2001). The things in this case are my body and my lungs and how I experience health in specific places. Most of the time, I have no problems of asthma, but when I get sick, it can stick to my lungs for days and sometimes weeks. Then, it is as if the places that I visit change, and I start to categorise them as good or bad for my health. A good place, for the most time, is my *home*. This is a place where I can relax, take medicine and be sure that the air that I breathe is clean and healthy for me. The places that are not good for me can be many, very varied and uncategorised. Some of these places I have needed to visit, for shorter or longer periods of time, include, when I was younger, my school and, now as an adult, my workplace. I have to go to my workplace to schedule meetings or take classes with my students. In doing so I take a risk, knowing that I am in places where my asthma may deteriorate, causing the period of recovery to lengthen or, if it is particularly bad, turn into a full asthma attack. In these places, my regular patterns of movement change. I go a little bit slower on the stairs, and I try to avoid those rooms that may have irritating scents produced by flowers or people wearing perfume. At the same time, I feel that I am, therefore, not part of the situation and instead I have a feeling of being *out of place* (cf. Cresswell 1996; Gleeson 1999; Dyck 2010).

Risk-taking is central for me when having asthma. Sometimes I am very aware of the risks I am taking in relation to my health and of how I am considering my decisions very carefully. At other times, I take risks without reflecting on it or having any concrete arguments why I am doing things as I do. For example, when I was young and there were parties at friends' homes that I really wanted to attend, it was not that unusual for party-goers to smoke cigarettes indoors. Cigarette smoke was and remains bad for my health, and, most often, if I had been at a party, my lungs would feel sour for a couple of days after an evening inhaling smoke. At the same time, I wanted to hang out with my friends. So the choice, most often, was to take the risk of getting worse and attend the party. Trying to say something about my friends' smoking was not always an option, so I just tried to blend in and keep my breathing difficulties to myself.

So, how can we use this example of going to the "things themselves"? My argument is that the moment of the party is usefully seen not only as a specific situation of social relations but as a specific place of social, material and temporal relations

[2] In the end of the 1980s and the beginning of the 1990s, ethnology, and more broadly research in the humanities and social sciences, became more and more aware of the researcher's own position to the fieldwork and the analysis of the study object. It became central to scrutinise the researcher's position and make this clear. In the 2000s, this changed to not only reflecting upon the researcher's own position but to the researcher using their own feelings and bodily sensations to gather new empirical material as part of their fieldwork. This methodological standpoint is central in this chapter to develop a methodological and a theoretical perspective to understand health through medical and geo-humanities.

that are all of interest. It was the place that pulled me to the party, because my friends were going there. But this was also the place where the cigarette smoke was going to be, together with a range of other social relations that I would need to negotiate in its presence. The place, then, demanded decisions in relation to a range of questions: should I say something about my asthma or should I say something about their smoking? Should I take the risk and hopefully have a good time, despite the likelihood of worsening health the day after? To understand what is involved in negotiating these kinds of questions when living with a chronic health condition, we need to go beyond understanding the example of the party in terms of a social situation and, instead, see it as a place where health *comes alive*.

It was the specific places that had a pull on me, very often because these places held within them all the fun – friends, alcohol, good music, dancing and so on. As I became older, they were also connected to places where I needed to be, for example, my workplace. Frequently, these places have started out as being relatively unproblematic for my health and as places in which I can be just like everyone else – hanging out, doing my thing, talking to people and so on. The places, however, change for me when something happens in and to my body as I start feeling the onset of breathing problems. The expression of my asthma is usually rather slow, so these breathing problems are not an asthma attack as such. Rather, the problems start with a feeling of something or a sensation that is not quite right. It may be the feeling of the air, or it may be the feeling of my lungs, but in either case, it is an experiencing of some minor sensation that is slowly getting worse. In this, my body, and specifically my lungs, comes forward to remind my consciousness of their presence (cf. Leder 1990). When this happens, I need to consider my relation to the specific place and ask myself if I should stay or if I should leave. What had been easy – just being in place – becomes troublesome and demands that I start to deal with my health in a different way. This chapter is how we can understand such places and what they can say about people's relation to health. But it is also a political project through raising my own voice to show how people are disadvantaged and left behind, to challenge the acceptability of this and to question whether the rest of the society might change or adapt their behaviour.[3]

Bodily Sensations and Place

Place has, for a long time, been theoretically and methodologically considered as something related to identity – places do something with people (cf. McDowell 1999; Chouinard et al. 2010; Soldatic et al. 2014). They may afford the possibilities through which to express a specific identity, or an identity may be attributed through association with a specific place. Therefore, building on the insights from my

[3] In my academic career as a researcher meeting other people with asthma, allergy and diabetes or using wheelchairs, disadvantages have become experiences that I share with them.

self-reflexive approach to the becoming of my asthma experiences in place, I have explored with people living with long-term illness the ways in which this interacts with specific places in creating and testing identities and practices. How do people take place? If we engage this question empirically, we can also ask how a person, or a group, may simultaneously be discouraged or excluded from a specific place. What kind of obstacles are there in a specific place?

The empirical material consists of 16 semi-structured interviews that were undertaken with young people aged between 15 and 20 years old.[4] The semi-structured interviews drew on an interview topic guide but also enabled the interviewees to introduce other themes for discussion. I recorded the conversations and subsequently transcribed them. The interview guide was informed by "The Explanatory Models Approach" of Kleinman and Benson (2006). Examples of questions that aimed to prompt conversation within an explanatory model framing included: What do you believe is the cause of our asthma? How serious is our asthma? How does our asthma affect your body? After the first interview, I equipped those participating with a camera to document aspects of their everyday experiences of living with asthma that were of importance to them (Harrison 2002). A second interview then took place based on these pictures.

Common examples of obstacles to full participation in specific places as identified by young people living with asthma included when the best friend's home becomes problematic because the family has a cat causing allergic reactions or when a party with friends becomes problematic because they are smoking cigarettes causing asthmatic reactions. What became apparent in the interviews is that the young people participating in the study often defied their bodily sensations and, instead, took the risk of being in these specific places. As with my own self-reflexive narratives, the friend's house or the party may be more important than the risk of worsening their asthma or allergy. The party and the friendship can be understood as not only an opportunity to socialise with friends but also the possibility to assert the wish to just do what you want. Nonetheless, these social situations create an accumulation of risks.

Social settings entail those living with asthma and allergies to trade risks and perceived gains. There are places, however, where there is limited choice in being there, in that attendance is obligatory, for example, at school. One of the boys interviewed – we can call him Emil – borrowed a camera from me and photographed[5] the places that were problematic for his asthma at a time when his asthma was particularly impacting upon his life. When we met the second time, Emil was no longer able to stay in school much because of his health problems which had deteriorated rapidly. While we discussed one of the pictures illustrating a dark corridor in Emil's School, he said:

---

[4]The interviews were part of my PhD project that resulted in the book *In a breath: A cultural analysis of asthma as limitation and possibility* (2007b).

[5]This method was inspired by Harrison's article "Seeing health and illness worlds – using visual methodologies in a sociology of health and illness: a methodological review" (2002).

> Where there is bad air I usually – if I really have to be there – do it as fast as possible. So I can leave it. If I stay for a longer time, I get problems. So I want to avoid these places at school. But where it's healthier, I prefer the air. It feels more comfortable breathing.

In many ways, I could recognise myself in what Emil said, even though it was a long time since I went to school. I did not have this kind of problem at school myself, but I have a body – and lungs – that has experienced similar sensations of *bad air* in specific places. This experience is as if the sensations are coming from the body and cannot be stopped (cf. Leder 1990). These bodily sensations, however, need also to be seen as part of creating the experience of the place and of the person's health. By *being there*, the person is experiencing both the air and the body, together with their interaction (cf. Frykman and Gilje 2003). At the same time, this bodily sensation often requires an action or reaction. I will come back to what kind of action this can be, but first I will discuss some basic phenomenological perspectives on bodily sensations.

Merleau-Ponty offers the insight that it is through our bodies that we are experiencing, in no specific order, the air, the place and our health (Merleau-Ponty 2002). It is through the body that Emil comprehends and effectively constructs the specific place. As an individual, we not only become aware of the bodily symptoms, but our awareness and construction of the specific place also change. Merleau-Ponty writes: "Our natural attitude is not to experience our own feelings or to adhere to our own pleasures, but to live in accordance with the emotional categories of the environment" (ibid., p. 442). This is what Emil *is doing;* similar to Merleau-Ponty, he "make(s) the discovery that" our lungs react[6] and he *acts* through his lungs. Sometimes this doing and acting creates a conscious thought, as I found in my own self-reflexive engagements, where Emil and I ask similar questions: do I need to be here or can I leave this place?

The bodily responses in these specific situations are not only bodily sensations, but they should also be understood as bodily habits that the person has learned to handle through living with asthma. As Merleau-Ponty points out, these bodily habits are not always articulated (ibid.) but are, rather, habits that create feelings or actions without the individual being aware of them. "We are our body" as Merleau-Ponty points out and continues: "we are in the world through our body, and in so far as we perceive the world with our body" (ibid., p. 239). Emil is not talking about how he lives his life through his body, but, over years of experience of living with asthma, he has developed bodily habits that lie in the background in his everyday life and that can generate strong potential feelings affecting how he perceives the world. These can be prompted by a change in the air, like a new smell, or a particular reaction in the body. There can be apparently insignificant changes that nonetheless awaken these habits of emotion and create a sense of fear. What is happening to my body? At the same time, this bodily sensation does not need to be in relation to asthma or an asthma attack. It can be comparable to Proust's memories when he feels the taste of the French madeleine cakes. In a moment, the body and the place

[6] Merleau-Ponty writes: "make the discovery that I am in love" (2002: 442), a good example of how our bodily feelings are in relation to the emotional categories that surround us.

prompt feelings that are not verbal but that the person relates to in the shape of memories.

So long as the lungs breathe as they should, they are in the background of the body's encounters with the world, and the range of conscious bodily perceptions do not include and are not influenced by the lungs. But when the person starts to feel that there is something wrong with their breathing, then the lungs come to the fore and affect how the person perceives the world. This reflects Leder's account of the *absent body*, a body in the background and effectively silent and invisible, until that moment when the body comes forward and the person has to deal with both the body's demands and the situation in which this occurs (Leder 1990). But the lungs can also "step forward" and affect perception not only when there is asthma or some form of breathing problem (cf. Merleau-Ponty 2002). The bodily habit of knowing what can happen when, as in the example with Emil, one is too long in places with bad air can make the absent body visible, even when there are no direct breathing problems at that moment. It is as if the bodily habits of knowing what can happen in different places also affect the person's health and force the person to act in relation to what can happen as much as what does happen. This is something that Emil points out in emphasising his practices; thus: "So I want to avoid the places at school".

The embodiment as discussed in this chapter has a methodological and a substantive focus on the body not as something that *we have*, but something that *we are* (Merleau-Ponty 2002; Frykman and Gilje 2003). We are our bodies, and it is with the body that we meet and perceive our environments, as well as experience health. To understand critical places, we therefore need to start with the body and how bodily perception and place are intertwined in complex ways. In a metaphorical way, we can say that a person with asthma thinks and acts with the lungs, but this is only part of the picture. We also need to understand risk-taking in relation to body perceptions, to experiences of health and to social threats.

Health Risks and Social Risks

Having a long-term illness such as asthma means taking different kinds of risks in relation to one's health. The nature of these risks will differ greatly depending on the long-term illness and so are not always comparable. In relation to the discussion so far, two different risks will be explored in what follows. These have emerged both as core findings from several of my and my colleagues' research projects on long-term illness and disability (Hansson 2007a; Hagen 2012; Alftberg et al. 2016; Cridland 2017) and as central when reflecting on my own experiences of having asthma. The first risk that I will discuss is connected to the long-term illness in more biological ways and, as such, can be seen as a form of health risk. The second risk is related to a different form of social threats associated with living with a chronic condition of ill-health and, as such, can be seen as a form of social risk.

Asthma, as discussed earlier, is experienced through the individual's perception, and the body, therefore, also becomes central in experiencing what kinds of health risks the individual is prepared to take. Risk, from this perspective, can be seen as an interpretation of past experiences that the individual has in relation to particular types of context (cf. Boholm 2009). For Emil his previous experiences made him classify different places as more or less dangerous for his health, and he pointed out a specific place in which: "If I stay for a long time, I get a problem". But it was not only that he calculated for how long he could stay in a specific place but that sometimes he avoided the risk of getting asthma altogether by not visiting the specific place. Through these previous experiences, the person can minimise what is known as an uncertainty for the health. The risks to bodily health in this way are negotiated explicitly in relation to particular places.

My own experiences suggest that these health risks include both those that can be known in advance, and are calculable, and those that appear unpredictably. Similarly to Emil, I can mediate the risks through an internal argument with myself that although I cannot stay for a long period of time in a specific place, a shorter period of time can be safe. The need to present an argument to oneself and negotiate the risk results from the benefits and pleasures of being in a particular place and which, on occasions, outweigh the risks, even high risks, of experiencing worse health. Emil, myself and others all have places where we really want to be despite knowing the risk to our breathing that these places pose. Taking this risk relates to timing and often comes down to a decision regarding how long is too long. The health risk, however, can come on much more quickly than expected. This is the case in the examples earlier of smoking at parties, of strong perfume coming into a room or of the unexpected presence of a pet animal. In these examples, the place is transformed very rapidly from a safe place to a highly risky place, and such transformation can be hard for the person at risk to cope with.[7]

The second risk that is central is a *social risk*. Living with a long-term illness is already a challenge in terms of the individual bodily experience of health, but an additional set of risks come from expressing the illness or talking about it. Ola, one of the young people interviewed, said the following when I asked him why he did not want other people to see that he had asthma and therefore hid his medication:

Ola: If there are people who do not know that you have asthma. If it is only me and my friends, I'll take the medicine. But if there are people who do not know me, it can be too dramatic. So you might as well go away.
Kristofer: In what way dramatic?
Ola: No, people think that you … you can collapse any second.

[7] Some of the risk that an individual takes can be calculated and more or less easy to cope with. One knows the risk but the traction to the place is greater, so the individual make a rational consideration. Other times there are no rational considerations, and the individual just *throws* themselves into the situation.

The relationship with other young people becomes problematic. The place turns into a social risk where Ola feels that too much tension is directed to him. This is problematic because it pulls this tension from others onto him and his body. For Ola, it is when he needs to take his medicine that the situation of the *unproblematic place* turns into something else. Most often, Ola tells me, he leaves the group and goes to a bathroom or something similar and takes his medicine in private, and, in effect, he camouflages his medication (Hansson 2005, 2007b). Then, nobody can tell that he has asthma, and he does not need to answer questions, or be stared at, from people he does not know.[8]

Ola was only around 20 years old when I held the interview with him, and we might expect that this camouflaging behaviour would be less marked as the importance of peer approval in youth declines with age. However, I, as an adult in my forties and a father, do exactly the same when I have problems with my asthma in public and need to take my medicine. Similar to Ola, I dislike it when people that I do not know look at me and wonder what is going on with my body and my health. I do not want to reveal a body that draws attention to itself because it is "sick" and effectively "out of place" in terms of what are appropriate bodily behaviours there. The place, then, becomes critical in such a situation through the relationship to other people and the normative bodily expectations that are part of the setting. The expression of asthma does not even need to be that acute; indeed, it can be under full control, but the reactions to chronic asthma that I experience if I reveal my condition become in many ways deeply problematic social relations. For this reason, I too, like Ola, tend to hide my condition, smuggle my medication into any given social setting and sneak off to a private space if I need to self-medicate; it is a form of passing (cf. Goffman 1963).

The concept of critical places combines both bodily health risks and situated social risks that the individual encounters and negotiates. While these two kinds of risk can be separated methodologically and analytically, experientially they occur simultaneously to create an accumulation of risks. Moreover, this accumulation of risks creates a practice, or set of practices, through which health becomes possible to study. For example, what kinds of emplaced health risk are confronted when a young man such as Ola does not want to reveal a "collapsing" body? What emplaced health risks are negotiated when this body is revealed to others? If we can understand this kind of accumulation of risks, we can also, methodologically and theoretically, understand how obstacles are constructed symbolically and materially.

[8] This discussion and example is further elaborated in Hansson (2005): "In this kind of *passing* a person can conceal a stigmatized identity (Goffman 1963). It is a kind of reflexive camouflage. At the same time this reflexive passing is not always done through individual choice. A person can be forced to act in accordance with a dominated discourse of what is accepted and what is not (Foucault 1987)" (ibid., p. 138).

In and Out of Places

So far critical places have been described as capturing an accumulation of bodily health risks, situated social risks and bodily sensations. These elements are adequate considerations in studying how health is not only experienced, but also how it is negotiated and challenged by the individual. We can, however, build in a further element if we see health experienced and constructed in relation to symbolic and material aspects of place. In this, I draw on perspectives on disability by Soldatic et al. (2014) who state that "[…] the spatial is not confined to the material and structural alone […] disability and spatial relations are constructed symbolically, culturally and materially" (p. 2). In this, obstacles, or in our case risks, are not objective but instead are the result of historical and cultural processes through which bodies and spatial relations are in constant creation over time (Gleeson 1996, 1999).

The specific histories of spatial change are thus central for investigating and understanding how critical places emerge over time. In the examples in which bodily sensations were associated with alienation from school, a party or similar, the obstacles came into being within a specific historical and environmental setting. When Emil's asthma began to worsen in the school facilities, this could have been from bad ventilation, poor cleaning or something else. But whatever the specific cause, the specific environment created the obstacles that Emil was fighting against. When we *go-along* with young people living with asthma and allergies into different locations, the spatial conditions are clarified (Kusenbach 2003). At the same time, it is important to not focus too much on the obstacles because it can mean that the complexity is missed in terms of the places' symbolic, cultural and material variability.

This methodological perspective intersects with Kitchin's (1998) argument in relation to disability studies that places exclude people in at least two ways: places are organised so that people with long-term illness know their place; and the place conveys a message that people with long-term illness experience being "out of place" (see also Cresswell 1996). Kitchin himself summarises this in the following way: "The spatialities of disability are configured to convince disabled people that they are 'out of place' and to keep them 'in their place'" (1998: 354).The concept of being out of place fits well with the approach of critical places. We can therefore ask whether people experiencing difficulties of being in a specific place come to self-identify as "out of place" and self-exclude themselves from those places. Are people instead seeking only those limited places that define them as "in their place"?

The decisions of the young people participating in my research to take risks, both health risks and social risks, suggest a certain resistance to being confined only to where they are "in place" and being excluded from where they may experience being "out of place". From the interviews, it is obvious that the young people want to take place and be a part of what is happening. At the same time and perhaps paradoxically, it is in part exactly this *risk-taking* that can produce the feeling of being "out of place". Returning to my own experiences, being aware of taking risks and being in a specific place almost always create a feeling of not belonging in that place. This is not because the potential obstacles would be absolute if manifest or because the social and material specificities make it impossible to perform an expected normality. It is, rather, from a much more diffuse and not-articulated feeling that this is not my place, partly as no-one else seems to take any risks in being at this place where I feel that my health may be lacking. However, I cannot know if there are other people apparently "in place" feeling similar feelings of being "out of place", and so this diffuse feeling results as I get *convinced*, using Kitchin's vocabulary, that this is not my place.

Kitchin's "out of place" cannot only be seen as a critical place but is also an emotional category of feeling *out of place*. It is when a "feeling makes its appearance", following Merleau-Ponty (2002: 442), that the individual is faced with the critical place in this respect. The term therefore can help us methodologically to study where and when these feelings arise and change a "normal" place into something else for the individual. But Kitchin's perspective in combination with critical places can also help us to see that the individual can have earlier experiences of the critical places and therefore avoid them. "Knowing one's place" can be understood as a concept that refers to the, consciously or unconsciously, decisions that a person enacts in considering whether to "take place" or not.[9]

Conclusion

The concept of critical places analyses the duality of both risks and benefits and how they can collide in one specific place and create a bodily experience that demands an active response from the person. In this way, the concept focuses attention on the phenomenological aspects of *doing* and what happens in specific places, but the concept can also be drawn on as an ethnographic method. The concept of critical places affords the ethnographic researcher the methodological tools for focusing on those places where not only the critical but also the health emerges. It is in critical places that health becomes visible and "alive" as the person needs to relate to sensations from the body. This gives the ethnographic researcher the opportunity to study how the person relates to their health. In a self-reflexive perspective, this can be the start for a political project that can help us to analyse those places and health phenomena that most affect people with long-term illness. It can give a voice to what is happening at and within the critical places.

[9] This perspective is not developed further in the chapter, but "knowing one's place" shall be understood as the person sees himself with the eyes of the other – how to see oneself as object (cf. de Beauvoir 1997).

References

Alftberg, Å., Apelmo, E., & Hansson, K. (2016). Inledning: Ljudmiljöer, kulturella praktiker och hörselnedsättning. [Introduction: Sound environments, cultural practise and hearing impairment]. In Å. Alftberg, E. Apelmo, & K. Hansson (Eds.), *Ljud tar plats: Funktionshinderperspektiv på ljudmiljöer [Sound takes place: Disability perspective on audio environments] 11*. Lund: Department of Arts and Cultural Sciences, Lund University.

Boholm, Å. (2009). Speaking of risk: Matters of context. *Environmental Communication: A Journal of Nature and Culture, 3*, 335–354.

Chouinard, V., Hall, E. C., & Wilton, R. (2010). Introduction: Towards enabling geographies. In V. Chouinard, E. C. Hall, & R. Wilton (Eds.), *Towards enabling geographies: "Disabled" bodies and minds in society and space*. Farnham: Ashgate.

Cresswell, T. (1996). *In place/out of place: Geography, ideology and transgression*. Minneapolis: University of Minnesota Press.

Cridland, M. (2017). *"May contain traces of": An ethnographic study of eating communities and the gluten free diet*. Lund: Department of Arts and Cultural Sciences, Lund University.

de Beauvoir, S. (1997). *The second sex*. London: Vintage.

Dyck, I. (2010). Geographies of disability: Reflections on new body knowledges. In V. Chouinard, E. C. Hall, & R. Wilton (Eds.), *Towards enabling geographies: "Disab-led" bodies and minds in society and space*. London: Ashgate.

Foucault, M. (1987). *The history of sexuality, vol. 2: The use of pleasure*. London: Penguin.

Frykman, J., & Gilje, N. (Eds.). (2003). *Being there: New perspectives on phenomenology and the analysis of culture*. Lund: Nordic Academic Press.

Gleeson, B. J. (1996). A geography for disabled people? *Transactions of the Institute of British Geographers, 21*, 387–396.

Gleeson, B. J. (1999). *Geographies of disability*. London: Routledge.

Goffman, E. (1963). *Stigma: Notes on the management of spoiled identity*. Englewood Cliffs: Prentice-Hall.

Hagen, N. (2012). A molecular body in a digital society: From practical biosociality to online biosociality. In M. Liljefors, S. Lundin, & A. Wiszmed (Eds.), *The atomized body – the cultural life of stem cells, genes and neurons*. Lund: Nordic Academic Press.

Hansson, K. (2005). Camouflage. *Ethnologia Europaea, 35*(1), 136–140.

Hansson, K. (2007a). Unga viljor – kritiska platser: Platsens betydelse för unga män med astma och allergi. [Young wills – critical places: The importance of the place for young men with asthma and allergy]. In S. Olin Lauritzen (Ed.), *Leva med allergi: Samhällsvetenskapliga och humanistiska perspektiv. [Living with allergy: Social science and humanistic perspective]*. Stockholm: Carlsson Bokförlag.

Hansson, K. (2007b). *I ett andetag: En kulturanalys av astma som begränsning och möjlighet [In a breath: A cultural analysis of asthma as limitation and possibility]*. Stockholm: Critical Ethnography Press.

Harrison, B. (2002). Seeing health and illness worlds – using visual methodologies in a sociology of health and illness: A methodological review. *Sociology of Health & Illness, 24*, 856–872.

Husserl, E. (2001/1900/1901). *Logical investigations* (Vol. 2). London: Routledge.

Kitchin, R. (1998). "Out of place", "knowing onés place": Space, power and the exclusion of disabled people. *Disability & Society, 13*, 343–356.

Kleinman, A., & Benson, P. (2006). Anthropology in the clinic: The problem of cultural competency and how to fix it. *PLoS Medicine, 3*(10), e294.

Kusenbach, M. (2003). Street phenomenology: The go-along as ethnographic research tool. *Ethnography, 4*, 455–485.

Leder, D. (1990). *The absent body*. Chicago: University of Chicago Press.

McDowell, L. (1999). *Gender, identity and place: Understanding feminist geographies*. Cambridge: Polity.

Merleau-Ponty, M. (2002/1945). *Phenomenology of perception*. London: Routledge.
Soldatic, K., Morgan, H., & Roulstone, A. (Eds.). (2014). *Disability, spaces and places of policy exclusion*. New York: Routledge.
Wendell, S. (1996). *The rejected body: Feminist philosophical reflections on disability*. New York/Oxon: Routledge.

Chapter 6
Sensing Nature: Unravelling Metanarratives of Nature and Blindness

Sarah Bell

Abstract Understandings of nature, landscape and the senses have seen important shifts over time, from early medieval ideas of land 'shaping' to more detached contemplative notions of land 'scoping' since the seventeenth century. Despite a growing body of research demonstrating the richness of people's diverse multisensory nature encounters, certain interpretations and embodiments of nature are persistently prioritised over others in both policy and practice. This essay explores how long-standing preoccupations with the visual in people's understandings of nature and landscape have intersected with overly simplistic metanarratives of 'blindness' and disability in ways that are constraining contemporary opportunities for genuinely inclusive nature experiences, experiences that could instead be promoting health and wellbeing through inspiring curiosity, embodied connection, movement and challenge. To do so, I draw on the findings of a 2-year in-depth qualitative study, 'Sensing Nature', exploring how people with varying forms and severities of sight impairment experience a sense of wellbeing (or otherwise) in diverse types of nature during the life course. The essay examines how study participants had come to embody and experience different nature encounters, pursuing a 'zest to curiosity' in 'mundane everyday' as well as 'riskier' natures. In doing so, it seeks to counter overly simplistic sensory stereotypes of people with sight impairment as 'sub-' or 'super-humans', understanding people first and foremost as idiosyncratic individuals with diverse nature interests, embodied knowledge and skills.

Keywords Sight impairment · Nature · Disability · Landscape · Embodiment · Wellbeing

S. Bell (✉)
University of Exeter, European Centre for Environment and Human Health, Exeter, UK
e-mail: sarah.bell@exeter.ac.uk

S. Atkinson, R. Hunt (eds.), *GeoHumanities and Health*, Global Perspectives on Health Geography, https://doi.org/10.1007/978-3-030-21406-7_6

Introduction

> The picturesque in nature, it is true, is shut out from me, and works of art are to me mere outlines of beauty, accessible to only one sense; but perhaps this very circumstance affords a stronger zest to curiosity, which is thus impelled to a more close and searching examination of details than would be considered necessary to a traveller who might satisfy himself with the superficial view. (Holman 1834: 4)

In this opening quotation, James Holman, a celebrated 'blind traveller' of the early nineteenth century, highlights his 'zest to curiosity' in nature, particularly since the picturesque is seemingly 'shut out' from him as a consequence of sight impairment. Later in his journals, he expresses a sense of relief that his encounters with nature are not romanticised by such 'purple prose' or wrapped in the 'fashionable language' of the sublime and picturesque (Paterson 2014). In this chapter, I explore this 'zest to curiosity' amongst participants of a 2-year study, 'Sensing Nature', which examined how (if at all) people with varying forms and severities of sight impairment experience a sense of wellbeing with diverse types of nature in the context of their whole and everyday lives. After reflecting on the role of the visual in past and contemporary metanarratives of nature and blindness, I draw on participant accounts to discuss the importance of more-than-visual nature encounters with sight impairment, and the implications of this for how inclusive nature access is currently understood and promoted. These accounts emphasise the need to progress from thinking about people's needs solely in terms of disability to understanding people as idiosyncratic individuals with diverse nature interests, knowledges and skills.

The Place of the Visual in Metanarratives of Nature and Blindness

The imagination of Western society has long been captured by the aesthetic qualities of nature and so-called 'natural' landscapes: from the Persian tradition of 'paradise gardens' and the monastic gardens of the Medieval Period to the English Landscape Garden movement of the eighteenth century, the Urban Parks and Open Air movements of the nineteenth century and the contemporary drive to 'reconnect' people with nature (Walton 2000; Ward Thompson 2011; Carpenter 2013; Cleary et al. 2017). Through this time, focus has shifted from the myriad multisensory qualities of nature (e.g. in hospital cloister gardens in the twelfth century) to what have since been critiqued as overly ocular-centric notions of nature and landscape (Rodaway 1994; Macpherson 2008, 2017). Ingold (2011) attributes this shift to a conflation in the seventeenth century of the original notion of landscape – linked to early medieval practices of working and living with the land (with the etymology of 'scape' traced back to the old English word 'sceppan' or 'skyppan', meaning 'to shape') – with the Greek term 'skopos' (to look), therefore prioritising a particular 'scopic regime of detailed and disinterested observation' of land, landscape and nature (Nettleton 2015: 764).

Such visual pre-occupations gained further momentum in the eighteenth century, as three aesthetic ideals emerged with the rise of Romanticism: beauty, the picturesque and the sublime (Macnaghten and Urry 1998). Coupled with the Industrial Revolution, when land was increasingly 'commandeered' for industrial development, such aesthetic connotations catalysed new social norms about how to appreciate, preserve and be in nature (Matless 1998), alongside a growing tendency to conflate 'seeing' landscape with 'knowing' landscape (Wylie 2007). The very meaning of the term 'aesthetic' was re-appropriated during this time; derived from the Greek word for 'sense perception', the term was originally used to explore 'the plenitude and complexity of sensations', only later narrowing in application to more visual forms of detached 'contemplation and judgement of beauty' (Howes and Classen 2014: 19). Contemplation of so-called 'natural' landscapes was reified during this period for moral, spiritual, mental and physical 'improvement', aspects of improvement that are not dissimilar to those apparent within contemporary public health messages advocating nature 'on prescription' (Bragg and Leck 2017; Bell et al. 2019), be it for mental health purposes, social or spiritual wellbeing or meeting physical activity guidelines (Carpenter 2013).

Yet, as noted by Rodaway (1994: 123):

> A dominance of the visual reduces the body to a surface and marginalises the sensuality of the body. The same argument can also be extended to spaces in general, that is, the environment becomes first and foremost visually experienced and other sensuous geographies are marginalised.

Recognising this argument, there have been efforts to counter dominant visual, romantic and pastoral tropes of nature (Braun 2005), alongside growing literatures in both geography and anthropology exploring how people 'see in and with the landscape and are co-constituted by it' (Macpherson 2008: 265). This work moves beyond notions of fixed, detached landscape encounters, instead defining landscape in terms of relational engagement, immersion, connection and immediacy, 'present and alive in and of its on-going animation and becoming' (Wylie 2009: 278). However, as noted by Morris (2011), an emphasis on the visual in such becomings tends to persist, and certain interpretations and embodiments of nature continue to be promoted over others (Spinney 2006; Brown 2017). For example, 'scenic' routes and 'view points' are often still prioritised over the 'constellation of tiny, individual and collective experiential moments and emotions which build into an overall impression' in and of nature (Richards et al. 2010: 1100). This highlights a 'friction between subjectivity and sociality' (Saerberg 2010: 366), influencing how people perceive and experience the world in varied ways at different times.

Informed by these ideas, this chapter explores how nature and landscape come into being amongst people living with varied forms and severities of sight impairment, emphasising that 'to be without sight is not necessarily to be without landscape' (Macpherson 2005: 98) or to be without nature. In doing so, it seeks to counter such persistent tendencies to prioritise the visual in how we understand and promote access to nature, but also to challenge the dominance of overly simplistic tropes of what it means to be sight impaired. As demonstrated by Bolt (2016), symbolic and literary ideas – or 'metanarratives' of blindness (e.g. as darkness, ignorance or concealment) – are frequently confused with reality, 'over-coding' blindness with negative emotion (Paterson 2014: 97). These metanarratives not only displace the embodied agency of people with sight impairment but also fail to engage with or recognise the importance of people's broader physical, cultural and social life circumstances and trajectories in shaping perception and experience. Focusing solely on one's visual impairment in this way reduces 'the complex person to a single attribute' (Bolt 2016: 10), rather than recognising them to be as idiosyncratic as anyone else (Richards et al. 2010).

The sensory experiences of people with sight impairment, as with other disabilities, often fluctuate in the public imagination from that of *sub*-human to *super*-human (Bolt 2016: 72), oscillating from what Saerberg (2010) describes as the tendency to devalue that which deviates from the dominant societal mode of perception, to lauding what Kleege (1999: 3) refers to as 'the ancient myth of compensatory powers'. In reality, whilst people may learn to *use* their senses more effectively with compromised sight – including their residual sight, hearing, touch, smell, etc. – this does not happen overnight and requires attention, work, time and patience (Macpherson 2009). As reflected in the wider literature on disability and impairment, such stereotypes (positive and negative) – and the exclusionary or overprotective social attitudes that stem from them – are often more disabling than the direct physical experience of living with impairment itself (Thomas 2007; Richards et al. 2010). In this way, impairment of a single sense – be it someone's sight or otherwise – catalyses a broader form of 'cultural impairment' across multiple intersecting domains (Keating and Hadder 2010), shaping and re-shaping people's opportunities to experience a sense of health, wellbeing or otherwise whilst negotiating their everyday geographies.

After briefly introducing the methods used within the Sensing Nature study, I subsequently examine how the intersection of these metanarratives of nature, landscape, blindness and sight impairment influences contemporary opportunities for genuinely inclusive nature experiences. Specifically, I discuss the role of the more-than-visual senses in enticing and sustaining what Holman describes as a 'zest to curiosity' in nature amongst participants, and the skillscapes (Hunt 2018) that nurture such opportunities. In doing so, I emphasise the need to recognise and respect the varied skills and approaches developed to move in and with nature in a meaningful way over time, and to prioritise people's interests and *capacities* rather than reproducing disabling experiences of segregation or exclusion.

The Sensing Nature Study

Conducted from February to December 2017 in England, the Sensing Nature fieldwork comprised two overlapping phases. In the first phase, I participated in both sight loss awareness and sighted guiding training, before volunteering with a range of activity groups around the country. These included both indoor and outdoor groups (e.g. walking groups, social groups) designed to bring together sight-impaired individuals with common interests. Time spent volunteering with these groups (attending over 15 activity sessions in total) provided a valuable opportunity to build an initial awareness of people's diverse sensory worlds. Rooted in these discussions, Phase 2 involved a series of in-depth interviews with 31 people, recruited from around the country with the assistance of the Royal National Institute for the Blind *RNIB Connect* magazine, the Thomas Pocklington Trust, Blind Veterans UK, the Macular Society, and local and regional sight loss organisations who offered to raise awareness of the study through their member networks. The Phase 2 sample included 15 men and 16 women, aging from mid-20s to mid-80s, living in rural areas, towns and cities, with a range of eye conditions (including glaucoma, age-related macular degeneration, retinitis pigmentosa, diabetic retinopathy, congenital cataracts, retinopathy of prematurity, Leber's congenital amaurosis, Leber's hereditary optic neuropathy, coloboma, retinal detachment, rod-cone dystrophy and sight loss caused by accident and/or brain injury). Fifteen participants were born with at least one eye condition, with 12 of these experiencing further sight changes later in life.

Of the 31 people who took part in Phase 2, everyone participated in an initial nature-themed life history interview, lasting between 1 and 3 hours, examining what nature is to them; how they perceive, experience and negotiate different types of nature in the context of their day-to-day lives; how this had changed through key life transitions and 'chapters' of their lives; and how they feel about existing efforts to support more inclusive multisensory nature encounters, based on their experiences both within the UK and beyond. Twenty-five of the thirty-one participants took part in a second 'go-along' interview (Carpiano 2009) within a setting they valued for encountering nature in their day-to-day lives. These in situ – or 'emplaced' (Riley 2010) – interviews offered more subtle insights into the strategies used by participants to negotiate varied forms of nature. Settings included participant gardens, local residential road/path networks, urban parks, woodland, coast and countryside areas, with interviews lasting from 20 minutes to 4 hours (according to participant preference). All interviews were transcribed and subject to in-depth inductive thematic narrative analysis (Riessman 2008). In this chapter, I discuss one strand of the emerging findings, sharing participants' 'zest to curiosity' in nature, and how opportunities to explore nature in this way are influenced by intersecting metanarratives of nature, landscape, blindness and sight impairment.

Reclaiming Nature, Landscape and Movement with Sight Impairment

Seeking More-Than-Visual Encounters with Nature

In the opening quotation of this chapter, James Holman asserts that the 'picturesque' in nature is 'shut out' from him as a result of life with sight impairment. I explore this contention in more detail here, drawing on the experiences of the Sensing Nature participants. As demonstrated by many of these participants, this assertion is largely based on a concept of blindness as total, when in reality most people with sight impairment do still have, and *use*, some degree of residual vision, whether it's light, colour or form perception. As such, curiosity for the picturesque in nature was not necessarily absent with sight impairment, but rather, participants highlighted a shift in how they explored such qualities, including changes in the *relative* importance of the visual alongside their wider sensory attunements.

One participant in his 70s, for example, explained how the 'big visuals' – rather than fine-grained visual details – in nature had become more important with the onset of age-related macular degeneration as he could no longer see the small-scale details that he used to enjoy. Reflecting on some of his most memorable nature experiences, he commented:

> We went up to the Argentine borders, the Iguazu Falls. Huuuge waterfalls… It's like an enormous valley you stand at the end of, and right round this whole valley, both sides, it's cascading in. It's quite unbelievable… And, it's sort of something you can still see. It's so big, and the bigger and the greater the landscape you look at, the less (pause), significance it [sight loss] has, as it were, you know? The more it comes down to an individual plant, the plant gets smaller, or to a flower, and the flower gets smaller and smaller, so the difficulties increase… because if you look at things up close, it is distorted, you know. But the bigger the block, the less affect that has. (Eric, male, 70s, dry age-related macular generation)

Being able to engage with these dramatic views, and to appreciate nature at such a large scale, reduced Eric's sense of impairment, whereas the pastimes he had dedicated much of his life to – gardening and botany – required more adaptation and adjustment due to the loss of colour and detail that comes with macular degeneration.

Several other participants, including those born with eye conditions, demonstrated a sense of awe through encountering such dramatic visuals, with one participant explaining that many of her 'biggest wows' had been visual: 'the sky and sunrises, sunsets, those would be really big wows, the way that the light plays on things' (Annie, female, 60s, congenital cataracts and linked complications). However, life with sight impairment had also inspired curiosity for a plethora of *non-visual* sensory pleasures in nature. As Annie, continued, for example:

> But sense of smell, whether it's wood smoke or people having garden bonfires where they're burning greenery, I love that smell, absolutely. So I can go into reveries and raptures over that sort of thing… And sounds, it will be the dawn chorus, owls hooting, all that sort of thing. Oh, see now, I've gone on to music, and I'm thinking, you know (laughs), how do you separate nature and music? 'Cause there's an awful lot that's in nature that's melodic, you know, heartbeat to drumbeat, wind playing in the trees and all that sort of thing - all interwoven, interconnected'.

The importance of such multisensory awe-inspiring moments in helping Annie to navigate emotions unfolding through different – and often deeply challenging – life transitions (including divorce and bereavement) was apparent throughout her interview. One spot along her daily walking loop (where we did the go-along interview) was particularly important to her; this was a place she had been able to walk to relatively safely with her children when they were young, and later, when she had needed to get out of the house during the more tumultuous moments of her previous marriage. Whilst visual qualities of the setting did play an important role in stirring and working through such emotions, this was accompanied by specific scents, sounds and atmospheric sensations that enriched and deepened Annie's sense of nature connection, peace and presence:

> My favourite bit is the lane here and, the times that I get really in tune and in touch with it would be, well it just happens, like sometimes the sky and the atmosphere can be really, really pregnant… I just feel very in touch and in tune. It's almost like I could reach up and touch the sky… and I think that has more to do with, just the feel of it… If we're walking of an evening, I quite often stop and just have five minutes listening to the birds, listening to the birds going to bed… It's still and peaceful and tranquil, and there's nobody else around. And, yeah, it's one of my favourite places for just coming and *being*.

Nonetheless, curiosity for nature's visual qualities was still apparent amongst participants without residual vision. One participant, Chris, even suggested value in working with poets to translate the affective force of nature's visual 'wows' in a way that people could relate to and engage with at a non-visual level:

> Lots of it, I think is aimed at the visual, 'Oh what a lovely view'. And when people say that, you want to say, 'And? Describe it. We want description!' And not everybody, lots of people can't do it… and it's a mix of both the practical, but also getting a sense of how the person *feels* when you see a wonderful view. And maybe, only a poet can do that, for example. (Chris, male, 50s, advanced congenital glaucoma, no sight)

In this way, Chris emphasised the importance of conveying how the visual is *felt*, both physically and emotionally (Saerberg 2015), in order to participate fully in shared nature experiences. The assumption that the visual can be put into words reflects 'how deeply embedded visual metaphors are in everyday language' (Wylie 2007: 145) and the perceived challenge in *sharing* experiences that go beyond the visual. Indeed, other participants without residual vision often used visual language to share and describe the sense of awe or perspective they experienced through non-visual sensations unfolding at the location of certain viewpoints. This was apparent, for example, in the narratives shared by Miles, a participant in his 70s who had lost all sight by the age of 7. Whilst discussing his love of being outside in the landscape, he commented:

> Broadly speaking, I enjoy interactions with water, and also with hills… I think with that there's a, partly it's freshness of air and the way that you tend to get bigger vistas than when it's flat. I mean I know (hesitates), visually, flat landscapes you get big skies, and all that sort of thing. But that's less obvious to me. Whereas if I'm standing at the top of a hill, then it gives me, usually, more of an impression of the surrounding countryside than if I'm at the bottom – I mean that is a bit of a generalisation, but you know, broadly speaking, when you're up you've got a bigger perspective and that's true for both vision and, and sound. (Miles, male, 70s, advanced congenital glaucoma, no sight)

For Miles, the big 'vistas' were still important, not for their visual effects, but for (as he later described it) the clearer 'sound pictures' he was able to create in such settings, and the different qualities of air and microclimate at such heights. In this way, although described using visual language, enjoyment was derived from alternative aspects of nature embodiment (Small et al. 2012), be it through particular aural or haptic experiences or, as others noted, the physicality and kinaesthetically felt sensations of progressing towards particular viewpoints (Spinney 2006; Saerberg 2010). Several participants therefore highlighted the visual as a relatively small part of the overall sense of achievement and release experienced through having the freedom to move through, explore and engage with nature in varied ways and at different scales:

> When I'm seeing a view, obviously I'm not seeing what you're seeing, 'cause I don't have the level of vision that you have. But I'm pretty sure that in some ways, I'm trying to compensate for that, either consciously or subconsciously, to some degree by imagining what *could* be there based on what I *can* see is there, or what other people tell me is there… So even though I can't see a view (pause), it's still relevant to experience it. I wouldn't want to over-egg that. I mean, I wouldn't want to go on a purely sightseeing holiday, for example. That would be entirely pointless. But if a view was one component of a set of experiences that I could appreciate, like climbing a mountain, you know, it's physical, it's tactile, it's sensory, it's hard work, and there's a sense of achievement at the end of it. And then you get to the top, and one of the things you experience is the view, then that would still be relevant. But I wouldn't do something, I wouldn't *usually* do something solely for a visual outcome'. (Brett, male, 40s, congenital glaucoma)

Sustaining a 'Zest to Curiosity': The Importance of Skillscapes

Many participants felt that nature still has much to offer in terms of health, wellbeing and pleasure, even without a visual interaction, and emphasised the importance of being attentive with all the senses. Importantly, and countering myths and stereotypes of compensatory 'super sense' development (Kleege 1999), this multisensory awareness did not 'miraculously' appear with sight loss. Rather, it was described as a quality that participants had to *work* at, developing new embodied dispositions and skills to find alternative ways of tuning into, navigating and deriving pleasure and meaning from different environments:

> I mean they say when your eyesight goes, your other senses improve – they don't. You *use* them more to listen to what's happening around you, so that you end up hearing the things that you never bothered to hear before. And they sort of, they rise in importance… the two main ones are touch and hearing. The hearing, to me, is enhanced by the background noise … you'll be walking down through the woods, you'll hear the birds in the background… and it fills that sort of (pause), space where there'd be nothing. Not that you have to fill it, but it's nice to have something to fall back on. (Quinn, male, 60s, retinitis pigmentosa)

Visual interactions with the world can generate a sense of permanence and perspective, with other sensory inputs often experienced as more ephemeral or fleeting (Rodaway 1994). As such, with the loss of sight, one's stability and continuous stream of sensory information can feel compromised, until, like Quinn, people learn to tune into the wider multisensory world of 'happenings' or movement (Hull 1990). Several participants therefore highlighted a 'zest to curiosity' in the dynamics of everyday nature, noting the value of 'observing' – or immersing oneself within – for example, the weather, local wildlife and plant growth, rather than relegating nature to the large swathes of coast or countryside that people tend to associate with 'getting out into nature'. As one participant explained, this realisation can take time:

> Your depth perception, where it goes fuzzy, is first to go. So I mean, it's bad enough to find the kerb edge, and that's only a drop, whereas unknown countryside and pitfalls etc., you know – therefore nature was something that you just thought was going to be an area you wouldn't be able to go to anymore. So that was a kind of bereavement of, of losing nature. Until you realise, actually there's nature all around us, you know, which was good, once you realise that…. So it's something you can still experience, in a different way. It's not better, it's not worse, but you can still experience it… And it might be, all you need to do is tend some plant pots on your windowsill, as opposed to saying, 'I need to climb Mount Everest'. And if you can imagine all the variations between a plant pot and Mount Everest, you will find something that you want to do and like and enjoy and get fulfilment from. (Chris, male, 50s, advanced congenital glaucoma, no sight)

Recognising that nature *can* be everything from a plant pot to climbing Mount Everest is important in light of current efforts to promote inclusive nature access. In what follows, I highlight how such efforts could potentially be enhanced through conceptualising people's nature encounters as 'skillscapes', that is, as fields of emplaced socio-material-affective relations that may induce 'certain capacities for skill' (Hunt 2018: 73), with skills developed 'ecologically, through bodily movements alongside environmental interactions' (Hunt 2018: 82).

'Skilling up' (Hunt 2018: 78) can happen at multiple scales of nature interaction. For example, for people who have lost confidence in tackling the more unpredictable, uneven, changeable qualities of nature, participants highlighted how supporting small-scale experiences, such as planting up a hanging basket, can be invaluable. This offers an opportunity to engage with the soil through the hands, to find new points of reference for identifying different plants – be it through scent, texture or phyllotaxis – and to gain a sense of achievement through producing something that *feels* right to them regardless of its visual impression. Over time, people may build more skills and confidence to tackle alternative forms of nature, be it the back garden, a local park or satisfying curiosities further afield. At this stage, participants emphasised the importance of accommodating the *kinaesthetic* sense; so often, the response of landscape architects, planners and environmental engagement organisations is to design a sensory garden for people with sight or other sensory impairments. Whilst such efforts are appreciated, they tend to privilege relatively simple forms of nature contact, via passive touch, hearing or smell, at the expense of more expansive, exploratory, multisensory experiences that nurture a sense of dimension, scale and orientation, spark curiosity and support people to move – and to move *freely* and safely – in and with nature:

> To be quite honest, most sight impaired people are completely bored by the sensory garden… People have a go at playing with it but it very often isn't particularly sensorially rich. You know, people think, oh you fill it with plants which have got lots of movement, so that the air moving through them provides lots of sounds, and lots of different textures and smells, and so on. But to some extent, once you've been in it, you've been in it. It's being able to *move* that I think is most enriching, and being able to move safely… One of the things that I like about the park near here is, it's got a number of different types of environments. And I think that's what needs to be considered. So you've got somewhere that is, you know, sort of cool and dark and damp. But you've also got somewhere that is high and light and airy, and somewhere where you can hear water, but also somewhere where you've got really lofty trees. So it's variety really. (Eve, female, 60s, congenital cataracts, later developing glaucoma)

As noted by Small (2015), the body senses as it moves, and does so most fluidly when movement and feeling are in unison. Participants, too, emphasised the need to recognise the whole body tactility of being in and with nature, and the importance of skilled movement in opening up such opportunities. Linked to this, many expressed frustration at the tendency to conflate the priorities and preferences of people with sight impairment with those reliant on wheelchair access. Two of the Sensing Nature participants did have additional mobility challenges and therefore appreciated efforts to enhance wheelchair access, with one participant in her 80s commenting:

> I think it's lovely if, if people don't expect too much, you know, it's lovely just to be able to sit in my wheelchair, sometimes, and just relax, and enjoy the pleasure of relaxing in the countryside. (Alva, female, 80s, advanced glaucoma)

However, many other participants explained how being limited to short sensory trails or small sensory gardens inadvertently created segregated experiences, treating them first and foremost as people with disability, rather than people with an interest in nature's varied forms and habitats. As one participant commented:

> The trouble is, access is seen as wheelchair users… But, sight loss is not really seen as a problem anywhere… So, I feel sight loss has been way down, you know, there's very little done to help people with sight loss out in the environment. So you can hire these buggies to go on. But, again, that's really motorised buggies, that's just for people who are in wheelchairs… And where they make paths to go across rural areas, areas of special interest with butterflies and things, sometimes that is, sight loss is not considered and there's pitfalls walking on that, you know? (Flo, female, 60s, congenital cataracts and myopia, detached retina, glaucoma)

By limiting people with sight impairment to relatively short wheelchair-friendly routes or worse still, by suggesting the use of unnecessary mobility aids, such as motorised buggies, such efforts undermine and overlook the dignity and varied capacities, interests and needs of people with sight impairment. As one participant commented in relation to the enforced use of airport buggies, for example:

> You are effectively saying my legs don't work, and you're actually now taking away my ability to walk independently, which might seem like a really small thing to you. But when I am already blind it's, it's totally unacceptable. (Brett, male, 40s, congenital glaucoma)

Whilst Brett's example does not relate to a nature encounter per se (beyond perhaps human nature), it captures an issue that recurred throughout the Sensing Nature interviews, whereby the *social* attitudes and misperceptions encountered in nature were almost more disabling than the physical risks posed by nature itself. This highlights the importance of countering 'ableist assumptions of vulnerability' (Porcelli et al. 2014: 873) that stifle and undermine people's 'zest to curiosity', failing to recognise the alternative skills and practical knowledges that people develop to negotiate and embody more complex environments:

> People's reactions to me make me want to stay away from a lot of things, and not get involved in a lot of things. So it affects my life quite, quite seriously – not the lack of sight, but more other people's interactions with me. And you know, that has led to serious depression over the years…I've never been on any of these walks, like the coastal path… and

> thought, 'Oh, I wish I could see that view'. Never have. And, that is part of me, somehow. So when people think, 'Oh poor thing, he can't see that view'. Yeah, fine, you think I'm missing out, maybe I am, but it doesn't bother me. That's not why I'm unhappy. It, it's people's reactions… They can be very overprotective… You know, you can never explore anything. (Seth, male, 50s, lost all sight in early life through injury)

This distinction is important; whilst participants often recognised that very little can be done to certain nature settings to improve access, safety and navigability (e.g. protected upland, coast or countryside areas), they felt that far more could be done to counter such debilitating negative social attitudes around sight impairment. To some extent, this relies on dismantling material, social and affective barriers to exploration within nature settings that *are* first and foremost 'built' and managed by and for people (e.g. urban parks, community gardens, coastal promenades). This may involve, for example, incorporating appropriate audio, tactile and clear-contrast visual markers to support independent navigation by people with sight impairment, rather than restricting their access to small segregated 'sensory' areas. By countering 'entrenched binary systems that differentiate and distance the world of the blind' (Bolt 2016: 125), members of the general public may start to notice, appreciate and respect the alternative skills and approaches adopted by people with sight impairment to move through such settings both smoothly and freely, recognising their *capacities* rather than fixating on disabilities. However, such transformations also rely on shifts in the skillscapes of outdoor activity providers, encouraging more openness amongst such professionals to adapt their specialist skill sets to the sensory priorities of people with sight impairment. In doing so, they could support people to tackle 'riskier' forms of nature in settings that cannot easily be modified or adapted, be it whilst caving, climbing, kayaking and so forth. As illustrated in the quote below, however, these opportunities were hard to find for many participants:

> It was going to be a parachute jump, and we were trying to raise money for it because we were doing it for charity… and we went to the local parachute place but they weren't very keen about it, they were really negative, and like, health and safety… and then my friend looked up a hang gliding place instead so we went over there and had a chat and they were really nice so we did hang gliding instead. (Annabelle, female, 20s, Leber's congenital amaurosis)

Since this participant – like any other novice parachute jumper – would have been strapped to a qualified tandem parachute jump instructor, there was no reason why sight impairment alone should have created barriers to taking part in the jump. As noted by another participant, one of the biggest barriers in being able to explore a full range of nature experiences is the lack of qualified people with the ability to appreciate and manage *sensible* risk with (in his words) 'grace and ease' (rather than *over*-perceiving risk), people who recognise the dignity of risk (Marsh et al. 2018) and base such judgements on people's individual parameters, skills and capabilities. Only with such transformations can people start to unravel overly restrictive metanarratives of sight impairment and blindness, and unsettle simplistic binary notions of 'able' or 'disabled' bodies in nature.

Concluding Remarks

In this chapter, I have examined how longstanding preoccupations with the visual in people's understandings of nature and landscape have intersected with overly simplistic metanarratives of 'blindness' and disability in ways that often constrain people's opportunities to pursue what 'blind traveller', James Holman, described as a 'zest to curiosity' in nature.

Whilst participants in the Sensing Nature study appreciated efforts made to create experiences that explicitly appeal to the senses, for example, via sensory gardens or sensory trails, and recognised the need to (re)build confidence in negotiating nature *gradually*, many expressed frustration at the tendency to delineate such segregated experiences, experiences that identify and construct people first and foremost by their impairments rather than as people from diverse backgrounds with an interest in the idiosyncrasies of nature. Many of these 'designated' experiences failed to cater for participants' desire to explore, to learn or to move freely and expansively in nature. Indeed, as noted by Spinney (2006: 729), encounters with landscape – and I would argue with nature in its broader sense – are often embraced not for passive sensory experiences but 'because of what it offers inwardly from an embodied and active participation within it'. Such participation may entail a degree of risk, yet opportunities to negotiate such risks in nature in the name of challenge and exploration are curtailed for many people through over-inflated perceptions of vulnerability and persistent stereotypes of what it means to live with sight impairment.

A recent report by the Sensory Trust in collaboration with Natural Resources Wales (2017: 9) argues that:

> Equality isn't about treating everyone the same, it is about respecting differences and finding ways to make everyone feel welcome and supported. At its root, what is most needed is a positive, open attitude and a welcoming approach combined with good communication and understanding.

In many ways, nurturing such positive, open attitudes relies on the unravelling of both negative and positive stereotypes that limit people's understanding of what varied bodies can do. It requires more widespread awareness of the diversity of nature encounters that support feelings of health and wellbeing, accepting nature in its *fullest* sense rather than continuing to reproduce narrow ideologies of the visual aesthetic, picturesque or sublime. Through 'exploring broader notions of health, wellbeing and human flourishing' and 'rejecting biomedical reductionism without abandoning the materiality of the body' (Atkinson et al. 2015: 74), future work in the medical humanities has an important role to play in promoting such awareness, a shift that is clearly evidenced by the current volume. Only through such critical approaches to health, wellbeing and embodiment will we start to move beyond embedded social norms, expectations and misinformed notions of embodied difference (Saerberg 2010) that fail to recognise, value or cater for a multiplicity of ways of sensing, being or moving through the world.

References

Atkinson, S., Evans, B., Woods, A., & Kearns, R. (2015). 'The medical' and 'health' in a critical medical humanities. *Journal of Medical Humanities, 36*, 71–81.

Bell, S. L., Leyshon, C., Foley, R., & Kearns, R. (2019). The 'healthy dose' of nature: A cautionary tale. *Geography Compass, 13(1), e12415.* https://doi.org/10.1111/gec3.12415.

Bolt, D. (2016). *The metanarrative of blindness. A re-reading of twentieth-century anglophone writing*. Ann Arbor: The University of Michigan Press.

Bragg, R., & Leck, C. (2017). *Good practice in social prescribing for mental health: The role of nature-based interventions*. York: Natural England Commissioned Reports, Number 228.

Braun, B. (2005). Environmental issues: Writing a more-than-human urban geography. *Progress in Human Geography, 29*, 635–650.

Brown, K. M. (2017). The haptic pleasures of ground-feel: The role of textured terrain in motivating regular exercise. *Health & Place, 46*, 307–314.

Carpenter, M. (2013). From 'healthful exercise' to 'nature on prescription': The politics of urban green spaces and walking for health. *Landscape and Urban Planning, 118*, 120–127.

Carpiano, R. (2009). Come take a walk with me: The 'go-along' interview as a novel method for studying the implications of place for health and wellbeing. *Health & Place, 15*, 263–272.

Cleary, A., Fielding, K. S., Bell, S. L., Murray, Z., & Roiko, A. (2017). Perspective essay: Nature connection and eudaimonic wellbeing: Exploring potential mechanisms. *Landscape and Urban Planning, 158*, 119–128.

Holman, J. (1834). *Voyage around the world, including travels in Africa, Asia, Australasia, America* (Vol. 1). London: Smith, Elder & Co.

Howes, D., & Classen, C. (2014). *Ways of sensing: Understanding the senses in society*. Abingdon: Routledge.

Hull, J. M. (1990). *Touching the rock: An experience of blindness*. London: Society for Promoting Christian Knowledge.

Hunt, R. (2018). On sawing a loaf: Living simply and skilfully in a hut and bothy. *Cultural Geographies, 25*, 71–89.

Ingold, T. (2011). *Being alive: Essays on movement, knowledge and description*. London: Routledge.

Keating, E., & Hadder, R. N. (2010). Sensory impairment. *Annual Review of Anthropology, 39*, 115–129.

Kleege, G. (1999). *Sight unseen*. New Haven: Yale University Press.

Macnaghten, P., & Urry, J. (1998). *Contested natures*. London: Sage Publications Ltd.

Macpherson, H. (2005). Landscape's ocular centrism – and beyond? In *Proceedings of the Frontis workshop from landscape research to landscape planning: Aspects of integration, education and application*, Wageningen, The Netherlands, 1–6 June 2004.

Macpherson, H. (2008). Between landscape and blindness: Some painting of an artist with macular degeneration. *Cultural Geographies, 15*, 261–269.

Macpherson, H. (2009). Articulating blind touch: Thinking through the feet. *The Senses and Society, 4*, 179–193.

Macpherson, H. (2017). Walking with visual impairments in the british countryside: Picturesque legacies, collective enjoyments and well-being benefits. *Journal of Rural Studies, 51*, 251–258.

Marsh, P., Courtney-Pratt, H., & Campbell, M. (2018). The landscape of dementia inclusivity. *Health & Place, 52*, 174–179.

Matless, D. (1998). *Landscape and englishness*. London: Reaktion Books.

Morris, N. J. (2011). Night walking: Darkness and sensory perception in a night-time landscape installation. *Cultural Geographies, 18*, 315–342.

Natural Resources Wales. (2017). *By all reasonable means: Least restrictive access to the outdoors.* A guide produced by the sensory trust in collaboration with, and on behalf of, Natural Resources Wales. Cardiff: Natural Resources Wales.

Nettleton, S. (2015). Fell runners and walking walls: Towards a sociology of living landscapes and aesthetic atmospheres as an alternative to a Lakeland picturesque. *The British Journal of Sociology, 66*, 759–778.

Paterson, M. (2014). Blindness, empathy, and 'feeling seeing': Literary and insider accounts of blind experience. *Emotion, Space and Society, 10*, 95–104.

Porcelli, P., Ungar, M., Liebenberg, L., & Trépanier, N. (2014). (Micro)mobility, disability and resilience: Exploring well-being among youth with physical disabilities. *Disability & Society, 29*, 863–876.

Richards, V., Pritchard, A., & Morgan, N. (2010). (Re)envisioning tourism and visual impairment. *Annals of Tourism Research, 37*, 1097–1116.

Riessman, C. K. (2008). *Narrative methods for the human sciences*. London: Sage.

Riley, M. (2010). Emplacing the research encounter: Exploring farm life histories. *Qualitative Inquiry, 16*, 651–662.

Rodaway, P. (1994). *Sensuous geographies: Body, sense and place*. London: Routledge Taylor and Francis Group.

Saerberg, S. (2010). Just go straight ahead. *The Senses and Society, 5*, 364–381.

Saerberg, S. (2015). Chewing accidents: A phenomenology of visible and invisible everyday accomplishments. *Journal of Contemporary Ethnography, 44*, 580–597.

Small, J. (2015). Interconnecting mobilities on tour: Tourists with vision impairment partnered with sighted tourists. *Tourism Geographies, 17*, 76–90.

Small, J., Darcy, D., & Packer, T. (2012). The embodied tourist experiences of people with vision impairment: Management implications beyond the visual gaze. *Tourism Management, 33*, 941–950.

Spinney, J. (2006). A place of sense: A kinaesthetic ethnography of cyclists on Mont Ventoux. *Environment and Planning D, 24*, 709–732.

Thomas, C. (2007). *Sociologies of disability and illness*. Basingstoke: Palgrave.

Walton, J. (2000). *The British seaside: Holidays and resorts in the twentieth century*. Manchester: Manchester University Press.

Ward Thompson, C. (2011). Linking landscape and health: The recurring theme. *Landscape and Urban Planning, 99*, 187–195.

Wylie, J. (2007). *Landscape*. Abingdon: Routledge.

Wylie, J. (2009). Landscape, absence and the geographies of love. *Transactions of the Institute of British Geographers, 34*, 275–289.

Part II
Voices

Chapter 7
Subjectivity, Experience and Evidence: Death Like Milk on the Doorstep

Hannah Bradby

Abstract Evidence-based medicine has been criticised as failing to include values and priorities as part of the evidence that determines treatment. Narrative-based approaches have developed to meet this lack and represent the subjective aspects of the experience of illness and treatment. But to inform medicine, patients' narratives should be reliable, stable and documentary such that contradictory, fantastic, metaphorical elements of experience are still excluded as legitimate evidence. This essay suggests that story offers a version of patient experience that encompasses wider aspects of the subjectivity of illness, suffering and loss. As the unreliable cousin of narrative, story's virtue is that it is harder to assimilate to the epistemological demands of biomedicine. Two unreliable stories representing experiences of pain, suffering, care and loss are presented. First, Jenny Diski's story of her double diagnosis with pulmonary cancer and fibrosis demonstrates ambivalent experiences that cannot easily be reduced to a stable, documentary narrative, relying as they do on memory and metaphor. Second, documentary narrative's limits to represent experience become apparent at the point where biomedicine's therapeutic methods fail and death comes into view. Sara Ryan's account of her son's death while in a 'care' unit illustrates how the truth of some experience may only be represented through imagined interactions that could never actually occur.

Introduction

Biomedicine has built its reputation by promising to extend the human lifespan and relieve the suffering attendant on illness and injury (Bradby 2016). Through the systematic use of scientific findings, evidence-based medicine has sought to improve health outcomes. The strong knowledge claims of bioscience have undercut the value of other sorts of evidence, especially those deemed subjective. Patients' values, their priorities for treatment and their experience of illness have consequently

H. Bradby (✉)
Uppsala University, Uppsala, Sweden
e-mail: hannah.bradby@soc.uu.se

S. Atkinson, R. Hunt (eds.), *GeoHumanities and Health*, Global Perspectives on Health Geography, https://doi.org/10.1007/978-3-030-21406-7_7

been lost from clinical decision-making. Narrative-based medicine has offered one means of adding experience and values back into the healthcare process.

In capturing patients' subjective experience, narrative approaches to medicine have established certain limitations to the form of narrative that can be admitted. Stable narratives of reliable subjects' experience may qualify as legitimate evidence to complement the evidence base that informs healthcare. The epistemological necessity of stable, documentable truths holds for narrative-based medicine just as for the evidence-based medicine that it supplements. Patients' experience and values must be presented as rational and stable narratives, particularly if they contradict professional priorities and guidelines with respect to treatment. Experience that cannot be assessed as documentary and as rational tends to be lost from view, leaving a range of experiences associated with illness lying outwith the definition of acceptable evidence on the grounds of being unstable, fantastic and/or irrelevant. A fatal diagnosis or injury is a moment when experiences cannot necessarily or easily be captured in stable, documentary, reliable narrative form. Without the prospect of cure, existential uncertainties which are not easily encompassed as stable evidence can become an important part of experience.

This chapter looks at how the end of life is written about in stories that would not be considered suitable as narrative evidence to inform healthcare decision-making. Narrative, in a stripped-down documentary form, has been subsumed to the medical project as a complement to the evidence base, but what has yet to be admitted are contradictory, fantastic metaphorical elements of experience. Story can offer a version of patient experience that encompasses wider aspects of the subjectivity of illness, suffering and loss. As the unreliable cousin of narrative, story's virtue is that it is harder to assimilate to the epistemological demands of biomedicine.

Medicine and Evidence

Medicine as knowledge-system, discipline and practice is a broad church that has managed to cohere as a professional project despite the ebb and flow of power and influence over time (Porter 1999). Medicine's social, economic and political influence is claimed on the basis of epistemological superiority and the efficacy of its therapeutic methods. The term **bio**medicine underlines how the medical professional project rests on the epistemological authority of science to support its knowledge claims. Biomedicine presents as a neutral form of knowledge, presumed to be universally applicable (Gordon 1988).

Developing out of clinical epidemiology, evidence-based medicine (EBM) epitomises the neutral power of biomedicine to apply scientific logic to address uncertainties around treating illness. EBM has been a powerful means of integrating the findings of systematic research into individual clinical decision-making to avoid idiosyncratic and precedent-driven choices through the logic of 'conscientious, explicit, and judicious use of current best evidence in making decisions about the care of individual patients' (Sackett et al. 1996). Evidence-based medicine is supplemented by narrative that represents patients' values and treatment priorities.

Evidence-based medicine defines guidelines and protocols of best practice through the meta-analysis of research results, inscribing a hierarchy of evidence, with the systematic review at its pinnacle as the most trustworthy form. The weight given to systematic reviews has earned EBM 'the reputation of dismissing the importance of the singular predicament of the patient and the individual judgment of the doctor' (Charon and Wyer 2008, p. 297). A normative focus on populations (Tonelli 1998) means that evidence-based medicine fails to account for marginalised population groups (Asthana and Halliday 2006) and has had the effect of undermining the value of evidence other than systematic reviews. The weight of experience, anecdote and intuition to inform clinical decision-making has been called into question, with EBM having 'standardizing and marginalizing effects … on other forms of knowledge' (Hanemaayer 2016). In particular EBM is seen as excluding patient views and narratives from consideration, one effect of which is to re-assert medical authority (Denny 1999).

The inability of evidence-based medicine to accommodate non-documentary knowledge has opened up space for narrative-based medicine to represent the subjective aspects of the experience of illness and treatment. While narrative-based insights may supplement EBM, the assimilationist tendencies of evidence-based thinking (Lambert 2006) imply a selective incorporation such that certain categories of experience remain inadmissible.

Narrative-Based Medicine and Its Limits

In order to counter-act the universalising tendencies of evidence-based medicine, erasing patients' (and also clinicians') values from consideration when planning treatment, attention has been focussed on narrative as a supplement to systematically reviewed evidence. Narrative (in combination with patient-centred care) is a means of countering the normative, standardising effects of evidence-based medicine:

The narrative provides meaning, context and perspective for the patient's predicament. It defines how, why and in what way he or she is ill. The study of narrative offers a possibility of developing an understanding that cannot be arrived at by any other means (Greenhalgh and Hurwitz 1998).

Narrative-based medicine offers a remedy for the excesses of evidence-based medicine – a means of reinstating 'the narrative singularity of both patients' and clinicians' lived experience' (Charon and Wyer 2008, p. 197). Good-quality decision-making for an individual patient requires an acquaintance with that person's context and clinical experience of combining best evidence with individual circumstances (CEBM Oxford 2015).

In soliciting narratives to supplement evidence-based clinical decisions, patients are assumed to be willing and able to produce coherent stable narratives to underpin cooperation with healthcare professionals who can access the evidence. This view depends on a set of assumptions about stable personhood, rationality and communication which may not be borne out in practice (Naldemirci et al. 2018). Suffering and existential uncertainty engendered by pain and/or curtailed life expectancy may call stable personhood and rational communication into question, such that a patient cannot produce a narrative that is usable in conjunction with other, rational evidence. The imminence of death might make it particularly difficult to make sense through rational narrative.

Potentially more problematic than patients' ability and willingness to produce suitable narratives is whether narrative, in the form deployed in narrative-based medicine, is capable of conveying the full range of subjective experience. Notwithstanding the debate over the extent to which narrative actually captures patient voices (Lambert 2006), some experience will never qualify as narrative evidence. The epistemological terms of biomedicine are such that contradictory, uncertain and/or absurd aspects of experience may not be rendered into appropriate evidence for analysis, even in narrative form. EBM has selectively incorporated narrative-based insights (Lambert 2006), and through this assimilation asserts the need for evidence to be stable and documentary, albeit narrative. Paradoxical, perverse and unstable experiences that cannot be easily rendered as documentary, and those narratives are lost to view. While a dichotomy between medical practice based on narrative and practice based on evidence is not clear cut,[1] narrative approaches nonetheless presume a documentary character.

The term narrative has, therefore, been used as a form of evidence which strips out and loses from view irrational aspects of all-too-human stories. As an alternative to narrative, the term 'story' that may include fictional, metaphorical and mythical elements, encompassing complex and contradictory aspects of experience, is proposed. In charting the limits of biomedicine and grasping for ways of representing the lived human experience of illness, injury and healthcare (including failed healthcare), the suggestion here is that we need to go beyond objective, documentary evidence (albeit in narrative form), to include stories which are not only subjective but also unreliable and fantastic.

The next section describes story, in contradistinction to narrative, before considering two unreliable stories representing experiences of pain, suffering, care and loss. Jenny Diski's story of her diagnosis and dying from pulmonary cancer and fibrosis shows ambivalent experiences that cannot easily be reduced to a stable, documentary narrative. Documentary narrative's limits to represent patient experience become apparent at the point where biomedicine's therapeutic methods fail and death comes into view. Sara Ryan's account of her son's death while in a 'care' unit illustrates how the truth of some experience may only be represented through imagined interactions.

---

[1] To avoid a false dichotomy between evidence and narrative, the term 'narrative evidence-based medicine' (NEBM) has been coined (Charon and Wyer 2008).

Story and Narrative

In the study of narrative (Riessman 1993) the terms story and narrative are used interchangeably, without defining their relationship to one another (Thornborrow and Coates 2005, p. 3). Narrative covers a wide range of forms and functions, including extended oral accounts of personal experience (Labov and Waletzky 1967), spontaneous casual conversation (Eggins and Slade 2006) and written and visual material (Squire 2012). Narrative has been defined in terms of its function and its structure, with a minimal definition being two events that are linked sequentially. The 'small story' of everyday chat has been contrasted with grand narrative (Georgakopoulou 2006) where 'big stories' are recognisable as the type of autobiographical accounts solicited in research interviews (Sools 2012). The 'particular and locatable life experiences' that are enfolded within 'small stories' attend to the process of working out concepts in local settings (Lorimer 2003, p. 214). In this chapter story is used as a term that encompasses and surpasses narrative: while both story and narrative can range from a minimal account of events to an epic tale, story has not been co-opted into the project of narrative-based medicine. While narrative-based medicine requires some form of reliable evidence, story retains the possibility of being unreliable and irrational.

Story

Story is a universal way of being human (Bruner 1990) through representing experience to ourselves and to others, rendering the everyday meanings by which we live (Bruner 2004). Story-telling allows different forms of knowledge to be connected (Bruner 1986) and organised into collective social representations of health and illness (Murray 2002). Story conveys personal as well as shared meaning, representing 'epistemic shifts on personal and intimate terms' (Lorimer 2003, p. 214). Stories are not only a way of 'organising knowledge, but they are constitutive rather than descriptive' (Game and Metcalfe 1996, p. 40).

While stories may have a recognisably Aristotelian beginning, middle and end, this structure may be acknowledged through its disruption or negation, with a timeline not necessarily proceeding in a linear fashion. The ordering of events, feelings and observations by narrative logic (Ricoeur et al. 1984) creates and conveys meaning from otherwise chaotic, disordered events. The disruption of pain, bodily sensations and secretions require stories to incorporate them into daily life (Bury 1982). Stories make sense of a life disrupted by distress and suffering, reconstructing and reconfiguring the world (Frank 1991; Kleinman 1988). A story could describe such disruptive sensations to allow their recognition as symptoms, facilitating medical management. But this would not be the only story available and might not be the most compelling. A compelling story entertains, enchants and even transports. Enchanting and entertaining stories include elements that the minimally defined narrative may not, such as metaphor, irony, exaggeration, satire and ridicule. The elements that render a story memorable and worth re-telling disqualify it from being reliable and valid evidence. The rehearsal of a story, polishing its timing for comic or dramatic effect, including new details and narrative intrigue, renders it inauthentic as narrative evidence admissible in a clinical encounter.

Stories represent an 'expression of yearning for the great escape from death' (Tolkien 1947), offering temporary escapism from the inevitability of mortality and suffering by transporting the audience beyond the space-time constraints of daily embodied life. Stories explain humans' short and limited lives and answer metaphysical questions around the nature of existence (Kearney 2002), bringing living generations into communication with forebears' meaning-making: stories are 'machines for the suppression of time' (Doja 2008).

Representing Experience: Writing Towards Death

The onset of a potentially fatal disease can be a powerful prompt to write, with the 'autopathography' (Aronson 2000) charting responses to a serious diagnosis. These experiences include the search for cure, rehabilitation (Conway 1997) and underlying causes (Burnside 2006), as well as decline and withdrawal from the social world as death encroaches (Diamond 1998; Kalanithi and Verghese 2016; Nobel 2005; Picardie 1998).

Jenny Diski was diagnosed with two lung diseases – fibrosis and cancer – both incurable. She wrote about her encounter with these 'two ways to die from the lungs' (Barber 2016) and with the healthcare professionals who diagnosed and treated her. Diski died 18 months after her diagnosis, as her book *In Gratitude* (Diski 2017) was being reviewed and published. Part memoir, part cancer diary, the book is made up of articles first published in the *London Review of Books*, documenting her encounters with healthcare providers and biomedical prognoses.

Diski's acerbic comments could be read as a guide for improving the experience of cancer services, as she charts, for instance, the slowing of time when waiting for the results of a scan (Diski 2014d), the deficits of the gowns provided for radiotherapy patients (Diski 2015b) and how to circumvent palliative care nurses' refusal to discharge a dying patient to her own home (Diski 2016).

On being given the news of her inoperable lung cancer, Diski addresses a 'Breaking Bad' joke to her partner: 'So – we'd better get cooking the meth'. The oncology specialists who have just given her the bad news do not react to the joke. Diski recounts her own embarrassed realisation that they may have heard similar 'crystal meth' jokes ever since the hit television show[2] aired in the UK and regrets

[2] An acclaimed American drama series, first broadcast in 2008, in which a terminal cancer diagnosis induces a previously law-abiding chemistry teacher to manufacture and sell crystal meth, in a bid to provide financial stability for his family after his death. https://en.wikipedia.org/wiki/Breaking_Bad

that she is already 'a predictable cancer patient' (Diski 2014b). Diski's forensic account of how she was given her grim diagnosis could be included in a 'Breaking Bad News' course which feature in clinical education (and which the show's title 'Breaking Bad' echoes). Over the course of her illness, Diski experienced compassionate, kindly care (Diski 2014d) and detached alienating treatment (Diski 2015b) from NHS staff. But her writing offers more than a series of pithy observations of the ups and downs of being a National Health Service patient.

In addition to reportage on receiving healthcare, Diski shows why we should take story seriously if we want to understand the experience of illness (Bradby 2017). She writes the story of her lung disease in the context of her unreliable memories, ambiguous emotions and apparently unrelated events, all of which have a bearing on what her illness and her treatment mean. Diski shows that her experience of illness and encounters with treatment cannot be understood without reference to the places, people and encounters that have made up her life. These include her troubled relationship with her parents (Diski 1997), being raped at the age of 14 (Diski 2009), suicide attempts and psychiatric treatment (Diski 2014a) as well as warm relationships with her partner and daughter. Prior to her diagnosis, Diski drew on her life experience in reviewing books and commenting on current political and cultural affairs, a practice that continued afterwards (Wilmers 2016).

Despite her horror at the prospect of committing cant or cliché in her final months, Diski wrote a cancer diary. She notes the competition from other 'survivors' diaries', but carries on because 'I'm a writer. I've got cancer. Am I going to write about it? How am I not?' (Diski 2014b). Diski's well-turned responses to the existential challenge of fatal disease and her evaluations of her care are impossible to disentangle from her life history. While narrative-based medicine seeks to strip out extraneous detail to represent the values and priorities that colour the experience of illness and so should guide treatment, the story of an illness is the story of a person embedded in her history and her network. Such a story is difficult to reduce to one dimension, one narrative segment.

As a troubled teenager, estranged from her parents and at the suggestion of her school classmate, Peter Lessing, Diski moved in and became the 'disputatious quasi-daughter of Doris Lessing' (Wilmers 2016). Despite knowing one another for 50 years (Diski 2014c), Doris Lessing had not featured in Diski's work because of a pact not to write about one another (broken by Lessing (Lessing 1988)). Doris Lessing died in 2013, a month after her son Peter, and Diski attended both death beds (Diski 2015d). In her last interview, Diski reports that after news of her cancer diagnosis was published, Lessing's agent contacted her to request that she write an intimate memoir of Lessing (Barber 2016). Diski had witnessed Lessing's private, domestic life as she became well known for 'The grass is singing' and as she 'won all the prizes in Europe – every bloody one' (Reuters 2007). Diski's inside-view of the Lessing household garnered considerable interest in London's literary journalism circles.

Diski's experience of fibrosis and cancer necessitates exploring Lessing's role in Diski's life. Diski's response to her illness circles around her sadness over 'the love she needed from Lessing, and did not get' (Enright 2017). Lessing had introduced Diski to literary London and had been a sort-of-mother for 50 years. Diski's pain of not being properly loved (Diski 2015a) and Lessing's refusal to acknowledge Diski's writerly success may be encoded in the book title *In Gratitude* sliding, as it does, so easily into 'Ingratitude'. Diski's experience of cancer/fibrosis cannot be understood without including Lessing's parenting of Peter, her quasi-parenting of Diski, Lessing's views on sex, writing, friendship, gender and her death, through which Diski refracts her own illness and dying.

Diski has been accused of narcissism and she refers to herself as a 'narcissistic writer': 'I start with me, and often enough end with me. I've never been apologetic about that' (Diski 2014b). Indeed, having described Peter and Doris Lessing's deaths as the context for her own, she writes: 'But back to me (it's not all about me, I know, but some of it is)' (Diski 2017, p. 240). Diski's willingness to train an analytic eye so unflinchingly on her own experience is not to everyone's taste and can be unnerving. But for understanding the complex, multi-layered and life-long nature of the experience of suffering and its implications for identity, her writing has a lot to offer. Diski is 'vividly in touch with her previous selves' (Harvey 2015) who discovered sex, literature and depression, with her experience of disease and dying intimately associated with those earlier moments.

Stories as a means of escaping (Tolkien 1947) or at least understanding death (Kearney 2002), suppressing time (Doja 2008) and allowing us to talk to our younger selves and those who have already died all of whom are on offer in Diski's memoir cum cancer diary. With cool rationality, she notes that having reached her late 60s she is not dying tragically young and that children burying their parents is the appropriate order of things (she has an adult daughter and two grandchildren born after the diagnosis). Nonetheless, the prospect of not knowing her grandchildren as they grow up induces in her a deep and lachrymose sadness. Her preoccupation with the deaths of Doris and Peter Lessing is an enquiry into how to approach her own.

The question of her own death and particularly its timing become more pressing as therapeutic options run out. Diski asks the questions of order and control (Kleinman 1988, p. 29) posing the inevitable 'how long?' question to the oncology specialist or 'Onc Doc'.

> When am I going to die? How long have I got? … As soon as you ask 'How long?', you can see Onc Doc's eyes shifting into eidetic motion to remember the figures for death in either case. He offers me statistics, smoothed to take in the doubleness of my disorders. He can say between 1 and 3 years, assuming I don't get an infection that my white cell system isn't strong enough to prevent from turning into fatal pneumonia rather than a common cold. The uncertainty is real. Onc Doc's skills are essentially statistical. Reading papers that include and conclude from other studies the mean average lifetime chances. But I'm not a statistic. As an individual, who knows (Diski 2015c)?

During chemotherapy Diski had already concluded that 'no one, Onc Doc, Onc Nurse, really knows very much, except in an academic way' and it is regretful that 'everything is presented to me statistically, as probabilities'. Diski smoked (Enright

2017, p. xi) as illustrated in her author photograph (Diski 2017), and an evidence-based response to 'Why me?' 'Why now?' questions about her diagnoses would take this into account. There is no mention of smoking as a context for her illness, except for an apparently unprompted message, tweeted on 27 Mar 2016: 'Hi guys. Just to say my cancer is not caused by smoking. Nor is my fibrosis. It's just cancer. Best wishes, Jenny'.

Diski laments being unable to find a way of breaking through the statistics-talk to ask questions about 'the cancer that is me and mine, what it is, how it is, how it and I are with each other' (Diski 2017, p. 117). The desire to discuss what her own cancer means is not satisfied by her clinicians' accounts of the available evidence, based on population findings.

Diski's account of her experience of cancer and dying avoids systematically reviewed evidence and does not constitute a documentary, stable narrative. The limitations of documentary forms of writing are apparent during her final months as she circles around the issue of death, trying out different approaches to the impending end. Reflecting on her vacillating fear of death, Diski notes she has likely seen her last Christmas and ponders what her grandchildren might remember of her. Her fear of death recedes but its closeness remains and she reports 'feeling that my death was waiting on the doorstep with the milk bottles'. There is a sense of resignation to an imminent end, which is portrayed as neither mystical nor religious, but in terms that are familiar and low key.

> For several days now I've been feeling as if I'm on a holiday, a short one coming to its end. Not an especially good one. Not sorry to be leaving, not sorry to have been here. No particular feeling one way or another. Not living in my place. Not familiar enough. As one might sit on the edge of a chair that is waiting for another occupant to take it over. It's the strangest of strange feelings. Best travelling clothes, a ticking of a clock that will go on ticking after you leave and after the next occupant too (Diski 2016).

This excerpt suggests that Diski may have achieved what palliative care terms a 'good death', coming to terms with the end of life. Diski's daughter, who was with her while she died, writes that during her mother's final days Diski 'was scared and didn't want to be alone. But who knows' (Diski 2017, p. 258). Diski's daughter underlines the 'not knowing' as the 'trouble with death', in that her mother's death was 'as mysterious to me as it was to her' (Diski 2017, p. 258). Diski's daughter describes the most difficult aspect of her mother's death and dying as 'her fury', seeing the fury, along with the writing, as part of a process of trying to understand and 'to make the unknown known or at least to create enough of something to observe and engage with' (Diski 2017, p. 259).

Seeking to understand or at least engage with the unknowable 'trouble with death' does not necessarily lead to a peaceable passing. Deborah Steinberg's blog[3] about living with breast cancer (of which she died in 2017) rails against pain and injustice to the end. Steinberg's friend Debbie Epstein described the blog as 'heartbreaking'.[4] Towards the end of her writing life and having been refused a

[3] https://darkcloudblog.wordpress.com/

[4] https://www.theguardian.com/world/2017/mar/06/deborah-lynn-steinberg-obituary

particular sedative by palliative care, Steinberg expresses her fury at her wishes being frustrated: 'i am angry beyond measure' (December 23, 2016 · 1:19 pm). A little later, after a friend had suggested that she might have been harsh, Steinberg writes:

> the problem is i don't want to die. but am anyway. i want to stop it. I'm fighting unconsciousness. i wont have peace or acceptance. there are bad ways to die. this is one of them. i resent dying. i resent having my day to day everything in other peoples hands. people like me make demands. we don't want to wait because waiting is itself a galling loss of selfhood. you fucking wait i want to say. but know that this is unfair and ugly. but also the truth. who wants to loose all the minutia that add up to autonomy. self respect. self definition.
>
> so in a way I'm not sorry. and am sorry about that.
>
> (posted December 23, 2016 · 6:14 pm)

For Steinberg the process of dependency arising from serious illness is infuriating because it infringes upon her subjectivity. In searching for ways to represent her dying process, Steinberg posts poems, as well as political commentary and mundane observation. Her last entry echoes Diski's anxiety about committing cliché.

> i thought i had something to say. but really, i don't. just the same things. bad worlds. goodbye cruel worlds cliches.
>
> (posted January 30, 2017 · 6:35 am)

The impossibility and unavoidability of death is a puzzle at the heart of accounts of fatal diagnoses. Writing towards one's own death does not necessarily bring peaceable resignation for the author, but the mystery of death is overcome when the narrative voice continues through the intervention of the dead author's partner (Diski 2017; Kalanithi and Verghese 2016; Picardie 1998). Writing about the person who has died is one means of bringing them back to life (Picardie 2001; Seaton 2002).

(Re)-imagining Life: Writing Back from Death

Sara Ryan's son Connor drowned in a bath at the age of 18, in a National Health Service unit for people with disabilities where he had been admitted in order to receive care. In her account of Connor's life and the pain of his avoidable death, Ryan (2018) brings her son back to life by remembering key moments from his life, including funny things that he said and did. Snippets of conversation are written out as dialogue, which give a flavour of the repetitive and insistent nature of Connor's questions, as well as their humour.

When he was primary school age, Connor asked his mother 'Mum? Are you my mum, Mum?'. Later he had a phase of calling his Mum 'Muv-aaar' which used to drive her potty. After his death she longed for that annoyance (Ryan 2018, p. 187). The conversations between mother and son become imaginary rifts on the sort of questions that Connor might have asked if he were participating in the process of making sense of his own death. The sense-making is done by imagining explaining death to Connor himself. The family pet, a dog called Chunky Stan, was alive at the time of Connor's death, and so Ryan imagines telling her son about the dog being put down.

> Mum, is Chunky Stan dead, Mum?
> Yes, he died after Christmas.
> Why, Mum?
> He was very old and the vet put him down.
> Did the vet put him down, Mum?
> Yes.
> Why, Mum?
> Because he was very uncomfortable and she thought it was kinder. We agreed.
> Why, Mum?
> Because he was so old.
> Mum, am I dead, Mum?
> Yes.
> Am I old, Mum?
> No.
> Why, Mum?
> You were never allowed the chance to grow old, matey.
> (Ryan 2018, p. 56)

The book describes the long, and eventually successful, campaign to get an inquest into Connor's death. The painful process of the inquest is documented, along with the public criticism of the Trust that had been responsible for the unit where Connor died. Ryan is highly critical of the individuals and organisations as well as the culture of neglect that pervades services for people with disabilities, not to mention the harrowing processes to which bereaved families are submitted in querying statutory care practices. Ryan is looking for answers to the 'why' question regarding her son's death by pursuing the formal and official channels. But after the inquest has finally noted failings of the process of care, there is still no good reason for the death. The legal and medical evidence does not explain why Connor died, at least not in human terms. And certainly not in terms that would have made sense to Connor himself.

> Mum. Mum?
> Yes?
> Mum. Was I born in a bath, Mum?
> Yes, you were. At home.
> Mum … Mum? Did I die in a bath, Mum?
> Yes.
> Why, Mum?
> I don't know.
> (Ryan 2018, p. 239)

Ryan's book grew out of a blog[5] that she had started about her family life, well prior to events that prompted Connor's admission to the unit where he died. In one post Ryan describes her fear of forgetting her son – that his death would obliterate him utterly, not just bodily.

---

[5] https://mydaftlife.com/

> I still remember LB though. […] I realised this yesterday when I imagined him on holiday with us. I could still see him, hear him and feel his presence strongly. His (constant) commentary, facial expressions, enjoyment, participation and humour. I could see him, sitting cross-legged on the beach, sifting sand through one hand, eyes half closed, basking in the sun like a contented cat.
>
> I wondered about this. On a windswept beach in Tenerife. One of my (many) fears was that I'd forget. That he'd lose shape, substance, being in time. His brutal and unexpected death would obliterate him. But it hasn't.
>
> (posted August 9, 2018).

One of the comforts of imaginary stories is that they overcome the finality of death.

Closing Thoughts

In order to represent the experience of illness and healthcare, forms of evidence beyond the documentary are implied. Representing the experience of death, dying and loss requires the imaginary and the metaphorical. Anthropology has considered its relationship with what fiction might mean in the context of factual accounts of the social and cultural world (Clifford and Marcus 1986; Strathern et al. 1987). Other social sciences have more regularly distanced themselves from the triviality of unreliable fictional accounts, in favour of the reliable validity of science. Fictional techniques illuminate the subjectivities of illness and of loss. How literary, journalistic, poetic and life-writing relate to and compete with the powerful truth-claims of scientific accounts of illness, treatment and loss should be part of a conversation with medicine.

This chapter makes a case for story as a way of knowing that has been underplayed, not only in evidence-based medicine but also in narrative-based approaches. 'The modern clinic is a densely narrated environment' (Atkinson 1997, p. 340) where stories do all sorts of work, including the socialisation of trainee practitioners (Becker 1961) and making sense of medical error (Bosk 2003). Sharing and constituting knowledge by story-telling has been harnessed by various health promotion initiatives (Frank et al. 2015; Stuttaford et al. 2006), and yet story is less often used as a method for representing subjectivity to inform healthcare. The term narrative has been subsumed into the medical project as a vehicle for stable knowledge of a subjective and qualitative nature that can complement the evidence base. With science asserting the universal truth of its knowledge claims, narrative approaches to medicine have taken on documentary expectations of knowledge as part of the wider biomedical project. People's experience of illness and of treatment is neither necessarily stable nor rational. How then can it be summarised in stable documentary narratives? 'Storytelling reveals meaning without committing the error of defining it' (Arendt 1968), and story offers ambiguous, paradoxical approaches to establishing the workings of the social world and the place of self, in times of health, illness, injury and madness. Evidence-based medicine is a means of managing 'the unruly social world in which medicine is practiced via objective scientific procedure'

(Goldenberg 2006, p. 2630) which has incorporated narrative-based medicine as a response to criticism of its limitations. The advantage of story is that it has not (yet) been subsumed to the biomedical project.

No more than its stripped down cousin narrative, can story give us 'untrammelled access to a realm of hyperauthenticity' (Atkinson 1997, p. 341). Story, when taken seriously, with the inclusion of its contradictory, metaphorical elements, can, however, bring insight into how illness, injury, treatment and loss are experienced. No more than narrative, is story a panacea for the limitations of healthcare, having distinct limitations as a method to represent experience: muted voices that cannot tell their own story and that have no champion can get lost (Bradby et al. 2009). Moreover, crafting a story is not necessarily in itself therapeutic, although it may be nonetheless informative.

While story has occlusions and limitations, it may explain troubling experiences, in ways that narrative cannot, and thereby offer comfort. Ryan's comfort comes not from the post-mortem evidence and inquest into her son's death but rather in being able to imagine his presence and his spirit. Ryan's imaginary dialogue with her son probes and perhaps even undoes the mystery of his absence from her life, albeit momentarily. Diski's story illustrates how the onset of mortal illness re-casts a life, and is experienced in contact with earlier selves, all of which may have a profound effect on expectations and experience of treatment. As the prospect of a biomedical cure recedes, the need to make sense of the onset of a disease in one's own terms may become more pressing and almost tangible, like milk on a doorstep.

References

Arendt, H. (1968). *Men in dark times*. Harcourt Brace Jovanovich: New York.

Aronson, J. K. (2000). Autopathography: The patient's tale. *British Medical Journal, 321*, 1599–1602.

Asthana, S., & Halliday, J. (2006). *What works in tackling health inequalities? Pathways, policies and practice through the lifecourse*. Bristol: Policy.

Atkinson, P. (1997). Narrative turn or blind alley? *Qualitative Health Research, 7*, 325–344.

Barber, L. (2016, April 17, 10, 11, 13, 15). Doris Lessing Took me in at 15. from the moment i got there, she talked about sex continuously. *The Sunday Times*.

Becker, H. S. (1961). *Boys in white: Student culture in medical school*. Chicago: University of Chicago Press.

Bosk, C. L. (2003). *Forgive and remember: Managing medical failure*. London/Chicago: University of Chicago Press.

Bradby, H. (2016). Research agenda in medical sociology. *Frontiers in Sociology, 1*. https://doi.org/10.3389/fsoc.2016.00014

Bradby, H. (2017). Taking story seriously. *Social Theory & Health; London, 15*, 206–222.

Bradby, H., Hargreaves, J., & Robson, M. (2009). Story in health and social care. *Health Care Analysis, 17*, 331–344.

Bruner, J. S. (1986). *Actual minds, possible worlds*. Cambridge, MA: Harvard University Press.

Bruner, J. S. (1990). *Acts of meaning*. Cambridge, MA: Harvard University Press.

Bruner, J. S. (2004). Life as narrative. *Social Research, 71*, 691–710.

Burnside, J. (2006). *A lie about my father*. London: Jonathan Cape.

Bury, M. (1982). Chronic illness as biographical disruption. *Sociology of Health & Illness, 4*, 167–182.
CEBM Oxford. (2015). *Trish Greenhalgh – 'real v rubbish EBM'*. https://www.youtube.com/watch?v=qYvdhA697jI
Charon, R., & Wyer, P. (2008). Narrative evidence based medicine. *The Lancet, 371*(9609), 296–297.
Clifford, J., & Marcus, G. E. (Eds.). (1986). *Writing culture: The poetics and politics of ethnography: A school of American research advanced seminar*. Berkeley: University of California Press.
Conway, K. (1997). *Ordinary life. A memoir of illness*. New York: W.H.Freeman and Company.
Denny, K. (1999). Evidence-based medicine and medical authority. *Journal of Medical Humanities, 20*, 247–263.
Diamond, J. (1998). *C: Because cowards get cancer too*. London: Vermillion.
Diski, J. (1997). *Skating to Antarctica* (First Printing edition). London: Granta.
Diski, J. (2009, November). Diary. *London Review of Books,* 52–53. https://www.lrb.co.uk/v31/n21/jenny-diski/diary
Diski, J. (2014a, September 11). A diagnosis. *London Review of Books*, 7–10. http://www.lrb.co.uk/v36/n17/jenny-diski/a-diagnosis
Diski, J. (2014b, February 6). I haven't been nearly mad enough. *London Review of Books*, 3–6. https://www.lrb.co.uk/v36/n03/jenny-diski/i-havent-been-nearly-mad-enough
Diski, J. (2014c, October 9). What to call her? *London Review of Books*, 11–13. https://www.lrb.co.uk/v36/n19/jenny-diski/what-to-call-her
Diski, J. (2014d, November 6). Schrödinger's tumour. *London Review of Books*, 23–24. https://www.lrb.co.uk/v36/n21/jenny-diski/schrodingers-tumour
Diski, J. (2015a, January 8). Doris and me. *London Review of Books*, 21–23. https://www.lrb.co.uk/v37/n01/jenny-diski/doris-and-me
Diski, J. (2015b, February 5). Was that when it was beaming me? *London Review of Books*, 29–31. http://www.lrb.co.uk/v37/n03/jenny-diski/was-that-when-it-was-beaming-me
Diski, J. (2015c, July 2). Spray it silver. *London Review of Books*, 13–14. https://www.lrb.co.uk/v37/n13/jenny-diski/spray-it-silver
Diski, J. (2015d, July 30). Why can't people just be sensible? *London Review of Books*, 13–16. https://www.lrb.co.uk/v37/n15/jenny-diski/why-cant-people-just-be-sensible
Diski, J. (2016, February 4). Mother's Prettiest thing. *London Review of Books,* 21–23. https://www.lrb.co.uk/v38/n03/jenny-diski/mothers-prettiest-thing
Diski, J. (2017). *In gratitude*. London: Bloomsbury.
Doja, A. (2008). Claude Levi-Strauss at his centennial: Toward a future anthropology. *Theory, Culture & Society, 25*, 321–340.
Eggins, S., & Slade, D. (2006). *Analysing casual conversation*. London: Equinox.
Enright, A. (2017). Foreword. In J. Diski (Ed.), *In gratitude* (pp. xi–xvii). London: Bloomsbury.
Frank, A. W. (1991). *At the will of my body*. Boston: Houghton-Mifflin.
Frank, L. B., Murphy, S. T., Chatterjee, J. S., Moran, M. B., & Baezconde-Garbanati, L. (2015). Telling stories, saving lives: Creating narrative health messages. *Health Communication, 30*, 154–163.
Game, A., & Metcalfe, A. W. (1996). *Passionate sociology*. London: Sage Publications.
Georgakopoulou, A. (2006). Thinking big with small stories in narrative and identity analysis. *Narrative Inquiry, 16*, 122–130.
Goldenberg, M. J. (2006). On evidence and evidence-based medicine: Lessons from the philosophy of science. *Social Science & Medicine, 62*, 2621–2632.
Gordon, D. R. (1988). Tenacious assumptions in western medicine. In M. Lock (Ed.), *Biomedicine examined* (pp. 19–56). Dordrecht: Springer.
Greenhalgh, T., & Hurwitz, B. (1998). *Narrative based medicine: Dialogue and discourse in clinical practice*. London: BMJ Books.
Hanemaayer, A. (2016). Evidence-based medicine: A genealogy of the dominant science of medical education. *Journal of Medical Humanities, 37*, 449–473.

Harvey, G. (2015, June 14). Jenny Diski's end notes. *The New York Times Magazine*, 34. https://www.nytimes.com/2015/06/14/magazine/jenny-diskis-end-notes.html
Kalanithi, P., & Verghese, A. (2016). *When breath becomes air*. London: Bodley Head.
Kearney, R. (2002). *On stories*. London: Routledge.
Kleinman, A. (1988). *The illness narratives: Suffering, healing, and the human condition*. New York: Basic Books.
Labov, W., & Waletzky, J. (1967). Narrative analysis: Oral versions of personal experience. In J. Helm (Ed.), *Essays on the verbal and visual arts* (Vol. 1966, pp. 12–44). Seattle: The American Ethnological Society.
Lambert, H. (2006). Accounting for EBM: Notions of evidence in medicine. *Social Science & Medicine, 62*, 2633–2645.
Lessing, D. (1988). *The memoirs of a survivor*. New York: Vintage Books.
Lorimer, H. (2003). Telling small stories: Spaces of knowledge and the practice of geography. *Transactions of the Institute of British Geographers, 28*, 197–217.
Murray, M. (2002). Connecting narrative and social representation theory in health research. *Social Science Information, 41*, 653–673.
Naldemirci, Ö., Lydahl, D., Britten, N., Elam, M., Moore, L., & Wolf, A. (2018). Tenacious assumptions of person-centred care? Exploring tensions and variations in practice. *Health, 21*, 54–71.
Nobel, I. (2005). *Like a hole in the head: Living with a brain tumour*. London: Hodder and Stoughton.
Picardie, R. (1998). *Before I say goodbye*. London: Penguin.
Picardie, J. (2001). *If the spirit moves you*. London: Macmillan.
Porter, R. (1999). *The greatest benefit to mankind : A medical history of humanity from antiquity to the present*. London: FontanaPress.
Reuters, L. C. (2007). *British author Doris Lessing reacts to Nobel win*. https://www.youtube.com/watch?v=vuBODHFBZ8k
Ricoeur, P., McLaughlin, K., & Pellauer, D. (1984). *Time and narrative*. Chicago: University of Chicago Press.
Riessman, C. K. (1993). *Narrative analysis*. Newbury Park: Sage.
Ryan, S. (2018). *Justice for laughing boy. Connor Sparrowhawk – A death by indifference*. London: Jessica Kingsley.
Sackett, D. L., Rosenberg, W. M. C., Gray, J. A. M., Haynes, R. B., & Richardson, W. S. (1996). Evidence based medicine: What it is and what it isn't. *BMJ, 312*(7023), 71–72.
Seaton, M. (2002). *The escape artist: Life from the saddle*. London: Fourth Estate.
Sools, A. (2012). Narrative health research: Exploring big and small stories as analytical tools. *Health, 17*, 93–110.
Squire, C. (2012). Narratives and the gift of the future. *Narrative Works: Issues, Investigations & Interventions, 1*, 67–82.
Strathern, M., Crick, M. R., Fardon, R., Hatch, E., Jarvie, I. C., Pinxten, R., et al. (1987). Out of context: The persuasive fictions of anthropology [and comments and reply]. *Current Anthropology, 28*, 251–281.
Stuttaford, M., Bryanston, C., Hundt, G. L., Connor, M., Thorogood, M., & Tollman, S. (2006). Use of applied theatre in health research dissemination and data validation: A pilot study from South Africa. *Health, 10*, 31–45.
Thornborrow, J., & Coates, J. (Eds.). (2005). *Sociolinguistics of narrative*. Philadelphia: John Benjamins Publishing Company.
Tolkien, J. R. R. (1947). *On fairy-stories*. Oxford: Oxford University Press. http://www.excelsior-classical.org/wp-content/uploads/2011/04/Tolkien_On-Fairy-Stories.doc.
Tonelli, M. R. (1998). The philosophical limits of evidence-based medicine. *Academic Medicine: Journal of the Association of American Medical Colleges, 73*, 1234–1240.
Wilmers, M-K. (2016, May 19). Diary. *London Review of Books*, 39. https://www.lrb.co.uk/v38/n10/mary-kay-wilmers/diary

Chapter 8
Borders of Blame: Histories and Geographies of HIV and AIDS in South Africa, 1980–1995

Carla Tsampiras

Abstract AIDS transcends, and is implicated in, spatial and temporal boundaries that we have only begun to explore and that require in-depth interdisciplinary and collaborative engagements. This chapter explores the overlap between disciplinary interests in history and geography and considers how they facilitate analyses of various boundaries constructed or reinforced by the State in relation to the emergence of, and responses to, HIV and AIDS in South Africa in the early years of the epidemic. From the framing and blaming associated with the first officially recognised cases of AIDS in the country to the maligning of Malawian migrant mineworkers, the creation and maintenance of borders of all kinds - from the physical to the socio-political, has had a direct impact on life and death in South Africa and in neighbouring countries like Malawi. In this chapter the intersections of historical and geographical perspectives reveals borders of blame applied to people but representative of more complex and interrelated systems of power and inequities. It argues for ongoing epistemic generosity and collaboration across disciplinary boundaries to subvert increasing barriers and restrictions related to health and humanity.

Major health epidemics are intimately linked to geographies of space, place, time and personhood. In both past and current epidemics, one of the most common socio-political events that occurs is the linking of illness to geographic markers or particular stereotypical embodiments of 'infected others'. The global influenza pandemic of the late 1910s became colloquially known as 'the Spanish 'Flu'. In the 1980s, AIDS was associated first with 'white, gay men' in the USA and UK and then 'black men' specifically geographically linked to Haiti. More contemporaneously, influenza in the English-speaking world is still commonly 'Asian' in origin - referring to a fairly loosely defined area. On one level the accurate linking of epidemics to places and people is epidemiologically important, but associations are also likely to be inaccurate and linked to socio-political bias rather than 'objective' scientific methodology. The bounding of diseases to places facilitates a sense of control and knowing, allows for the assignation of blame, and alleviates fear of the unknown in the

C. Tsampiras (✉)
University of Cape Town, Cape Town, South Africa
e-mail: carla.tsampiras@uct.ac.za

S. Atkinson, R. Hunt (eds.), *GeoHumanities and Health*, Global Perspectives on Health Geography, https://doi.org/10.1007/978-3-030-21406-7_8

same way maps (or nowadays GPS) alleviate concerns for people travelling unfamiliar routes.

Being conscious of the socio-political, economic and historical influences on these geographies of space, place, time and identities linked to health are important, as is revealing them, challenging them or addressing them. Doing so, however, may take us away from comfortable, known, controllable disciplinary or methodological spheres.

Some of us travelling disciplinary routes linked to understanding health, especially routes that challenge or traverse conventionally located epistemic and pedagogic boundaries, have had to navigate uncharted terrain. We may have felt fear in the face of unanticipated methodological obstacles or unknown responses to pedagogical experimentation. We may also have realised that a lack of control and acknowledgement of different ways of knowing can be invigorating and allow us to navigate by means other than the most conventional ones. By reflecting on geographies of health epidemics and disciplinary practices associated with understanding epidemics, we can begin to generate new dialogues between different intellectual traditions. Respectfully sharing tales of our adventures and critical evaluations of our findings allows us to undertake new journeys or undertake our own journeys differently.

This chapter is one such respectful sharing of experiences and findings linked to history research into the early years of the AIDS epidemic in South Africa (1980–1995). It is offered by an historian interested in social histories of health and in the intersections between 'race', gender, sexuality and class. One who is also interested in deepening conversations about health in and between the fields of Medical and Health Humanities (MHH) and GeoHumanities and has been influenced by health historians and geographers.

When it came to crossing borders, reconfiguring landscapes and dealing with sites of malevolent geographies, the late political geographer Glen Elder insisted on new engagements with hostels and migrant labourers, centring gender and weaving complex interactions between all those involved in migrant labour (2003). His analysis of hostels for municipal workers as opposed to miners, his focus on womyn[1] who were not even meant to be part of the hostels, and use of feminist and queer analyses of space showed how research and analysis could be undertaken differently.

Historian Shula Marks has shaped, formed, explored and mapped numerous aspects of southern and South African histories. Her work on apartheid and health formed the basis of academic articles and reports that would go to international organisations, and she worked with others outside her discipline long before inter-,

[1] I concur with Walker's argument that 'the boundaries of the category "woman" are rendered extremely elusive by the operation of other significant markers of social power' and have different meanings at different moments in history (Walker, C. (1990). *Women and Gender in Southern Africa to 1945.* Cape Town – RSA: David Philip, p. 26). For this reason, I assert my preference for using the term 'womyn' (as inclusive of transpeople who identify as womyn) in my work and publications when possible.

trans- and multidisciplinary work was *de rigueur* (Marks & Andersson 1983, 1986, 1989). When she brought her formidable knowledge and scholarship to examining AIDS (2002, 2007), she allowed for new spatial and temporal topographies of understanding and enquiry. The work and advice of Elder and Marks allowed a mature history postgraduate starting out on her PhD to think differently and take chances in relation to methodology and practice and in challenging and holding boundaries.

Overlaid on these works, and the shifts in enquiry and methodology they facilitate, are more recent works by social science/geographers like Hunter (2010) and historians of pharmacopolitics like Parle et al. (2018). The burgeoning of fields like GeoHumanities and MHH allows for explorations of, and experimentation in, methods, means, relationships and reasons for engaging in discussions and research about health. This chapter explores the connections between AIDS histories and geographies in southern Africa and considers how future areas of exploration can be facilitated by GeoHumanities.

Geographies of AIDS

AIDS transcends and is implicated in spatial and temporal boundaries that we have only begun to explore. There are complex geographies linked to AIDS and understanding AIDS that range from the cellular to the societal and require increasingly in-depth collaborative engagements.

The series of AIDS-related complexes that result in death mean that from a biomedical perspective, one virus (the human immunodeficiency virus, known commonly as HIV) is implicated with a host of co-infections and illnesses such as tuberculosis, hepatitis B and hepatitis C. Diagnosis and treatment methods are complicated and may need to address immediate opportunistic infections such as oesophageal candidiasis, as well as addressing co-infections and providing lifelong ART. The virus is of course not really only one virus but one virus group with multiple sub-types or clades. As a result of recombination, the virus group perpetually evolves and at the cellular level spatial boundaries between HIV and CD4 T lymphocytes (the white blood cells so vital for the immune system) are continually removed and reformed.

Moving from the microscopic and cellular level to the levels of regional and global public health, AIDS as a health crisis also straddles time and space differently, occupying zones of temporal dissonance depending on context and the types of conversations taking place. Contingent on location, and personal or professional background, AIDS is a contemporary or re-emerging health crisis; a health concern that is now a chronic, manageable condition; or one that is sufficiently under control to welcome 'post-AIDS' theorising (Tsampiras 2017).

There is a geography of knowledge about HIV and AIDS in South and southern Africa that has its origins in ideas shaped by recent colonial world orders and global extractive economies. To date, the hierarchies of AIDS-related academic disciplines have favoured what is seen and written by medical and health sciences and overshadowed work in humanities, arts and social sciences (Tsampiras 2017). The knowledge hierarchies extend beyond disciplinary boundaries to include socio-economic ones demarcated by global political powers and privileges, resources and access to publishing and research agenda setting.

Exploring various geographies of AIDS across disciplinary boundaries through GeoHumanities or MHH-inspired research can increase our understandings of, and engagements with, the geographies of AIDS. However, before we consider how these engagements might shape AIDS research in the future, let us foray into the overlap between history, geography and other boundaries in relation to AIDS in South Africa in the past.

Mapping the Early Years

The first two people to be officially recognised as dying from AIDS in South Africa were flight stewards Ralph Kretzen who died on 26 August 1982 and Pieter Daniël (Charles) Steyn who died on 1 January 1983. Their deaths from the newly named Acquired Immunodeficiency Syndrome were reported in South African newspapers in early 1983 (Grundlingh 2001a, 87). Government responses in the first few months of 1983 were aimed primarily, as Grundlingh notes, at 'quell[ing] any possible panic reactions amongst the white public and … assur[ing] them that they were in no danger' (Grundlingh 2001b, 125).

In July 1983, the prestigious *South African Medical Journal (SAMJ)* published two articles by medical professionals dealing with AIDS in the country. One discussed immunological abnormalities amongst 'ten homosexual men' and, alongside information on test results, contained a number of comments about the nebulous notion of 'promiscuity' (Anderson et al. 1983). The other article detailed the first two 'cases' (Ras et al. 1983) defining the men as homosexual and indicating that they had spent time in the USA. Between the evidence of types of infection and radiography reports, is information on 'case one's' repeated admission to hospital, increasingly laboured breathing and news that '[t]he patient died shortly after admission to the intensive care unit' (Ras et al. 1983, 141). About the second 'case', we learn that '[a] week before admission to a private nursing home he had collapsed while showering, and although he recovered he complained of weakness, fever and sweating attacks' slightly further down the page, '[h]e died 11 days after admission in hypoxic respiratory failure and acute renal failure with acute marrow hypoplasia' (Ras et al. 1983, 141). These sentences act as absent referents to the everyday lives of Ralph and Pieter and permeate the boundaries of written scientific academic text and tone with messy, everyday details of life and death. They remind us of the everyday geography of health and disease that take place outside of medical geographies of health in everyday places and actions – showering, breathing, being conscious, travelling, caring, loving and feeling.

Initial responses to the first officially recognised AIDS cases in South Africa were characterised (as they were globally) by fear and reflected existing stigma and 'moral panic', heightened by the conservative morality that pervaded South Africa in the 1980s (Grundlingh 2001a, b; Hodes 2018). As Phillips (2001, 13) noted, the initial identification of HIV positive people 'evoked a cascade of finger-pointing along the fissures of existing prejudices, a process turned into a veritable avalanche from the late 1980s'.

The hegemonic narrative about AIDS and HIV in South Africa constructed by medico-scientific and political elites in the early years of the epidemic followed a well-known trajectory constructed and created in the English-speaking 'north'. It included 'high-risk groups' and the use of AIDS avatars like 'the promiscuous homosexual', 'the infected foreigner', 'the infecting prostitute' and 'the sexual black body'. These avatars reduced complex people living in complex times to stereotypical 'types' that were unquestioningly researched, shaped and created out of existing moral and political conservatism. The narratives in South Africa were influenced by the socio-political upheavals in the country at the time. Although they did evolve to narratives more closely related to human rights and hope by the 1994 transition, for the most part the responses by elites who wrote about, and responded to the emerging crises, generally revealed varying degrees of prejudice including racism, sexism, classism and homophobia (Tsampiras 2013).

The hegemonic narratives were mediated by moral boundaries initially particularly those associated with sexual orientation, 'race', gender and class (Tsampiras 2008, 2015, 2019). AIDS was identified in the USA, but in the English-speaking narrative, its geographic emphasis soon shifted from 'white homosexual men' in a part of the Americas (or similar counterparts in the UK) to 'black' Haitians to 'black heterosexual people' associated with most of the African continent but particularly an undefined 'Central/central Africa'. The same spatial twists and turns occurred in South Africa with variations that reflected the state of turmoil in the country and the influence of conservative apartheid ideologies (Tsampiras 2015).

Ralph and Pieter transgressed the established boundaries of 'white', heterosexual masculinity and their professions allowed them to move beyond the country's borders and the repressive confines of the homophobic apartheid government. They benefitted from their class and 'race' privilege but their departure from an idealised straight 'white' masculinity put them outside the heteropatriarchal hegemony that shaped (and often still shapes) identities in South Africa (Gevisser and Cameron 1995; Reddy 1998; Tsampiras 2014). The apparent 'transgressions' of dominant socio-political or moral borders allowed the arbiters and creators of those dominant borders to easily apportion blame.

Writing about the constructions of 'risk groups' in the USA, Glick Schillers et al. (1994, 1339) noted that from the outset of the epidemic, 'authorities, researchers, and writers often reported the disease as originating from outside, wherever that was, and brought by that which was alien'. In South Africa, fear of infection by outsiders or infection from across borders surfaced not only in relation to homosexual men in South Africa but also in relation to almost everyone sharing the continent. Amongst political and medico-scientific elites shaping and framing the dominant AIDS narrative, the rest of the continent became 'alien' and South Africa was constructed as apart from it.

South Africa was separated by an imaginary *cordon sanitaire* and fuelled by ideas of the 'total onslaught'. Maylam (2001, 204) explains that in the face of mounting internal resistance and external pressure, official propaganda encouraged the idea of a 'total onslaught' – a concerted, communist-inspired, international assault on South Africa's political and economic order. As Visser (2004) has shown, ideas of specific 'perils' or dangers were reinforced in literature and were also present in government-influenced film and media. Beyond popular and visual culture, Lawson has drawn attention to the South African government's attempts to create a physical *cordon sanitaire* around its northern borders during this period, to prevent individuals with HIV or AIDS from entering the country (Lawson 2008, 32–35). The propagation of fear and threat under the guise of the 'total onslaught' allowed for the concomitant 'total national strategy', which Davies and O'Meara (1985, 183) explained included 'a sophisticated matrix of economic and other "incentives" applied together with military and other "disincentives"' to allow South Africa to 'assert its hegemony over the states of the region and to become the internationally recognised "regional power" in Southern Africa'. The South African government frequently breached other countries' borders or drew labour temporarily over them.

The *cordon sanitaire* had symbolic, imaginary, psychological and physical aspects to it and the very real anxieties felt about the future of South Africa could be exploited by suggesting the cordon could be crossed. The borders around South Africa were spaces of conflict – they were places that some young people crossed over to join military training camps in order to fight as part of the liberation struggle, or went into exile to plan and prepare for post-liberation South Africa. Borders were also places where other young people, conscripted into the army, were sent to 'defend' those at home from attack or destabilise neighbouring countries supporting the liberation struggle. As African countries 'north of the border' fought for and received independence, politicians and others also played on the fears and anxieties of minority communities or privileged elites and much of that fear was centred around the breaching of the country's 'borders'. In relation specifically to AIDS, the idea of the ever-encroaching external threat crossing borders was absorbed into the medical imaginary and is best represented by the following model published in 1988 depicting the spread of 'heterosexual HIV infection' into South Africa (Fig. 8.1).

Schoub et al.'s (1988) model is replete with boundaries and borders and border crossings that happen in only one direction and compartmentalise people and places. In this understanding of transmission, male homosexuals are at the opposite end of the page from heterosexuals and alongside intravenous drug users and the rest of Africa. Other articles published in the *SAMJ* reinforced the notion of infectious migrant labourers. Becker, for example, conjectured in 1986 that the lack of information on 'heterosexually' spread AIDS in the country would soon change as '... some of the several hundred thousand migrant workers in the RSA may bring infection from neighbouring countries where heterosexual spread is already established' (1986, 27).

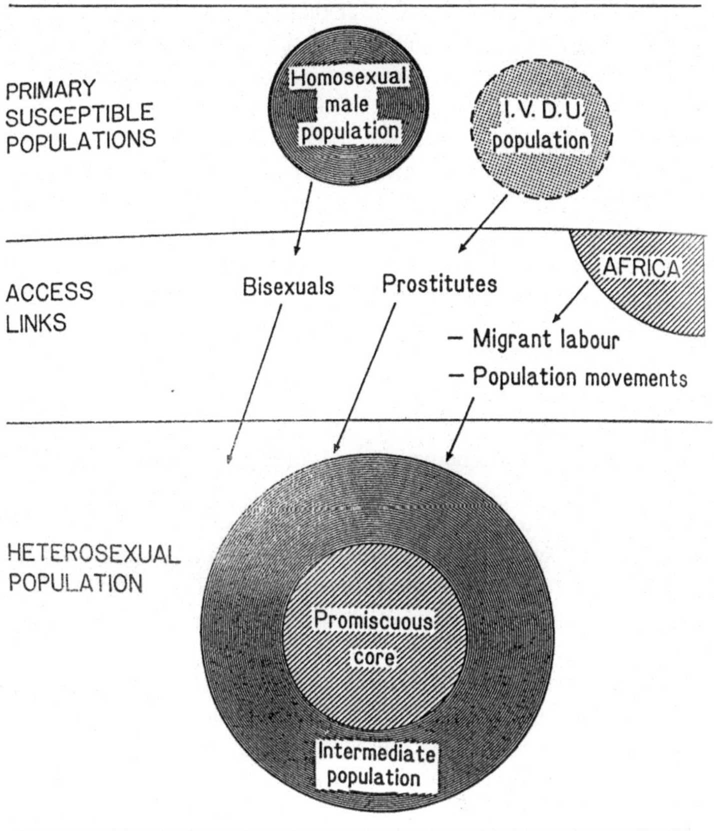

Fig. 8.1 Schoub et al.'s 1988 model showing how AIDS might infect 'the heterosexual population' in South Africa

In a climate of fear-inducing propaganda, it is unsurprising that the AIDS epidemic was framed as an ever-encroaching menace carried by 'alien' bodies from 'outside'. In times of political uncertainty, when regimes are in decline, or when state or corporate actions or inactions need to be obfuscated, 'others from outside' are a useful distraction. Despite people dying of AIDS within the borders of the country and evidence of increasing domestic infection rates, AIDS was presented as an external menace making its way from beyond the borders.

Not every articulation of this idea was necessarily linked to deliberate, nefarious propaganda nor deliberately ill-intended; nonetheless, its influence (conscious or unconscious) was sufficient to manifest in the pages of a medical journal and in speeches in parliament. The government did, however, deliberately deploy the idea of external infected vectors as a means of distracting attention away from its initial inaction in responding to AIDS within the country. As will be shown below, it also deliberately deployed the idea in relation to certain migrant workers.

On 7 January 1983, less than a week after Pieter Steyn's death, the Department of Health 'called a meeting of representative specialists to formulate a surveillance and containment strategy' for AIDS ('Acquired immunodeficiency syndrome' 1983, 98). The AIDS Advisory Group (AAG – also known as the Advisory Group on AIDS or AGA) was established and set about gathering epidemiological data, addressing the safety of the blood supply, and providing information to the government in a period lasting until 1986, that was primarily concerned with information gathering and advice giving. Members of the group included medical experts and researchers, government officials, and medical representatives of the military and the mining industry. Despite the apparently hasty convening of the AAG, the government's subsequent lack of concern about AIDS would suggest that this was an exercise in allaying public panic. After the initial meeting, the AAG was only officially formed and recognised as an advice-giving body to the government in 1985 (AIDS Advisory Group 1986).

It is the next period of time between 1987 and 1989, overseen by Minster of Health Dr. Willie van Niekerk (a member of the ruling National Party or NP), there were two transitions in what can be considered a period of 'facts and fear'. The first transition was from a process of information gathering to one of information distribution with an emphasis on public health education on AIDS aimed at adults. It is the second transition, however, that is of particular interest in relation to geographies and borders of blame. During this transition, narratives about the threat of AIDS and the infected outsider moved from being a political device to justify inaction, to being actively employed to justify practical, legislative responses that identified and blamed people from beyond the country's borders and took measures to remove them. In a long history linked to the violent seizure of land, coercive wage labour practices and the damaging extraction of minerals, political elites were able to use AIDS and geographies of blame to address concerns with migrant mineworkers.

Minerals and Malawian Miners: Selective Fears from Beyond the Border

Even as international and local hegemonic AIDS narratives began to move away from the notions of 'high-risk groups', mineworkers, particularly migrant labourers from beyond South Africa's borders, were turned into a new 'high-risk group'. The idea of, and anxiety about, infection being associated with mineworkers and migrant labourers has long featured in South Africa's history and was evoked during discussions about AIDS in parliament.

The close association between mining corporations and successive South African governments, as well as the intimate link between living and working conditions on the mines and the (ill)health of mineworkers (migrant and others) and the friends, families and colleagues of mineworkers, has long been researched (Burke and Richardson 1978; Jochelson 2001; Katz 1994; McCulloch 2002, 2009; Packard

1990, 1993). As much of the existing scholarship shows, the 'alien infected other' has provided an easy (absent) referent summoned by establishment figures to evoke fear, or further the political or economic goals of a variety of ruling elites (Grove and Zwi 2006).

To examine how the 'alien infected other' was evoked in parliamentary discussions by political elites, it is necessary to be aware of the creation of the tricameral parliament. In 1983, a 'whites only' referendum had shown support for parliamentary 'reform' and resulted in the creation of a tricameral parliament. This model resulted in two new parliamentary chambers operating alongside the 'white' House of Assembly (HA) in which real power remained vested: the House of Representatives (HR) to which 'coloured' people could elect 'coloured' representatives, and the House of Delegates (HD) to which 'Asian/Indian' people could elect 'Asian/Indian' representatives (Beinart 2001). It is in these houses that discussions about AIDS occurred and policies, plans and laws were debated and passed.

During the first five years of the AIDS epidemic, little mention was made of threats from outside South Africa in parliamentary discussions recounted in *Hansard*. In 1987, however, Mr. F.M. Khan (HD), after quoting figures for AIDS deaths and possible infection rates in other African countries, cautioned 'we cannot rely on the fact that we have boundaries. [AIDS] will reach South Africa; as a matter of fact it already has ... How will one stop infected people from outside our borders coming into South Africa?' (*Hansard* HD 1987a, col. 997).

In this discussion and those in the other chambers of parliament, 'black' mineworkers and migrant labourers, more generally, became a new 'high-risk' group – 'the infected foreigner' – by virtue of their nationalities, 'race' and status as infected outside vectors 'carrying the disease prior to entering the country' (*Hansard* HR 1987b, col. 2712; *Hansard* HA 1987c, col. 5161).

In June 1987, W.J.D. van Wyk (HA) presented a series of questions in parliament to Van Niekerk inquiring whether workers 'from other African countries' were 'medically examined with a view to Aids (sic)', how many were seropositive, and what was being done about it? (*Hansard* HA 1987d, col. 265–266). Van Niekerk reported that seven cases of AIDS had been diagnosed, and 940 blood tests had returned positive results. He declared that South Africa's response was in line with international practice in that 'persons with positive blood tests' were hospitalised, examined and 'rigorously counselled on an individual basis' about preventing further infection and, where possible, sexual contacts were traced. He emphasised that the same procedure was applied to 'foreign as well as South African miners' and commented that the mines had 'launched an extensive educational campaign on Aids' (*Hansard* HA 1987d, col. 265).

Van Niekerk's reply did not indicate the extent of the government's involvement, if any, with the mines' overall procedures or education campaign. However, 'the Cabinet [had] requested the Chamber of Mines to repatriate all foreign miners with positive blood tests', and the 'responsibility for medical tests on mineworkers lies with the mines' (*Hansard* HA 1987d, col. 266). Despite the historically close association between the government and the Chamber of Mines (a body established to represent the interests of the large mining companies), the burden of responsibility for HIV positive miners was to be borne by the industry. As the year proceeded though, Van Niekerk, in consultation with other government departments, passed legislation to facilitate the removal of 'foreign' HIV positive mineworkers.

In a number of instances, discussions in parliament raised concerns about AIDS and HIV infection as a pretext to consider the migrant labour system more broadly. In some of these instances it is the transgressed borders of the notion of 'family' and sanctioned heterosexual marriage that are discussed. In the House of Delegates, in 1987, A.E. Arbee highlighted living and working conditions in the mining industry:

> The hostel concept is an outdated one. If the hon. Minister is concerned, as we all are, about the danger of Aids spreading in such conditions, I believe he has a moral obligation to allow the development of family life of miners. He should not allow the Chamber of Mines to get away with cheap working conditions by allowing people from the neighbouring states to live there and work under such conditions. We have unemployment in our own country. We should allow our own workers the opportunity to work in the mining industry and to live within the proximity of the mine in a healthy environment, otherwise this deadly disease will not only cripple the mining industry, but also pose a danger to all South Africans ... [I want to ask the Minister] at least to deal with this matter very seriously, namely in getting the Chamber of Mines to encourage families to live with the people working in their mines. That is very important. (*Hansard* HD 1987e, col. 1767)

Arbee's comment links hostel life to ill-health and the breakdown of families and calls for corporate accountability. It also reflects an argument that was gaining ground in South Africa – that 'local' unemployment problems should be solved by employing local labour rather than importing 'foreign' labour (Wilson 2001; Nattrass 1995). Arbee used apprehension about the spread of AIDS to articulate concerns about living and working conditions on the mines, and to make a moral argument about the sanctity of marriage and family. These arguments were also articulated in 'nationalistic' terms that differentiated between 'local' and 'foreign' mineworkers. In a reply which indicated an unwillingness to address the structural problems raised by Arbee, the Minister of Economic Affairs and Technology addressed neither AIDS nor local unemployment, instead referring to migrant workers who returned to 'their countries' (*Hansard* HD 1987f, col. 1770).

By late 1987 and into 1988 debates about migrant labourers and mines, framed within discussions about AIDS, were evoking images of migrant workers as potentially 'infected foreigners' (*Hansard* HR 1987g, col. 2710; *Hansard* HD 1988a, col. 7581; *Hansard* HR 1988b, col. 9509). On 1 September 1987, Van Niekerk stated in the House of Representatives that '... we have to come up with some plan regarding the labourers, the foreign workers whose tests are positive' (*Hansard* HR 1987i, col. 2752). He also linked HIV to broader security issues, commenting that the development of a vaccine was unlikely because the virus was 'like a terrorist' (*Hansard* HR 1987b, col. 2712).

Some members of the House of Representatives and the House of Delegates did not think Van Niekerk's suggestions went far enough (*Hansard* HD 1987h, 3036–3037). Mr. L.J. Hollander (HR) raised the issue of proof of 'AIDS status' as a requirement for everyone seeking entry into South Africa:

> It does not help us to combat the illness locally if people are still crossing our borders. Here I am referring particularly to carriers. Carriers can have the illness for years without being aware of the fact that they are Aids carriers. People who enter our country will have to prove that they are not ... we cannot try to find a cure for Aids within our country if carriers from outside are still entering our country ... This does not apply only to mineworkers. I understand that this disease is rife in some African countries. People from America – incidentally, this is where the illness had its origin – and all those entering the country by air or sea, should give proof of being free of Aids. (*Hansard* HR 1987b, col. 2712)

Despite the apparent lack of awareness about the difference between HIV and AIDS and any mention of how window periods would be accounted for, Van Niekerk's 'plan' was revealed in the House of Assembly on 3 September 1987. He emphasised the apparently high occurrence of HIV infection amongst 'workers from outside the borders' and from 'Central Africa', noting that entry into South Africa of 'workers from high-risk areas... require[d] special attention' (*Hansard* HA 1987j, col. 5148–5149). He went on to inform the house that discussions with 'other departments and [affected] organisations ... including the Chamber of Mines', had begun and the Department of Health (DoH) had 'already prepared draft regulations in terms of the Health Act' which would allow transmittable diseases including 'AIDS or HIV contamination' to be identified and isolated in 'special units' for compulsory medical treatment. The Department of Home Affairs had been tasked with facilitating the removal of 'such persons' from South Africa, using 'aliens control legislation and regulations' (*Hansard* HA 1987j, col. 5149). AIDS per se is not a transmittable disease yet, as Cameron and Swanson (1992, 211) later commented in their critique of the legislation that came out of this plan, the schedule of communicable diseases under the Health Act included AIDS but not HIV.

While the new measures were not meant to apply to all visitors, but to all 'labourers recruited elsewhere', Van Niekerk specifically drew attention to migrant labourers from Malawi. His department had 'sent two delegations to Malawi to negotiate with Government officials there', and they were 'available to help them with an epidemiological study, to help them with their guidance and to help them to combat that situation in that country' (*Hansard* HA 1987j, col. 5148–5149). In blood samples tested from blood transfusions, five per cent of Malawian mineworkers tested HIV positive compared to less than one per cent of workers from Mozambique, Lesotho and Botswana who were, therefore, not considered a 'problem' (*Hansard* HA 1987j, col. 5148–5149). As South Africa had yet to adopt any kind of comprehensive response to AIDS, it is unclear what guidance the delegation could have offered the Malawian government. It appears that the offer of help amounted to little more than monitoring and the distribution of basic public health information.

There was a lack of logic in the selective application of the legislation. If the theory was that no HIV-positive person should be allowed to work in the country for fear that they would spread HIV, then the legislation should have applied to all people working in or entering the country, and not be dependent on arbitrary percentage differences in national infection rates. Van Niekerk spoke about the plan in relation to migrant workers on mines, but not to foreign workers in other sectors. Migrant labourers from the various sectors of South Africa that the government had forcibly removed people to and set up on the basis of 'ethnicity' – the so-called homelands or Bantustans including the government-declared 'independent states' of Transkei, Bophuthatswana, Venda and Ciskei – were also absent from any of the discussions. Despite these contradictions, Van Niekerk was confident: 'with this measure the Republic of South Africa may rest assured that we are doing everything in our power to spare South Africa from this disease, which has a 100% mortality rate' (*Hansard* HA 1987j, col. 5150).

Responses to Van Niekerk's measures varied along party political lines. Dr. W.J. Snyman of the right-wing Conservative Party (CP) appeared to have noticed the contradictions in Van Niekerk's statement and suggested that testing and deportation of HIV-positive people should be considered for 'all visitors to South Africa' (*Hansard* HA 1987j, col. 5151). Dr. D.M.S. Barnard of the liberal Progressive Federal Party supported calls for isolation and deportation, but called on the Minister and the Chamber of Mines 'to treat these unfortunate people ... as human beings':

> I think it would be good for our relations with the rest of the world and for our labour relations if we could show that the mines and the South African people have compassion and will treat these unfortunate people in a way that is humane and a credit to our medical profession of South Africa. (*Hansard* HA 1987j, col. 5161–5162).

Replying to Snyman, Van Niekerk noted that testing all visitors to South Africa had been considered, but because of the delay in the formation of antibodies, the costs involved and the fact that tests were not '100 % accurate', it had been deemed 'impracticable to test all people entering SA' (*Hansard* HA 1987j, col. 5214). The same issues would apply to mineworkers, but no comment was made on this, indicating perhaps that the money to test mineworkers was available and there was less concern about accuracy for migrant labourers' tests. Van Niekerk took pains to 'make it quite clear' that his suggestions in relation to isolation did not refer to 'institutionalisation in the sense of quarantine' but rather 'a place where he [sic] can be treated to eliminate the risk of infection' (*Hansard* HA 1987j, col. 5217–5218). He reinforced this point: 'I wish to reiterate this very clearly so that the media do not think we intend placing them all in concentration camps'. These would be 'places where they will not contract other infections because they have no resistance to them and they die from other infections'. Deportations would 'be done with great circumspection and compassion'; South Africa had already sent two delegations to Malawi and were 'going to help them with their programme in handling these cases'; this would 'of course be done with great compassion in the true tradition of medical services in South Africa' (*Hansard* HA 1987j, col. 5218).

Noteworthy in Van Niekerk's reply is the desire to separate NP responses to AIDS from statements made by the CP about putting people with AIDS into quarantine; the evident concern with ensuring 'accurate' media coverage of the measures; a vague explanation of what these 'places' actually would be or do; and the ironic projection of the South African government as thoughtful and compassionate. There was, however, little substantive information about what the government understood by compassion, and there was no reflection on the effect these deportations would have on the livelihoods of migrant labourers and their families.

The legislation appeared in the *Government Gazette* of 30 October 1987. An amendment to the Admission of Persons to the Republic Regulation Act, 1972, now included 'acquired immune deficiency syndrome (AIDS); and infection with human immuno deficiency [sic] virus (HIV)' to the list of diseases which would render a person 'a prohibited person' if they were 'afflicted' with them (*Government Gazette*, 1987a, 1–2). New regulations for the Health Act, 1977, 'relating to communicable disease and the notification of notifiable medical conditions' now included AIDS amongst a list of communicable diseases (*Government Gazette*, 1987b, Health Act, 68–80). These two pieces of legislation allowed for people who were HIV positive or had AIDS to be quarantined, denied entry into the country or repatriated. While there is no evidence to date that people were quarantined, people were repatriated.

In April 1988, during a vote on national health, Dr. Snyman of the Conservative Party questioned why the DoH annual report had listed the number of AIDS cases in the country but did not mention mineworkers, 'particularly those from Malawi' (*Hansard* HA 1988c, col. 6143). Van Niekerk dismissed the omission: members were already 'aware of the Malawian mineworkers', and '[n]o worker from any high risk country will be allowed into South Africa without an Aids negative certificate'. Moreover, 'hundreds – almost all – of those who were already working in South Africa, are already back in their countries of origin' (*Hansard* HA 1988d, col. 6332). Van Niekerk appeared to suggest that Malawi was no longer considered the only high-risk country and proceeded to outline infection rates in Tanzania and Zaire (current day Democratic Republic of the Congo), reiterating the danger of AIDS and linking it specifically to the threat of infected persons in exile crossing over the border into South Africa:

> [AIDS] is a disease of which we must take cognisance because people who are being trained in certain areas of Africa, and who sneak across our borders, are Aids carriers. We must now take note of the possibility of positive Aids results among those people, because they are trained in camps in which, as we know, there is a high incidence of Aids. I am mentioning these things because it is a matter we shall take further on the security level. (*Hansard* HA 1988d, col. 6332)

In a single debate, Van Niekerk had linked AIDS to mineworkers, other countries and the 'terrorists' (freedom fighters and exiles who lived and trained in bases and camps in other countries) but had failed to recognise or acknowledge any 'real' threat from within South Africa. 'Africa' was not only associated with infection but also with sheltering 'terrorists'. The statement is replete with menacing images of borders being breached by terrorists carrying AIDS as well as AK-47s. A week later, in a similar evocation, Van Niekerk personified AIDS and described it as 'creeping into this country' (*Hansard* HD 1988e, col. 7520).

For the people who were not members of the liberation movements or soldiers of their armed wings, but mineworkers from Malawi, the changing legislation directly affected their livelihoods and security, as Chirwa's (1998) work shows. Over a 4-year period (1988–1992), the legislation was used to repatriate approximately 13,000 Malawian mineworkers, creating tensions between the South African and Malawian governments, between the Chamber of Mines and the South African and

Malawian governments, and between the repatriated miners and the Malawian government.

Cameron and Swanson argue that '[p]ublic fear and panic, as well as a desire on the part of the authorities to be seen to take strong and decisive action, make the temptation to resort to coercion and to invoke the law in applying it strong' (1992, 232). The deportation of Malawians served to both emphasise government action and to alleviate internal political and economic pressures, neither of which related to the alleged health crisis facing South Africa from Malawians nor offered any realistic means of addressing the alleged health crisis. Chirwa suggests that by this time researchers at the South African Institute for Medical Research had 'argued that AIDS was already entrenched in South Africa and the ban on the employment of foreigners with the disease would not do anything to curb its future development in the country', further undermining the government's stated reasons for deporting Malawians (1998, 72). Rather, the need to use regional labour more efficiently, the economic climate in South Africa combined with economic sanctions and disinvestment, the labour and social unrest (including strikes) in the country, rising unemployment and internal pressure to reduce dependency on 'foreign' labour influenced the government's decisions. The pervasive fear of AIDS, together with the idea of 'high-risk groups', was used to justify the exclusive repatriation of Malawian mineworkers. The government did not apply the legislation nor demand the repatriation of mineworkers with HIV infection to any other countries.

Horwitz's work has highlighted the complex association between infection rates (in her examples of TB, HIV and AIDS) and migration and has cautioned that nuanced understandings are needed to account for the 'whole context of political, economic and social circumstances, not merely the migrant nexus' (2001, 119). During the late 1980s, the South African government focused on a fear of AIDS (not on other issues related to health and migrant labour) and used it to implement policies more closely linked to economic and political dynamics than health concerns. After repatriations had begun, Van Niekerk did not again mention infection rates in Malawi nor the danger from infected foreign mineworkers in parliamentary discussions.

Malawian government officials complained that the mandatory testing procedure was degrading, undertaken without consent and contrary to international anti-discrimination policies relating to HIV status and employment. The Chamber of Mines, too, was reluctant to repatriate 'less militant' migrant labourers in the face of increasing militancy amongst South African miners. Despite these concerns, the testing and repatriation continued (Chirwa 1998, 67, 71). For 13,000 people and their families and dependents, association with a specific geographic location was sufficient to fundamentally change their lives under the pretext of health concerns with consequences that would last much longer than the legislation used to bring it about (Chirwa 1997, 1996; Mtika 2007).

At the end of 1989, when Van Niekerk was replaced by Dr. Rina Venter, the DoH's stance would change, but the issue of infection from beyond the *cordon sanitaire* would remain a feature of discussions about AIDS. The department's 1989 annual report noted that in terms of 'combating and controlling AIDS, the key responses were preventing spread through the screening of blood products, by 'informing and motivating the public toward safer sexual practices' and issuing free condoms on request. The report failed to refer to the new strategic plan that had been mentioned in other official documents. It referred instead to the AAG's plan from 1987 noting that it had been implemented by the department and 'all measures contained in that plan [were] strictly adhered to' (DNHPD Report 1989, 28). The annual report was silent on the repatriation of mineworkers. The inherent contradiction in introducing legislation to discriminate against those with HIV or AIDS, while simultaneously addressing 'high-risk behaviour rather than high-risk groups' in local AIDS programme, could not be sustained with any credibility.

Terrorists and Spectres: Ongoing Fear About Borders

In December 1990 official data indicated that 613 cases of AIDS had been identified, and 270 people had died from AIDS-related illnesses in South Africa ('Update AIDS' 1990). By December 1993 the number of identified AIDS cases had reached 2697 with 711 deaths; and the results of the annual national HIV surveys of women attending antenatal clinics (initiated in 1990) indicated growing HIV prevalence rates in this group. Between 1990 and 1993, AIDS featured more prominently in DoH annual reports, in official epidemiological reports and in parliamentary debates. A period of extended engagement and education took place and coincided with a shift in the health-care focus of the DoH that was influenced by the changing political situation in the country which saw F.W de Klerk becoming leader of the National Party and State President, the freeing of political prisoners (including Nelson Rolihlahla Mandela), the unbanning of anti-apartheid organisations and the start of negotiations that led to the 1994 elections (Beinart 2001, ch.10).

Soon after her appointment as minister of health, Venter declared: '[n]ot a single community is immune to this dreaded disease. All of us in South Africa are concerned about this problem. Aids is not coming – it is here' (*Hansard* HD 1990, col. 137). She framed AIDS not as an external spatial threat but an internal reality, the first health minister to do so. She highlighted the limitations of legislative interventions, but seemed to absolve the government of any responsibility to address AIDS:

> I would like to stress that one cannot control it [AIDS and HIV infection] by legislation or any measure that the State might take. I think it is necessary for private enterprise to have an employment policy on HIV carriers or Aids sufferers, based on realistic and sound scientific principles. It is also important that all workers should be educated on the basic aspects of HIV infection and how it should be prevented. Individuals could be identified for specific training at Aids training centres. (*Hansard* HD 1990, col. 137)

While narratives about 'infected outsiders' still featured in parliamentary discussions, specific legislation intended for select groups of people with AIDS was no longer considered. Some of the legislation enacted earlier was either revoked or, in the case of repatriation, 'relaxed after criticisms' (Chirwa 1998; Sadie and Van Aardt 1992). The powers given to officials to force people to have HIV tests and the listing of HIV as a health concern that would render a person 'prohibited', were rescinded in the Aliens Control Act 96 of 1991. As Cameron and Swanson note 'On 1 October 1991, the government without publicity or fanfare removed AIDS and HIV infection from the list' and 'the government's own AIDS Unit discouraged reliance on legislative measures' (1992, 217 & 221).

While Venter did not completely abandon assumptions about external vectors of infection, she did signal a change in official government discourse around border controls: '[m]igrant workers on the mines are not the only problem – airlines and hauliers are too. There is no practical way to seal these borders' (*Hansard* HD 1990, col. 137). From 1991 the focus on outside vectors shifted from migrant mineworkers to returning exiles, and drew on ideas of 'foreignness', which were contrasted with notions of 'belonging' and 'citizenship'. These assumptions were evident in discussions about a proposed amendment to the South African Citizenship Bill that would, among other things, allow exiles to return to South Africa and make access to South African citizenship easier.

In February 1991, Mr H.D.K van der Merwe (CP), voicing opposition to the amendments, warned against an uncontrolled influx of people at a time of increased public fears about personal safety. He drew attention to research that showed that both 'white' and 'black' South Africans felt increasingly unsafe and insecure (*Hansard* Extended 1991a, col. 1135–1136), an unsurprising assertion in the face of the political upheaval and violence being experienced in the country.

He insisted that the proposed amendments were dangerous and called on the minister 'not [to] allow individuals, or a larger number of people into South Africa, who make it unsafe for everyone, whatever one's skin pigment might be'. He asserted that 'a terrorist is a terrorist, whether he is against apartheid or whatever else', and because they might have AIDS: 'It is not only the insecure situation which prevails in South Africa today, but also the entire medical and health situation in South Africa which causes us great concern, and this is a result of people who have not been tested [for AIDS] coming into the country' (*Hansard* Extended 1991a, col. 1135). In reply, I.J. Pretorius (NP) dismissed Van der Merwe's comments: '…the hon. member is now conjuring up spectres. All he is trying to do is score a few political points', suggesting that the NP now did not share the view that the nation's health system was under threat from untested people arriving in South Africa (*Hansard* Extended 1991a, col. 1137).

The CP MPs tenaciously and repeatedly demanded HIV tests for immigrants, exiles and migrant labourers. Snyman made the point:

> I should very much like to repeat the insistence of the CP in respect of the so-called exiles who are now going to return in their thousands from lands to the north of the Limpopo, where the incidence of Aids has already assumed epidemic proportions …
>
> ... Why cannot it be required as a condition for the acquisition of citizenship or permanent residence that a negative blood test in respect of Aids should be produced? It is absolutely essential. (*Hansard* HA 1991b, col. 174)

This statement elicited interjections in the House. Venter did not deny the link between returning exiles and AIDS but raised a legal question:

> As far as the exiles are concerned, I think it is essential for me to ask the hon. member on what grounds one can refuse persons who have been identified as citizens of this country access to South Africa because they are Aids sufferers? (*Hansard* HA 1991b, col. 175)

In March 1991 Snyman, after extrapolating future AIDS figures for South Africa, argued:

> Because the incidence in neighbouring countries to the north of us, for example Uganda, Zambia, Malawi and Zaire – is far higher than that in South Africa, the CP feels very strongly that exiles who left this country more than 7 years ago – in other words, people who have already lost their South African citizenship in terms of the Aliens Act – should only be allowed to naturalise themselves if in addition to the normal requirements in respect of contagious diseases, for example tuberculosis, they also test negative for HIV, before they qualify to become citizens of this country.
>
> In light of the alarming figures in countries to the north of the Republic it is of absolutely urgent importance to protect the inhabitants of this country. When the CP requested this in January of this year, only the ANC reacted by refusing. I think the time has come for the hon. the Minister to give a clear answer to our representations. We want to know what the Government is planning to do in respect of the protection of our people. (*Hansard* HA 1991c, col. 3145)

This statement clearly uses multiple fears and threats for party-political purposes; it encodes 'white' fears about 'black' people and the changing political context in the country. It has as much to do with fears about AIDS and HIV infection as it does with challenging the policies of the ruling National Party.

By identifying those who needed to be protected as 'our people', the CP invoked a sense of 'white' nationalist identity that would not include the returning exiles. The exiles – most of whom would have been 'black' people and all of whom would previously have been considered terrorists and communists by the CP, its supporters, and many other conservative South Africans – were clearly not considered to be in need of 'protection' and did not belong to 'our people'. Snyman continued to raise fears about 'Aids sufferers freely crossing our northern borders' until at least April 1992 (*Hansard* HA 1992a, col. 5388).

The CP never explained how an influx of people (HIV positive or otherwise) could be a threat to security. Would the threat be in terms of bodily and physical integrity with a sub-text that evoked ideas of rape and/or 'the diseased black body'/ or 'the black peril or danger'? Part of the 'total onslaught' rhetoric included the notion of the 'white' minority in South Africa (and other elites) perpetually threatened by perils or dangers including the '*rooi gevaar'* (red peril) of communism and the '*swart gevaar'* (black danger) evocative of 'black' people swamping 'white' people. These and other threats featured prominently in the public and political imagination of sectors of the South African population (Brown 1987; Posel 2001; Visser 2004). Or would it be an economic or social threat to public health resources, or community resources required to care for people with HIV? It was the undefined, nebulous nature of the threat that made it powerful enough, or at least sufficiently adaptable enough, to camouflage the actual arguments or political points being made.

In her reply to Snyman, Venter merely referred to her previous responses:

> I gave him the answer then that on the basis of the fact that the exiles coming back to South Africa – if they are citizens of South Africa – have right of entry, we will have to assist them with their problem. (*Hansard* HA 1991c, col. 3159)

While this statement implied that 'their problems' were outside of 'our' problems, there was at least an acknowledgment that the returning exiles were not foreigners and had a right to health care on their return.

The issue of providing proof of HIV status continued to feature in parliamentary discussions until early 1993 and turned towards new immigrants and migrant labourers, but this time with a focus on Mozambique (*Hansard* HR 1991d; HA 1992a; Extended 1993a; HR 1993b). As a 'frontline state', a country that sent migrant labourers to South Africa, one in which the ANC had a mission (the Maputo mission) and within which the SA government had covertly funded 'resistance' groups, Mozambique had a long and fraught relationship with the apartheid government. While more nuanced research is needed into the detailed relationship between Mozambique and South Africa at this time, the reasons given in parliamentary discussion for this focus on Mozambicans were the high HIV infection rates in the province then known as Natal (now KwaZulu-Natal or KZN). KZN borders Mozambique and Mozambicans were suspected of being the vectors of infection in that area.

Venter brought discussions about proof of status to an end in parliament by observing:

> One of the reasons why it is very difficult to implement such a step is due to the window period of this infection. The person could already be a carrier. One cannot identify him [sic] as a carrier by testing because he has not yet developed anti-genes. On what basis can a state issue a certificate stating that a person is really free of the virus? I believe that this is actually a political argument. I believe South Africa should base its standpoint on sound scientific proof. One really needs factual information on which to base this issue. (*Hansard* Extended 1993a, col. 6734)

The link between Mozambicans and returning exiles requires more investigation, but what does not, is how the conversations and overtones evident in discussions about the HIV status of returning exiles cast the debate in a particularly racialised and racist light. The idea of ANC members 'bringing AIDS into South Africa' was also of concern to the ANC. In a post-1994 context the legacy of this racialised discourse was that research into the infection rates of returning MK soldiers created controversy (Shell 1999).

The racialised discourse meant that little attention was paid to the predominantly white South African Defence Force (SADF) soldiers who had been based beyond the borders to 'protect' apartheid South Africa – despite the fact that, as Lawson points out, many of them were based in areas with high HIV-prevalence rates (Lawson 2008, 45–47). Lawson notes that SADF soldiers were 'holed up' in northern Namibia with their army headquarters in the town of Katimo Mulilo which was considered the epicentre of the region's epidemic and suggests that '[w]hen apartheid's illegal wars came to an end in the late 1980s, thousands of SADF soldiers came home. With them, surely came HIV' (Lawson 2008, 46). Lawson also describes the Caprivi Strip – which borders Angola, Botswana Namibia, Zambia and Zimbabwe – as the military frontier of the region and further discusses the movement of ANC cadres through Uganda, Tanzania and Zambia. The established borders of secrecy associated with years of covert and overt fighting were maintained even after 1994. While the politics of race and blame played out from the 1980s into the new millennium, soldiers from all sides of the struggle received few appropriately devised anti-HIV messages, endured HIV-positive status with little support or access to treatment, and died of AIDS (Whiteside et al. 2006).

The language of fear, threat and danger that permeated discussions about AIDS and HIV between 1980 and 1995 provides a sense of the political and social mood during this period. Discussions in parliament between 1990 and 1995 shifted from a focus on 'AIDS-infected migrant labourers' to a broader consideration of migrant labour and the conditions experienced by migrant labourers. The notion of 'the infected other' was never totally banished, even in a post-apartheid parliament. In 1994 Mrs B.M. Nzimande (ANC) declared '[a]s we are talking now, our country is being inundated with refugees and illegal immigrants – several million from high-risk Aids areas' (*Hansard* NA 1994, col. 3459). During the same debate the new Minister of Health, Dr. N. Dlamini-Zuma referred to 'illegal aliens' and noted that 'there may be problems in that area ... and ... our health services may be inundated, but that is not within my jurisdiction. It is within the jurisdiction of the Minister of Home Affairs' (*Hansard* NA 1994, col. 3459). For all the human rights discourses that began to feature in the new parliament from 1994, and despite the fact that the ANC had been exiled in other African countries, refugees and illegal aliens from 'high-risk Aids areas' were apparently already 'inundating' the health service.

Conclusion: Malevolent and Benevolent Geographies

In the South African AIDS narrative, the use of borders and constant gazing across them drew attention away from key problems within the country's socio-political and economic borders. AIDS was present in South Africa in the early 1980s, and early case studies made it clear that people of all 'sexual orientations' were susceptible. Acknowledging and responding to that, however, would require the transgression of borders linked to morality, sexuality, sexual orientation, 'race', class, gender and other inequalities. A redirecting of attention to 'high-risk groups' and people from beyond the country's borders ensured that comprehensive strategies that recognised that the disease could affect everyone did not need to be conceived of.

This chapter has shown how the creation and maintenance of certain borders (psychological, physical, emotional or otherwise) had a direct impact on people's lives and deaths. Malawian migrant labourers were one such group of people and were part of a mineral-industrial complex closely associated to racialised, heteropatriarchal spatial and economic planning, and a long and problematic relationship with (ill)health. It is a contribution to a body of work around a variety of aspects of migration, migrant labour and AIDS that came, and continues to come, from non-academic organisations and from a variety of disciplines (Brummer 2002; Campbell and Williams 1999; Crush 1999; Crush et al. 2005; Lurie 2000). This body of work has helped configure the boundaries of my work but also allowed me to bridge, disregard, or recreate them.

Marks identified one of the mandates of history in her seminal piece on AIDS in South Africa emphasising 'the value of historical analogy in challenging dominant preconceptions, locating contemporary reactions in context and enabling a sense of perspective' (2002,13). The ability of history to reflect on, reveal and locate the origins of ideas or responses linked to health, analyse them and identify their trajectories is invaluable in creating an historical consciousness about health and in deepening understandings about health concerns and those affected by them. Elder noted though that '[t]he present is senseless without the past but similarly, the present is more than the past' (2003, 3). Both Marks and Elder are scholars whose works display exceptional disciplinary craft and innovation while also being accessible to those working in other disciplines. They combine critical and rigorous writing with a lived epistemic generosity and a commitment to academic activism. Future multi-, trans- and interdisciplinary work that emulated this could be challenging and invigorating, which is exactly why fields like GeoHumanities and MHH need to be developed – to bring the best of all disciplinary practices together, learn what is known or still needs to be known from them and then take steps into new futures of understanding.

How would our understandings of AIDS change if we worked deliberately on engaging with an established topic in a new way? What mutual benefits would be derived if geographers combined their disciplinary mandates, manifestos or curiosities with historians who make use of the strengths of history which Berridge explains lie in 'its sense of process and of change over time, in its generalizing capacity to interweave and assess different forms of source material and theoretical perspectives, above all, in its scepticism about received opinion and what appears to be the norm' (1994, 233).

This historical analysis required, and depended on, an engagement with the physical geography of the region; the socio-political landscape of the time; the geographies of knowledge; the temporal dissonance of the health concern under consideration; the archaeology of archives and sifting of sources; the contours of the body; the mapping of the psycho-social power of emotion; and shifts and ruptures of socio-economic and political fault lines. This brief examination of a certain point and type of AIDS history in South Africa reveals the geographies and borders of blame applied to people, but representative of more complex, interrelated systems of power and inequities which are revealed on closer inspection – it was a brief journey over numerous borders on a small chartered corner of a disciplinary map.

Discussing creativity and place in GeoHumanities, Dear notes that '[c]reative interdisciplinary exchanges occur when people venture beyond their disciplines and speak across boundaries in their respective tongues about matters of common concern' (2011, 12). There is a mandate here to think about our own intellectual and disciplinary borders and work out which to keep and which to cross. There is also an urgent need to examine a number of situations where borders are being built or breached (fortress Europe, Brexit, Mexico, Palestine, the planetary systems) that require the combination of complementary and critically challenging abilities to make more sense of the imminent social (ill)health and blame they imply. If we creatively spoke in our respective tongues about these as matters of common concern, we could disrupt the malevolent geographies of disease being suggested and help shape more benevolent or revolutionary geographies of collective, collaborative well-being.

References

Acquired ImmunoDeficiency Syndrome (AIDS). (1983, January 22). *South African Medical Journal (hereafter SAMJ), 63*(4), 98.

AIDS Advisory Group's first year. (1986, June 7). *SAMJ, 69*(12), 775.

Anderson, R., Prozesky, O. W., Eftychis, H. A., Van der Merwe, M. F., Swanevelder, C., & Simson, I. W. (1983). Immunological abnormalities in South African homosexual men. *SAMJ, 64*(4), 119–122.

Becker, W. B. (1986, October 11). HTLV-III infection in the RSA. *SAMJ* (Supplement), 27.

Beinart, W. (2001). *Twentieth century South Africa*. Oxford: Oxford University Press.

Berridge, V. (1994). Researching contemporary history: AIDS. *History Workshop, 38*, 228–234.

Brown, B. (1987). Facing the 'black peril': The politics of population control in South Africa. *Journal of Southern African Studies, 13*, 256–273.

Brummer, D. (2002). *Labour migration and HIV/AIDS in Southern Africa*. IOM (International Organization for Migration) Report. Retrieved from Southern African Regional Poverty Network (SARPN). www.sarpn.org/documents/d0000572/P534_Labour_Migration_HIV.pdf

Burke, G., & Richardson, P. (1978). The profits of death: A comparative study of miners' phthisis in Cornwall and the Transvaal, 1876–1918. *Journal of Southern African Studies, 4*, 147–171.

Cameron, E., & Swanson, E. (1992). Public health and human rights – The AIDS crisis in South Africa. *South African Journal on Human Rights, 8*, 200–233.

Campbell, C., & Williams, B. (1999). Beyond the biomedical and behavioural: Towards an integrated approach to HIV prevention in the southern African mining industry. *Social Science & Medicine, 48*, 1625–1639.

Chirwa, W. C. (1996). The Malawi government and South African labour recruiters, 1974–92. *The Journal of Modern African Studies, 34*, 623–642.

Chirwa, W. C. (1997). 'No TEBA...forget TEBA': The plight of Malawian ex-migrant workers to South Africa, 1988–1994. *International Migration Review, 31*, 628–654.

Chirwa, W. C. (1998). Aliens and AIDS in southern Africa: The Malawi-South Africa debate. *African Affairs, 97*(386), 53–79.

Crush, J. (1999). Fortress South Africa and the deconstruction of apartheid's migration regime. *Geoforum, 30*, 1–11.

Crush, J., Williams, B., Gouws, E., & Lurie, M. (2005). Migration and HIV/AIDS in South Africa. *Development Southern Africa, 22*, 293–318.

Davies, R., & O'Meara, D. (1985). Total strategy in southern Africa: An analysis of South African regional policy since 1978. *Journal of Southern African Studies, 11*, 183–211.

Dear, M. (2011). Creativity and place. In M. Dear, J. Ketchum, S. Luria, & D. Richardson (Eds.), *GeoHumanities* (pp. 9–18). New York: Routledge.

Department of National Health and Population Development. (1989). *Annual Report. Cape Town general collection, SA Reports*. National Library of South Africa, Cape Town, South Africa.

Elder, G. (2003). *Hostels, sexuality, and the apartheid legacy: Malevolent geographies*. Athens: Ohio University Press.

Gevisser, M., & Cameron, E. (Eds.). (1995). *Defiant desire: Gay and lesbian lives in South Africa*. New York/London: Routledge.

Glick Schillers, N., Crystal, S., & Lewellen, D. (1994). Risky business: The cultural construction of AIDS risk groups. *Social Science & Medicine, 38*, 1337–1346.

Government Gazette. (1987a). *Government Notices – Department of Home Affairs, No. R. 2439*, October 30, 268, 11014. Pretoria, South Africa.

Government Gazette. (1987b). Health Act, 1977, No. R. 2438, October 30, 268, 11014. Pretoria, South Africa (see particularly Annexure 1).

Grove, N. J., & Zwi, A. B. (2006). Our health and theirs: Forced migration, othering, and public health. *Social Science & Medicine, 62*, 1931–1942.

Grundlingh, L. (2001a). Early attitudes and responses to HIV/AIDS in South Africa as reflected in newspaper, 1983–1988. *Journal for Contemporary History, 26,* 86–103.

Grundlingh, L. (2001b). Government responses to HIV/AIDS in South Africa as reported in the media, 1983–1994. *South African Historical Journal, 45*, 124–153.

Hansard (1987a). House of Delegates, June 17, vol. 13, col. 997.

Hansard (1987b). House of Representatives, September 1, vol. 14, col. 2712.

Hansard (1987c). House of Assembly, September 3, vol. 19, col. 5161.

Hansard (1987d). House of Assembly, June 23, vol. 20, col. 265–266.

Hansard (1987e). House of Delegates, August 10, vol. 13, col. 1767.

Hansard (1987f). House of Delegates, August 10, vol. 13, col. 1770.

Hansard (1987g). House of Representatives, September 1, vol. 14, col. 2710–2713 and 2751–2752.

Hansard (1987h). House of Delegates, September 2, vol. 14, col. 3036–3037.

Hansard (1987i). House of Representatives, September 1, vol. 14, col. 2752.

Hansard (1987j). House of Assembly, September 3, vol. 19, col. 5148–5150; 5161–5162; 5214; 5217–5218.

Hansard (1988a). House of Delegates, April 26, vol. 3, col. 7581.

Hansard (1988b). House of Representatives, May 10, vol. 4, col. 9509.

Hansard (1988c). House of Assembly, April 18, vol. 3, col. 6143.

Hansard (1988d). House of Assembly, April 19, vol. 3, col. 6332.

Hansard (1988e). House of Delegates, April 26, vol. 3, col. 7520.

Hansard (1990). House of Delegates, February 22, vol. 21, col. 137.

Hansard (1991a). Extended Public Committees – Assembly, February 15, vol. 23, col. 1135–1137.

Hansard (1991b). House of Assembly, February 26, vol. 29, col. 174–175.

Hansard (1991c). House of Assembly, March 18, vol. 24, col. 3145–3159.

Hansard (1991d). House of Representatives, May 13, vol. 26.

Hansard (1992a). House of Assembly, April 27, vol. 33, col. 5388.

Hansard (1992b). House of Assembly, March 24, vol. 32.

Hansard (1993a). Extended Public Committees – Assembly, April 30, vol. 40, col. 6734.

Hansard (1993b). House of Representatives, May 24, vol. 41, col. 9015.

Hansard (1994). National Assembly (NA), October 20, vol. 2, col. 3459.

Hodes, R. (2018). *HIV/AIDS in South Africa in Oxford Research Encyclopedia of African History* [online version]. http://oxfordre.com/africanhistory/view/10.1093/acrefore/9780190277734.001.0001/acrefore-9780190277734-e-299. Accessed Feb 2019.

Horwitz, S. (2001). Migrancy and HIV/AIDS: A historical perspective. *South African Historical Journal, 45*, 103–123.

Hunter, M. (2010). *Love in the time of AIDS: Inequality, gender, and rights in South Africa*. Scottsville: University of KwaZulu-Natal Press.

Jochelson, K. (2001). *The colour of disease: Syphilis and racism in South Africa, 1880–1950*. Basingstoke: Palgrave.

Katz, E. (1994). *White death: Silicosis on the Witwatersrand gold mines, 1886–1910*. Johannesburg: Wits University Press.

Lawson, L. (2008). *Side effects: The story of AIDS in South Africa*. Cape Town: Double Story. See specifically chapter 1 'From "gay plague" to "African AIDS"'.

Lurie, M. (2000). Migration and AIDS in southern Africa: A review. *South African Journal of Science, 96*, 343–347.
Marks, S. (2002). An epidemic waiting to happen? The spread of HIV/AIDS in South Africa in social and historical perspective. *African Studies, 61*, 13–26.
Marks, S. (2007). Review: Science, social science and pseudo-science in the HIV/AIDS debate in southern Africa. *Journal of Southern African Studies, 33*, 861–874.
Marks, S., & Andersson, N. (1983). *Apartheid and health: Report to the World Health Organisation*. Geneva: WHO Publications.
Marks, S., & Andersson, N. (1986). Women and health in South Africa. *Canadian Women Studies, 7*, 93–97.
Marks, S., & Andersson, N. (1989). The state, class, and the allocation of health resources in southern Africa. *Social Science & Medicine, 28*, 515–530.
Maylam, P. (2001). *South Africa's racial past – The historiography of racism, segregation and apartheid*. Aldershot: Ashgate.
McCulloch, J. (2002). *Asbestos blues: Labour, capital, physicians and the state in South Africa*. Bloomington: Indiana University Press.
McCulloch, J. (2009). Counting the cost: Gold mining and occupational disease in contemporary South Africa. *African Affairs, 108*(431), 221–240.
Mtika, M. M. (2007). Political economy, labor migration, and the AIDS epidemic in rural Malawi. *Social Science & Medicine, 64*, 2454–2463.
Nattrass, N. (1995). The crisis in South African gold mining. *World Development, 23*, 857–868.
Packard, R. (1990). *White plague, black labor: Tuberculosis and the political economy of health and disease in South Africa*. Pietermaritzburg: University of Natal Press.
Packard, R. (1993). The invention of the tropical worker: Medicine research and the quest for Central African labour on the South African gold mines, 1903–36. *Journal of African History, 34*, 271–292.
Parle, J., Hodes, R., & Waetjen, T. (2018). Pharmaceuticals and modern statecraft in South Africa: The cases of opium, thalidomide and contraception. *Medical Humanities, 44*, 253–262.
Phillips, H. (2001). AIDS in the context of South Africa's epidemic history: Preliminary historical thoughts. *South African Historical Journal, 45*, 11–26.
Posel, D. (2001). What's in a name? Racial categorisations under apartheid and their afterlife. *Transformation, 47*, 50–74.
Ras, G. J., Simson, I. W., Anderson, R., Prozesky, O. W., & Hamersma, T. (1983). Acquired immunodeficiency syndrome – A report of two South African cases. *SAMJ, 64*(4), 140–142.
Reddy, V. (1998). Negotiating gay masculinities. *Agenda, 14*, 65–70.
Sadie, Y., & Van Aardt, M. (1992). VIGS-politiek in Suid-Afrika: 1987–1992 [AIDS politics in South Africa: 1987–1992]. *Politikon: South African Journal of Political Studies, 19*, 81–100.
Schoub, B. D., Smith, A. N., Lyons, S. F., Johnson, S., Martin, D. J., McGillivray, G., et al. (1988). Epidemiological considerations of the present status and future growth of the acquired immunodeficiency syndrome epidemic in South Africa. *SAMJ, 74*(4), 153–157.
Shell, R. C. H. (1999). *Trojan horses: HIV/AIDS and the military in Southern Africa, an epidemiological analysis with reference to Grahamstown and Port Elizabeth in the Eastern Cape*. Conference paper presented in East London, South Africa.
Tsampiras, C. (2008). "Not so 'gay' after all": Constructing (homo)sexuality in AIDS research in the *South African Medical Journal*, 1980–1990. *Southern African Historical Journal, 60*, 477–499.
Tsampiras, C. (2013). *Politics, polemics and practice: A history of narratives about, and responses to, AIDS in South Africa, 1980–1995*. PhD Thesis, Rhodes University, South Africa. http://vital.seals.ac.za:8080/vital/access/manager/Repository/vital:2517
Tsampiras, C. (2014). Two tales about illness, ideologies, and intimate identities: Sexuality politics and AIDS in South Africa, 1980–1995. *Medical History, 58*, 230–256.
Tsampiras, C. (2015). From "dark" country to "dark" continent: AIDS, "race", and medical research in the South African Medical Journal, 1980–1995. *Journal of Southern African Studies, 41*, 773–796.

Tsampiras, C. (2017). Temporal dissonance: South African historians and the post-AIDS dilemma. *Journal of Medical Humanities*, 1–17. https://doi.org/10.1007/s10912-017-9440-x. Accessed 21 Mar 2019.

Tsampiras, C. (2019). Removed from context and complexity: Women in early AIDS research in the *South African Medical Journal*, 1980–1995. *Journal of Southern African Studies, 45*(1). https://doi.org/10.1080/03057070.2019.1564869.

Update: AIDS in South Africa since 1982 (as on 14 December 1990). (1990). *Epidemiological Comments*, 17, 12.

Visser, W. (2004). The production of literature on the "red peril" and "total onslaught" in twentieth-century South Africa. *Historia, 49*, 105–128.

Whiteside, A., De Waal, A., & Gebre-Tensae, T. (2006). AIDS, security and the military in Africa: A sober appraisal. *African Affairs, 105*(419), 201–218.

Wilson, F. (2001). Minerals and migrants: How the mining industry has shaped South Africa. *Daedalus, 130*, 99–121.

Chapter 9
Which Patient Takes Centre Stage? Placing Patient Voices in Animal Research

Gail Davies, Richard Gorman, and Bentley Crudgington

Abstract The growth of personalised medicine and patient partnerships in biomedical research are reshaping both the emotional and material intersections between human patients and animal research. Through tracing the creative work of patients, publics, scientists, clinicians, artists, film-makers, and campaigning groups this chapter explores how 'patient voices' are being rearticulated and represented around animal research. The figure of 'the patient' has been a powerful actor in arguments around animal research, mostly 'spoken for' by formal organisations, especially in publicity material making ethical justifications for the need and funding of medical research. Here, patient voices make corporeal needs legible, gather expectations and resources, and provide the horizon for embodying future hopes. However, the accessibility of digital media, alongside local institutional experiments in openness, is creating alternative spaces for voicing patient interfaces with animal research. On research establishment websites, and elsewhere, patients' perspectives are emerging in short films, taking up positions as narrators, tour guides, and commentators, inviting the public to follow them into these previously inaccessible spaces. The embodied experience of patients, sometimes severely affected by the current absences in biomedical research, are used to authorise their presence in these places, and allow them to ask questions of animal researchers. The films are powerful and emotional vehicles for voicing patient experiences and opening up animal research. They also refigure the affective responsibilities around animal research, resituating a public debate around ethics within the body of the patient. The future expectations personified in the abstract figure of the patient, are rendered turbulent in the ambiguous corporeal encounter between human and animals undergoing similar experiences of suffering.

G. Davies (✉) · R. Gorman
University of Exeter, Exeter, UK
e-mail: g.f.davies@exeter.ac.uk; r.gorman@exeter.ac.uk

B. Crudgington
University of Manchester, Manchester, UK
e-mail: bentley.crudgington@manchester.ac.uk

S. Atkinson, R. Hunt (eds.), *GeoHumanities and Health*, Global Perspectives on Health Geography, https://doi.org/10.1007/978-3-030-21406-7_9

Dying for the other

We hear the patient before we see them. Beatrix De Costa is performing. There is purpose and poise. It is not clear if the patient is leading or being led.

The split screen tells us we shall be exploring options and experiences, as well as spaces and how to navigate them.

The screens fill with containers of knowledge, concrete ones, plastic ones. Hands, arms, bodies, controlling substances in controlled environments. The coordinated effort of Hope.

Three screens. Three choices. Look directly at what is affecting you. Look at something different. Look away entirely.

As we travel deeper into this patient journey, some decisions are taken for us and choices are taken away from us.

Time is counted down. Three fingers are splayed on the floor. Two fingers are held up for counting. One mouse is being prepared for death.

The patient decided what they wanted to confront, which patient takes centre stage, troubling the narrative of who is dying and who is other.

The triptych is culled, panel by panel. And so is the mouse. There is only one panel. There is only one voice.

Reflections on Beatriz Da Costa (2011) *Dying for the Other*[1]

This opening commentary describes a film by artist, activist and academic Beatriz da Costa, who died in 2012 after 'negotiating with cancer for many years' (Kelley 2017, p. 230). Her work, *Dying for the Other*, with its multiple screens projecting patient perspectives, clinician interventions and animal experiences, makes strikingly visible the companion species and questions of relation that connect 'suffering, remedy, and harm' (Kelley 2017, p. 233) in healthcare and research. Patients' experiential knowledge is increasingly emphasised by those seeking to calculate remedy and harm, improve healthcare delivery and translate biomedical research. However, the animals underpinning pre-clinical research, used in safety and efficacy trails, or standing in for patients as mouse avatars within personalised treatment rarely appear in these accounts. Laboratory animals are usually kept offstage, silent and unseen. Da Costa's film oscillates between triptych and single screen, giving us choices about where to look. And, as the animal is killed, it acknowledges we may choose not to watch: "the moment of looking away becomes an integral part of viewing the work" (Kelley 2017, p. 238). Her film introduces a series of questions we want to address in this chapter: How is the idea of patient voice articulated and transformed in relations with animal research? How does personally and publicly addressing the companion species whose suffering you share change patient experiences and responsibilities? And, how do subject and object positions shift when there is an enduring anxiety over which patient takes centre stage?

In what follows, we describe extracts from four further films, which involve listening to patients[2] talk about their encounters with biomedical research and labora-

[1] https://vimeo.com/33170755, last accessed 24/09/2018.

[2] 'Patient' is a complex, dynamic and relational category, contestable and temporally located, sharing boundaries and imbrications with other advocacy movements (Epstein 2008). Here, our use of 'patient' is as a gateway to consider the entanglements of a wide range of people with animal research, and consider how people's experiences of, and encounters with, disability and illness acts to influence attitudes to animal research. The growing transparency of animal research produces new entanglements and knowledges, producing an arena of debate amongst all those who are directly or indirectly involved in such networks (Callon and Rabeharisoa 2004).

tory animals.[3] These emerging interfaces present a challenge to scholars seeking to expand the inclusion of patient voices within science (Boenink et al. 2018) and animal voices within society (Srinivasan 2016), respectively and collectively. These encounters have the potential to shape research priorities, understand animal experiences and remake narratives around health. They also produce tensions. Moves towards making animal research more public through patient voices may also make it more intimate, as the 'somatic sensibilities' (Greenhough and Roe 2011) of patienthood are shared across sentient beings who have in common a 'physical vulnerability' and 'susceptibility to injury and illness'. We suggest that the interface of the GeoHumanities and health geographies offers an important space to explore how these novel expressions of patient voice are emerging and to reflect on how they augment, fragment or refract with ideas of giving voice for and about animal others in the production of health. Whilst health geographies are attentive to how health, medical research and care are situated and mobilised, the humanities offer an attentiveness to practice, which does not follow theory and holds onto the indeterminacy of voice, preceding the settlement of these meanings. This openness to "when voice happens, how voice happens, or where voice happens" (Mazzei 2008, p. 45) is essential for these often ambiguous encounters between patient experiences and animal research. In seeking to listen carefully, we draw on our embodied experiences, as researchers, patients and publics, to reflect on how patient voices are created and animals related, and how we are interpolated as audiences.[4]

We start and close with the work of artists and poets reflecting on the entanglements of human and animal health and the intermingling of patient perspectives with creative practice. In the next section, we introduce the policy developments creating interfaces between patients and animal research, before considering how films from patient groups, research advocates and research institutes connect people affected by health conditions with the animals that share their condition. In our accounts, we want to draw attention to the multiplicity of places in which patient voices are being crafted and the different ways these position who is responsible for whom and to whom in interspecies relations of care (Davies et al. 2018). We explore the different ways in which patients and animals are attributed rhetorical and political power in deploying and (re)signifying multiple discourses around health, research and embodiment (Ganchoff 2008). We do this with the aim of adding understanding to how different voices are mobilised in public debates around science (Michael 2009) and helping develop careful ways of speaking for and with

[3] This research is part of the wider Wellcome Trust collaborative award on the Animal Research Nexus (205,393/Z/16/Z). This programme explores the changing historical and social relations around animal research from different perspectives, including the growth of patient and public involvement and engagement within the practices of animal research. For further information see https://animalresearchnexus.org/ (last accessed 28/09/2018).

[4] Our professional and personal backgrounds are diverse: spanning veterinary science, art, human geography, policy, gender, sexuality and experiences of acute and chronic illness. We do not name the different descriptions of films that we have authored. Whilst patient voices have the potential to radically remake these interfaces around animal research, this will not be achieved through allocating meanings in ways that promise the authenticity, fixity or the singular truth of voice.

multiple others in health research (Dam and Svendsen 2017). We also seek to use the creative methods of the humanities to hold these spaces and narratives open and recognise that conversations between patient voice and animal research may start with not knowing what to say.

Patient Voices and Animal Research

The question of how to place patient voice in relation to animal research is not an easy one to answer, yet it is increasingly being asked, both directly and indirectly, through the conjunction of two developments that are opening up new interfaces within animal research. The first is the movement towards increasing patient and public involvement (PPI) within health and care research. Formal practices of PPI are increasingly embedded as an 'official ideology within legislation' and 'widely endorsed by both lay and professional groups within health and social care and in academia and medical research in the UK' (Gibson et al. 2012, p. 533). PPI seeks to enable 'patients' to apply their priorities to the development and delivery of health services (Tritter 2009) and improve the quality and relevance of health research by drawing on the lived expertise of patients, acquired through personal health journeys (Caron-Flinterman et al. 2005). The second is the growing commitment towards openness in animal research after several decades of secrecy and security, especially in the UK. The Concordat on Openness in Animal Research was launched in the UK in 2014 with signatories across universities, medical research charities and others agreeing to provide greater public clarity about when, how and why animals are used in research (Understanding Animal Research 2017). This move to increase transparency also has potentially 'transformative implications for the relationship between science and society' (McLeod and Hobson-West 2016, p. 791).

To move, at the same time, towards widening involvement in and increasing openness around animal research opens opportunities for reshaping research practice, science policy and public conversations around animal research. Powerful claims are made for patient voices across these domains. Rothwell (2006) describes how patients and their carers play an increasing role within debates on animal research, with significant impact and media interest. Reporting on emerging trends in animal research within neuroscience, Pankevich et al. (2012, pp. 57–58) note that 'people living with disease hold particular interest in learning about new research' and describe how 'individuals living with disease are knowledgeable about the use of animals in research and can also take part in public engagement efforts'. They suggest patients can help to 'personalise' the issue of animal research, shifting debates towards how animal experimentation can help people. However, these dual movements bring an irruption of uncertain new meanings around the place of patient voice in relation to animal research.

The more abstract figure of 'the patient' has long been a powerful actor in arguments around animal research, mostly 'spoken for' by formal organisations, especially in publicity material, making ethical justifications for medical research funding. Here, patient voices make corporeal needs legible, gather expectations and resources, and provide the horizon for embodying future hopes. However, being open about, and involved in speaking about, animal research may bring personal complexities for those patients who now find themselves speaking for and perhaps to animal others, as well as public responsibilities within an often controversial public debate. Individual patient voices address a more complex and uncertain future, with the potential to reconfigure narratives, and (re)gain, (re)possess and (re) claim moments and experiences of hope (Power et al. 2018). They demonstrate hope is not a universal experience or singular achievement; hopes for health become entangled and contested with hopes around animal lives. That which cures may harm (Derrida 1981), creating ambiguous and enigmatic 'questions of relations' around health (Stengers 2011, p. 30). Adding patient voices may not produce clarity or confirmation; meanings are added to the complex intersection between biomedical research, human health and animal welfare. The divergent expectations around what a patient voice is and does, and the inclination that many have to look away from animal research, may also result in meanings that are missed, silenced, eluded or excessive.

This complexity is recognised in institutional communications around animal research. The organisational steps towards public openness and patient involvement in basic biomedical research using animals have been hesitant. There are sporadic experiments that speak to the potential of these interfaces. The increasing accessibility of digital media, particularly short films on websites of patient advocacy groups, research institutions, artists and others, is creating spaces to explore expressions of patient voice in relation to animal research. Patients can be found taking up positions as narrators, tour guides and commentators, inviting the public to follow them into the previously inaccessible spaces of laboratory animal research. The embodied experience of patients, sometimes severely affected by current absences in biomedical research, authorise their presence in these places, and allow them to ask questions of animal researchers and explain their experiences to wider publics. They also potentially refigure the affective responsibilities around animal research, by connecting the 'to and fro' of public debate around the ethics of animal research with the personal experiences of patients. The future expectations personified in the abstract figure of the patient, are rendered turbulent in the ambiguous corporeal encounter between specific patients and their laboratory animal companions. These films show patients moving through the spaces of research, speaking in different configurations to, for and with the researchers using animals in their work.

Speaking to Research

There is no introduction. The film starts with a man talking, a headshot against a wall, with slightly blurred focus. Two people, a man and a woman affected by Parkinson's, talk about their experiences of being involved with research. Their dyskinesia acts as an affective reminder of the embodied stakes for them as people affected by this health condition.

The woman describes herself and her illness as a valuable commodity to researchers, but also that she enjoys the exchange. She feels empowered to give direction to a research project. She notes that 'patients can only benefit from that sort of input'. Later, she explains 'it's actually humanised the research, it's made them a person, someone I could communicate with'.

The man explains how he gets to have his say. 'If science wants to do something for science sake…' the man begins and the video zooms in '…then you have a voice, and you get the chance to be heard'. Patient voice has the potential to hold scientists to account.

The film ends with a sketch of a megaphone and the call to 'have your say in Parkinson's Research'. The address is to other people affected by Parkinson's, using patient voice to champion patient voice. The film emphasizes how patients can shape research direction, but also transform relations, meeting researchers and making new friends.

Reflections on Parkinson's UK (2017) *How Can You Have Your Say in Research*[5]

We want to contextualize emerging engagement around animal research by starting with the ways patient involvement itself creates and shapes a particular idea about patient voice. The short Parkinson's UK film exemplifies an opportunity for speaking in ways that are empowering. The film amplifies, and the megaphone underscores, this is an occasion to shape the production of scientific research, as well as deepen social relations around science and patient advocacy. The film recognises the transactional, as well as relational, aspects of patient-researcher connections, suggesting these exchanges have value to researchers, as well as being valued by patients. The chance to be heard means these are not necessarily in opposition. The film mobilises a version of 'patient voice' to champion democratic involvement of patient voices in speaking to research.

The Parkinson's UK film does not talk about animal research but is indicative of aspirations for work on public involvement. Involving patients in research has been a growing trend in health research, aiming to 'give voice' to the people affected by the health conditions that the research aims to benefit. As Callard et al. note, there is growing recognition that 'knowledge 'from the bedside' must feed back into the laboratory if the translational endeavour is to have any real success' (2012, pp. 390–391). They argue decisions over future drug design ought to take into account patients' lived experiences rather than solely drawing on scientific representations and forecasts of presumed therapeutic value. Involving patients in setting research priorities can ensure that research is most relevant to those who might be characterised as its 'end users'.

Patient involvement activities themselves are varied, ranging from the identification of research priorities and questions, the appraisal of study design, to the dissemination of research results. Patients can become involved at multiple points within the research pipeline and increasingly further 'upstream'. The accessibility of digital media has created additional and alternative spaces for voicing these patient interfaces with research, offering platforms that legitimise, stabilise and encourage patient involvement in research. These films serve many purposes: they are voice-pieces for medical research charities to highlight moves embedding patient centricity, for scientists to further demonstrate the importance of research, and for fundraisers to produce affective and emotional connections.

[5] https://www.youtube.com/watch?v=F8fDYCEEE0Q, last accessed 24/09/2018.

The format of the film authenticates an emerging genre of giving voice through patient involvement. People affected by health conditions sit facing the camera and voice their opinions about the researchers they encounter and the research that shapes their lived experiences of embodiment. The presentation conveys why it matters what they say. The talking head format underlines the authenticity and authority of the spoken patient voice: the direct camera angle locates the voice in the affected body whilst simultaneously emphasising the value of articulate speech in having your say in the practices and relations of research.

The next film echoes this format. Yet, something is different. Introducing animals as another participant in this conversation shifts the relations between patients, researchers and publics. The ways in which patient voice patterns the subject and object of conversation shifts, as further subjectivities are introduced.

Speaking for Research

The voice-over starts with footage of a white mouse in a cage nibbling a cardboard tube. We're told that Genetic Alliance (an umbrella patients association) and Understanding Animal Research (a group promoting openness in animal research) took patients to a UK teaching hospital to learn more about the animals used to study their conditions. The film cuts from the mouse to a headshot of a man against a wall.

The man begins to talk, the other voices in the background give his voice an informal authenticity. None of the people are named, nor are the health conditions they are affected by. The man explains he was interested in finding out about the high animal welfare standards, the day to day routines, and the researcher's relationships with the animals. He says he is reassured by this, having previously taken on faith the animal care within research. He values being able to see this in practice and in person.

We switch to a woman, sat in the same spot. She echoes it was her first time seeing laboratory animals and her impression was the animals seemed cared for and happy. She articulates her concern from the standpoint of her condition, that it does involve and require research using mice. Whilst it was an eye-opener to see what this involved, 'they all looked fine'.

The next person launches into an energetic discussion about using animals as models for deafness research, noting their advantages in basic research and the later stage of a drug development and regulatory testing. He speaks less about visiting the facility and mounts a robust defence of the necessity of animal research. He concludes by stating how 'well-husbanded' the animals are, laughing that the animals 'have better living conditions than I do at home'.

The final man, younger than the others, starts from a different position, talking about enjoying meeting the researchers themselves, rather than the animals, though inexorably the conversations comes around to the mice in the room. He says researchers were enthusiastic about the welfare of the animals and explains the enrichment strategies they used. He notes the 'respect' the researchers showed the animals, 'rather than just being, this is an experiment, this is something living as well'.

Reflections on UAR & Genetic Alliance (2012) *What Do Patients Think of a University Animal Research Lab?*[6]

[6] https://www.youtube.com/watch?v=j2RRnwtnsjw, last accessed 24/09/2018.

In this film, as before, the patients appear before the viewer as a sequence of perspectives, their voices conveyed by the commanding conventions of the talking head. Many of the roles in this conversation about biomedical research are the ones encountered before: around the potential of patients as spokespersons, around roles of researchers, and the economics of drug development. But there are also mice: as research models, as something living, as happy and well husbanded. Everyone mentions the mice. These are animal bodies that matter, to patients, to researchers and to the public too. Yet in talking about these other animal bodies, something changes in the way that patients are granted authority to challenge the nature and direction of scientific research that uses animals. The speaking voice of the patient is transformed to give recognition to the animal subject that cannot speak.

The patients in this film similarly explain why they appreciate the opportunity to have their say: they value talking to researchers, listening to researchers and talking about research. But, in the final cut of voices, there is less here on how patient voice might be empowered to contribute to health research, through the embodied experience of living with disease and challenging the science. Rather, the focus is on who is speaking for the animals and how they account for the ways ethical research is enacted in practice. By entering into the animal research facility, and reporting back, the patient gains a privileged position as witness to the experiences of animals. They are reassured from their own personal positions, as someone potentially in a position to gain from this research. Yet, they also gain an uneasy responsibility, in public and in private. The responsibility taken for animal care that lies with the research establishment is conveyed, via the patient, to a public audience. The patient bears witness and provides a public warrant that what they encounter is acceptable care. Responsibility may also have a more personal dimension, as the potential to understand their own condition through research and through animal's bodies opens the potential for shared suffering (Haraway 2008). Their voices convey some anxiety as they make frequent and specific reference to how well looked after the animals are, and the necessity of doing animal research. The patient voices become one of speaking for, rather than to, animal research.

It is possible to speculate about this redirection of patient voice, drawing on the two dimensions of responsibility above. The first is the recognition of animal research as a matter of public concern and debate. Public discussions around animal research are conventionally framed in relation to an ethical debate between pro- and anti-positions. This polarised debate leads to the expectation that patients should take sides. It is necessary to add your voice to one perspective or another rather, to speak for animal research in the pursuit of further understanding of disease and therapeutic possibilities, or to speak against, as in the case of Animal Aid campaigns where people affected by health conditions stress animal research not be done 'in their name' (Animal Aid 2017). The UAR film is a powerful and public counter to this, through which patients act as advocates for research in general and offer ethical assurance around animal research in particular. However, the second question of personal responsibility is more uncertain. The encounter with the animal model may also bring to mind the responsibilities that come with having a body that is vulnerable and to consider how suffering is shared not just with other patients but also with animals too. There is no easily accessible vocabulary or socially acceptable script to articulate these aspects of experience. They may be expressed as either hesitancy or a move quickly into the conventional frameworks of ethical debate. It is also possible to speculate that this absence of language leads to a further absence, this time around the potential for patient voices to challenge researchers about the direction and nature of animal research in the same way they might question other aspects of health and care research.

This film thus changes the content and address of the patient voice. Patients are empowered to speak for research in public, though not necessarily to shape research in practice. The patient voice is a powerful actor in public arguments around animal research, providing assurance as a privileged witness for the experience of the animals and humanity of the researchers they have met. This is a distinctive contribution to ethical justifications for animal research, but it may come at the cost of the power of patient voices to reshape research directions, even as personal encounters with research animals may become increasingly common.

Speaking to the Other

"Tom, this is one of the alco-mice", a man we take to be a scientist explains, as he pulls a cage from a rack of individually ventilated mouse cages. "Ah this one? This is our man?" replies Tom, his voice clear. The scientist launches into a description of 'the blueprint of life, our DNA', his gloved hands gesticulating as Tom looks into the cage. We cut to a close up of a mouse stretching and climbing on the roof bars. The scientist explains the experimental design in everyday language. It hinges on a "very simple choice, very much akin to if you or I went into a pub and said to you 'what would you like to drink?'". The shot switches to the Tom's face staring into the cage, as the scientist explains that one bottle of liquid contains water and one "10% alcohol, so the equivalent of a strong beer".

Tom's face is hidden by his facemask. Only his eyes are visible as he glances to affirm the scientist's continued description, then back to the AlcoMouse. The scientist tells us most mice won't touch the alcohol, whilst the AlcoMouse will take 85% of their daily fluid intake from the alcohol containing bottle. We see a mouse drinking. The scientist stresses the importance of choice. "He chooses the ethanol all the time basically", Tom comments. "85% of the time, yeah", the scientist confirms and corrects. He explains this experimental set up reveals the choice is largely driven by genetics.

Tom finds his voice: "So that gene is basically, that's the one that's saying this isn't socially driven, it's not driven by peer group pressure, it is basically, that's their make-up, that's the way they were designed, and that's what they are going to choose". The camera pans around. Tom's face is clear for the first time, directly in focus. "From my point as an alcoholic, that's something that is great for me to hear, that if there is a similar gene in adults, or humans, this gene would say that it's not just my peer group pressure, it is the fact that I need to drink, and I want to drink, and I actively seek to drink, rather than seeking water". It's the first time Tom acknowledges this condition, though it's been implied throughout. The voice-over picks up this connection to explain the AlcoMouse gene is also present in humans and this is a small, but important step towards understanding "the disease that blighted Tom's life for 15 years".

We see Tom again, facemask down, his voice full of emotion, "I came in here thinking I was just going to look at a mouse that had been fed alcohol". His voice starts to break. "And this one mouse, has given me in 15 minutes, a better understanding of my own illness, than 15 years of trying to search for answers. To be told, that there's a possibility that there's a link to a signal in my brain that was making me crave the alcohol more. For me it's …" He's emotional now. "… I can't get it through how, how, both upsetting, that it's never been told to me before, but also liberating that I'm getting answers. Just from that mouse." Tom's eyes fill with tears. "One mouse." The camera moves from the mouse and back to Tom. Off screen we hear some ask if Tom is okay. "Nah I'm alright. I'm fine. Just got that off my chest. I'm sound. Happy. Happy. That's the thing. Happy." We see a final shot of Tom, his face a mix of different emotions.

Reflections on BBC Horizon (2010) *Miracle Cure? A Decade of the Human Genome, Alcomouse and MRC Harwell*[7]

Within the animal facility, the nature of the encounter, the address to the other and expressions of voice around animal research are radically transformed. The exchange is not directed to the public, as an outward expression of support for research, though that may be implied. Mediated through film, facemasks and the cage face, this is a deeply private and personal conversation, which changes the corporeal and social understanding of what it means to live with a condition and to live with another with that condition. Ganchoff (2008) suggests that patienthood can act like a collective identity. Animal research opens up this collective identity beyond human socialities to remake patienthood through interspecies relations. These entanglements are multi-faceted, reflected in the affective sometimes faltering position of patient voice in this encounter.

The animal other becomes a companion to the unfolding understanding of a condition for scientists and for patients. These animals are not simply 'furry test tubes' (Garner et al. 2017), even and perhaps especially whilst inserted into experiments designed to reduce environmental complexities to bring out genetic cause. Animals are responsive beings, bred to share some biological similarities to human disorders whose validity is always located within the experimental situation (Ankeny et al. 2014). Animals are also powerful figures in cultural narratives that contribute to reallocating responsibilities around health. The alcoholic mouse model offers a new narrative to reframe the 'biographical disruptions' (Bury 1982) experienced around illness like alcoholism by anchoring them in biology. The unworldly innocence of the experimental animal, removed from social interactions and peer pressure, yet still repeatedly needing to drink, offers a different account of the person's experience of their own condition. The animal gives the patient a new kind of voice, in the placing of a craving within biology, in the authority to speak as a patient, in the feeling of it being 'liberating'. Tom speaks directly to the mouse, mediated by experimental situation and scientific interpretation: "this is our man".

The conclusion of the film reassures us that having processed this complex journey, Tom is happy. We are also told, by the scientist this time, that the animals are also happy. The film leaves us with this note of ease, despite acknowledging there is a long way to go before this animal model will yield results with clinical implications. For the patient, the experience of seeing their illness reflected, embedded and embodied in another living being opens up a position to place their voice within the

[7] https://vimeo.com/118265337 – last accessed 24/09/2018.

discourses of biomedical research. This is a new conversation in which there is the potential for different kinds of answers and for finding hope.

Yet hope can be elusive, and identifications can also involve estrangements. The final film we introduce is an animation by Nate Milton of Poem to My Litter by Max Ritvo. The film of Ritvo's poem was completed posthumously, after his death from cancer in 2016. In the poem, Ritvo writes of an experiment in which cells cloned from his tumours were placed in mice in the hope of finding more promising and personalised treatments. The development of these co-clinical trials introduces a direct relationship between patients and animal research. This unexpected, and likely for some unspeakable, intimacy is articulated through the haunting lines of animation and poetry.

Embodying the Other

On a medicinally beige background, handwritten in black, all caps.

POEM TO MY LITTER.

A scratchy outline of a Max Rivto appears, looking directly at the camera, addressing the viewer personally, there is no difference between their skin tone and the background, flattening the image, this is only an outline. Max is only one dimension of this story, but the contrasting white rims of their spectacles frame their eyes, telling us this is from their perspective.

Max shrugs and touches their chest at the mention of "men's" and "my tumours" gently signalling that Max is human and Max is sick. Tactile ownership is consistent. The perspective shifts to a profile.

Max is replaced, or refracted into 12 white mice, nestled together. The narration explains "My doctors split my tumours up and scattered them into the bones of twelve mice".

A purple gloved human hand holds a mouse by the scruff of the neck for inspection, its body hangs, orb like, one dark eye marking the centre of mass, because, as Max explains, they watch each mouse like a crystal ball, unreadable in this telling.

Death; "I wish it was perfect". Flesh is replaced with a skeleton, explored by wriggling white lines, the animacy of tumour cells, darkly yet honestly foreshadowing the path of the narration.

A white dot starts in the right femur and then to the lungs, down the legs and up to the throat, hovering and trembling with potential momentum, like a laser pointer in medical lecture, charting the history of the colony. The anatomy and autonomy of narration.

Max, shirtless, in denims and floating. A shadow under their feet confirms they have a physical presence, even if they have begun to transcend the weight of their story.

Reflectively, we are watching fireworks with Max. Exploding from some central foci and radiating away, an optical metastasise, not exactly Brownian motion, pure white, but the path feels familiar.

As the background colour changes, it dawns on us that the shadow is blood, the skin flayed, a mouse exsanguinated, quite dead. One more breath here.

You are the soft point, 4 or 5 feet away that Max, in the Vrksasana position, is gazing at. Voice and posture are centred, their eye contact with you is tethering. Two more breaths here. Contemplative.

"To complicate the story" another mouse is held by the scruff and injected, defences have to be lowered to allow the full story to develop.

We leave the violence of the last two stanza and look down on a personalised scene. All the mice cradled in Max's arms. "No playing favourites". Clumped together, seemingly identical, held just above heart level, a macro manifestation projected from the thyroid beneath. Max the Human, Max One, Max Two. United, divided, dividing over several scales.

"They are like children you have traumatised and tortured so won't let you visit" is narrated over strobed images of Maxes, flickering lighting, anxious, like Shelly. Monsters are not the only things we are creating.

Rubbing their head Max confesses a hope. That, even when transplanted and grown in another, good will come of suffering.

A red mark dances over the words, a bouncing ball to reveal the cadence of this multi-species empathy.

If this is not familiar to you then this is how you voice this rage. Rubor, tumor, calor, but no hubris, that has been shed along with all the fur.

"That's peace. Which is what we want. Trust me." Trust Maxes.

Reflections on Nate Milton's animation of Max Ritvo (2016) *Poem to My Litter*[8]

Conclusion

We opened and end this chapter with voices produced through artistic practice, poetry and animation to underscore the current uncertainty around the place of patient voice in animal research. Voice can 'elucidate, clarify, confirm, and pronounce meaning' (Mazzei 2008, p. 47). Voice can provide truth, fixity, knowledge and authenticity. Voice can also challenge and remap truths, stability and ideas of authentic meanings. Voices slip, escape, refuse to become pinned down or perceptible. Voices intrude into discourse, capable of bringing discomfort and uncertainty (Mazzei 2008). The work of da Costa questions how we speak about illness, who leads the conversation, who has choices and who does not, as she journeys through treatment. Ritvo interrogates who 'we' are, in our relations of living and dying together, resituating a public debate around balancing harms to animals and future benefits to patients within a complex and specific present where disease outcomes are affected by the simultaneous trajectory of tumours in one person and their personalised mice. The artists voice both connection and disconnection. Da Costa presents a narrowing frame as options and tissue are removed from both patient and animal on their journey together. For Ritvo, narratives of kinship show mice as traumatised children who won't let him visit. They both signal the ambiguous relations of personal experiences around animal research and the uncertainties that come with communicating these in public.

These discussions bracket three more conventional presentations of patient voice in relation to biomedical research where meanings are seemingly more organised. In these, patients voice their experiences to guide research practices. They speak publicly about their support for animal research. They find new ways of understanding the conditions that affect them from within the animal research facility. They

[8] https://www.youtube.com/watch?v=NGr44gjkoxc last accessed 24/09/2018.

speak to, for and with animal research. The preposition alters as perspectives shift and experiences, assurances and authorities are exchanged. Yet here too, we hear echoes of the ambiguities that art is able to express within the complex and changing patterns around patient voice. Who guides and frames the conversation and to whom is the discussion addressed? How does speaking for and with animals in research remake personal subjectivities and identities? And what are the absences and erasures as subject positions shift around the central uncertainty of how to give voice to changing relations with animals? This is an unresolved conversation that merits further consideration in the spaces of the GeoHumanities, drawing as it does on the construction of meanings across species, the changing location of health and knowledge, and the possibilities of an altered horizon for political practices.

In concluding, we indicate some of the openings and closings associated with these possibilities. We want to acknowledge first what is positive in the way patient voices are being mobilised, framed and most critically listened to, as part of an evolving relationship between people affected by health conditions and biomedical research. The encounters between patients and animals can alleviate concerns and worries patients may have about being reliant on research involving the use of animals; witnessing the care deployed for animals gives confidence and reassurance around their intimate entanglements with the worlds of laboratory animals. The animal model may offer new commonalities to speak about a health condition formerly associated with individual stigma, offering novel resources through which to narrate biography, as well as opening up future horizons in which biological trajectories may be reversed. With appropriate care and caution, we suggest there is the potential for further conversations here in which patient voices may have a direct role in speaking to the research priorities, protocols and particularities of animal models that might best represent what is important to them about their health.[9]

However, we also want to acknowledge the significance of these cautions. It is often implied that patient voice helps to 'personalise' the issue of animal research, shifting the focus of debate from animal suffering towards how animal experimentation can help and benefit both patients and publics. Such claims imagine a certain kind of patient and actively enrol patients as having public responsibilities and duties around articulating animal research. There are cautions here about how patients are made to stand in as public witnesses, when Fredriksson and Tritter (2017, p. 96) and others argue it is inappropriate to use 'patients as proxies for the public' in healthcare decisions. There is also careful consideration needed around how the situated act of speaking about animal research changes expressions of patient voice. These do at times align easily with the economies of hope and humanisation that go to support the pursuit of more research, better treatment and pharmaceutical value (Sunder Rajan 2017). Yet, the dominance of affective tropes around

[9] Davies and Gorman are doing further in-depth interviews and ethnographic work, as part of the Animal Research Nexus programme, with engagement professionals, researchers, patient groups and publics to understand how PPI may be able to engage meaningfully around animal research. We would like to thank those medical research charities, communications organizations and research institutions whose conversations have informed our reflections in this chapter here.

hope in relation to patient groups (Coyle and Atkinson 2018), can make it more difficult to express ambivalence or uncertainty. Past histories of protest against animal research seem to amplify these tendencies. The distinctive and disruptive voice of the artists comes from outside. Their experiments with the expressive qualities of film and language outlive their experimental encounters with the biological sciences. As the ongoing work of translation between patient, public and animal bodies in the biosciences continues, we argue these should be joined by increasing attention to the translation of meanings across species and spaces in the GeoHumanities.

References

Animal Aid. (2017). We, the patients, say no to animal experiments! *Animal Aid.* https://www.animalaid.org.uk/the-issues/our-campaigns/animal-experiments/victims-charity-campaign/patients-say-no-animal-experiments/. Accessed 1 Oct 2018.

Ankeny, R. A., Leonelli, S., Nelson, N. C., & Ramsden, E. (2014). Making organisms model human behavior: Situated models in North-American alcohol research, since 1950. *Science in Context, 27*, 485–509.

Boenink, M., van der Scheer, L., Garcia, E., & Burg, S. v. d. (2018). Giving voice to patients: Developing a discussion method to involve patients in translational research. *NanoEthics, 12*, 1–17.

Bury, M. (1982). Chronic illness as biographical disruption. *Sociology of Health & Illness, 4*, 167–182.

Callard, F., Rose, D., & Wykes, T. (2012). Close to the bench as well as at the bedside: Involving service users in all phases of translational research. *Health Expectations, 15*, 389–400.

Callon, M., & Rabeharisoa, V. (2004). Gino's lesson on humanity: Genetics, mutual entanglements and the sociologist's role. *Economy and Society, 33*, 1–27.

Caron-Flinterman, J. F., Broerse, J. E. W., & Bunders, J. F. G. (2005). The experiential knowledge of patients: A new resource for biomedical research? *Social Science & Medicine, 60*, 2575–2584.

Coyle, L.-A., & Atkinson, S. (2018). Imagined futures in living with multiple conditions: Positivity, relationality and hopelessness. *Social Science & Medicine, 198*, 53–60.

Dam, M. S., & Svendsen, M. N. (2017). Treating pigs: Balancing standardisation and individual treatments in translational neonatology research. *BioSocieties, 13*, 1–19.

Davies, G., Greenhough, B., Hobson-West, P., & Kirk, R. G. W. (2018). Science, culture, and care in laboratory animal research: Interdisciplinary perspectives on the history and future of the 3Rs. *Science, Technology, & Human Values, 43*, 603–621. https://doi.org/10.1177/0162243918757034.

Derrida, J. (1981). *Dissemination*. New York: Continuum.

Epstein, S. (2008). Patient groups and health movements. In *The handbook of science and technology studies* (Vol. 3, pp. 499–539). Cambridge, MA: MIT Press.

Fredriksson, M., & Tritter, J. Q. (2017). Disentangling patient and public involvement in healthcare decisions: Why the difference matters. *Sociology of Health & Illness, 39*, 95–111.

Ganchoff, C. (2008). Speaking for stem cells: Biomedical activism and emerging forms of patienthood. In S. Chambré & M. Goldner (Eds.), *Patients, consumers and civil society* (Vols. 1-0, Vol. 10, pp. 225–245). Bingley: Emerald Group Publishing Limited.

Garner, J., Gaskill, B., Weber, E., Ahloy Dallaire, J., & Pritchett-Corning, K. (2017). Introducing Therioepistemology: The study of how knowledge is gained from animal research. *Lab Animal, 46*, 103–113.

Gibson, A., Britten, N., & Lynch, J. (2012). Theoretical directions for an emancipatory concept of patient and public involvement. *Health, 16*, 531–547.

Greenhough, B., & Roe, E. (2011). Ethics, space, and somatic sensibilities: Comparing relationships between scientific researchers and their human and animal experimental subjects. *Environment and Planning D, 29*, 47–66.

Haraway, D. J. (2008). *When species meet*. Minneapolis: University of Minnesota Press.

Kelley, L. (2017). The political life of cancer: Beatriz da Costa's dying for the other and anti-cancer survival kit. *Environmental Humanities, 9*, 230–254.

Mazzei, L. A. (2008). An impossibly full voice. In *Voice in qualitative inquiry* (pp. 57–74). London: Routledge.

McLeod, C., & Hobson-West, P. (2016). Opening up animal research and science–society relations? A thematic analysis of transparency discourses in the United Kingdom. *Public Understanding of Science, 25*, 791–806.

Michael, M. (2009). Publics performing publics: Of PiGs, PiPs and politics. *Public Understanding of Science, 18*, 617–631.

Pankevich, D., Wizemann, T., Mazza, A.-M., & Altevogt, B. (Eds.). (2012). *International animal research regulations: Impact on neuroscience research*. Washington: National Academies Press (US).

Power, A., Bell, S. L., Kyle, R. G., & Andrews, G. J. (2018). Hopeful adaptation' in health geographies: Seeking health and wellbeing in times of adversity. *Social Science & Medicine*. https://doi.org/10.1016/j.socscimed.2018.09.021. Accessed 22 Mar 2019.

Rothwell, N. (2006). Public engagement on the use of animals in biomedical research. In *Engaging science: Thoughts, deeds, analysis and action* (pp. 38–43). London: Wellcome Trust Publication.

Srinivasan, K. (2016). Towards a political animal geography? *Political Geography, 50*, 76–78. https://doi.org/10.1016/j.polgeo.2015.08.002.

Stengers, I. (2011). *Cosmopolitics I*. Minneapolis: Minnesota University Press.

Sunder Rajan, K. (2017). *Pharmocracy: Value, politics, and knowledge in global biomedicine*. Durham: Duke University Press.

Tritter, J. Q. (2009). Revolution or evolution: The challenges of conceptualizing patient and public involvement in a consumerist world. *Health Expectations, 12*, 275–287.

Understanding Animal Research. (2017). Concordat on openness on animal research in the UK. *Concordat on Openness on Animal Research in the UK*. http://concordatopenness.org.uk/. Accessed 1 Oct 2018.

Chapter 10
Surviving Homelessness in Melbourne: The Niching of Care

Cameron Duff

Abstract Most studies of homelessness emphasise the suffering and social exclusion experienced by people living without secure housing. It is only more recently that scholars have turned their attention to the means by which homeless people struggle to maintain their health and wellbeing despite their exclusion. This interest has been a strong feature of recent affective and performative geographies of homelessness. This chapter adds to this literature by offering insights into the affective and performative aspects of health-in-the-making, along with novel empirical reports of the lived experience of health and wellbeing in specific homeless communities in Melbourne. I address the first aim by extending recent geographical work on 'niching practices' in urban settings. I argue that select urban sites support the health-in-the-making of homeless people by affording the creation of enabling niches. Turning to my ethnographic data, I indicate how these niches enable the expression of novel durational agencies and affective capacities. These reports help to elucidate the means by which precarious life is managed in cities. I conclude by briefly considering the implications of my analysis for ongoing efforts to articulate a novel politics of homelessness grounded in an affective right to the city.

Surviving Homelessness in Melbourne: The Niching of Care

City life inevitably mediates the health and wellbeing of urban dwellers, even though the nature of this mediation continues to arouse contention. Mostly, cities are characterised in terms of the risks they pose to health (Rhodes 2002; Rydin et al. 2012), with a lingering contrast between a more rural and salubrious 'state of nature' and the hazards of the urban form (Galea and Vlahov 2005). That cities and urban life pose significant threats to the health and wellbeing of urban dwellers is even more entrenched in debates concerning the health of marginalised communities including the homeless and those living with mental health problems. Of course, the

C. Duff (✉)
RMIT University, Melbourne, VIC, Australia
e-mail: cameron.duff@rmit.edu.au

S. Atkinson, R. Hunt (eds.), *GeoHumanities and Health*, Global Perspectives on Health Geography, https://doi.org/10.1007/978-3-030-21406-7_10

existence of health inequalities in urban settings is beyond contention, and the fact that marginalised communities in particular face disproportionate risks to their health and wellbeing can scarcely be gainsaid. What this consensus has inadvertently engendered, however, is a near prohibition on discussions of the enabling or health-promoting aspects of urban life for people living with housing or mental health problems in cities. Concern for these aspects is at once dismissed for overlooking the myriad hazards that disadvantaged groups face in cities, whilst apparently endorsing the neo-liberal view that civil society ought to bear the primary responsibility for responding to housing and mental health problems in cities, rather than the state (Jacobs et al. 2003; Lancione 2016). Mindful of these admonitions, housing researchers have mostly sought to catalogue the risks and harms experienced by people living with mental health or housing problems in cities, such that understandings of the enabling or health-promoting features of city life in and for these communities have rarely emerged (see Lancione 2013; Duff 2017).

The cost of this prohibition is enduring indifference to the characteristic features of what I will call, after Marcel Hénaff (2016), *health-in-the-making*. Hénaff (2016: 38–40) treats the city as an always-incomplete 'machine' of flows, exchanges and transitions. This 'city in the making' (Hénaff 2016: ix–x) organises, produces and transforms resources, modes of sociality, populations, styles and aesthetic preferences, social identifications and forms of subjectivation. It is a place where 'culture itself becomes dense' as practices, techniques, crafts and forms of know-how mingle and evolve in a 'crucible where all knowledge converges' (Hénaff 2016: 42). The city is always unfinished, requiring the ongoing practical attention of any number of urban dwellers for its 'making', maintenance and repair. Hénaff (2016) insists, of course, that only some of these efforts are organised or planned. While the great infrastructural projects that undergird cities are the subject of meticulous organisation and direction, most of the city's social, affective and aesthetic features do not conform to this model of urban planning. Much like de Certeau's (1984) celebrated account of the everyday tactics and strategies by which a dizzying array of practices emerge in cities, Hénaff's urban imaginary emphasises the 'making-do' or muddling through by which cities are maintained in a bricolage of practices, routines, techniques, discourses and political strategies. What this logic helps us to understand is how urban inhabitants must also invent, compose and maintain varied practices of making-do by which their health and wellbeing may be sustained, protected, recovered or restored in everyday urban life (see Andrews 2017; MacLeod 2017).

The aim of this chapter, therefore, is to risk a concern for the experience of health-in-the-making among people experiencing housing and/or mental health problems in cities in order to illuminate the means by which individuals struggle to maintain their health and wellbeing despite their exclusion. Without ignoring the enormous everyday health challenges that people living with housing and/or mental health problems face in cities, it is deeply concerning that scholarly, policy and legal debates so routinely reduce individuals to these problems (Cloke et al. 2008; Lancione 2017; Vasudevan 2015). Reducing homelessness and mental illness to discrete health problems emphasises the utility of, and the urgent need for, welfare interventions in cities, even as it occludes understanding of the everyday tactics and strategies, practices and routines, by which individuals struggle to maintain their health-in-the-making. One can surely endorse the former even as one insists on the urgency of the latter. Prioritising the latter does, however, require that one move beyond the usual focus on findings in public health, housing and homelessness studies, social work and psychology to explore contributions across the GeoHumanities – in studies of ecologies, contexts and dwelling practices – where important insights are emerging into the experience and promotion of health in sites of significant social and material disadvantage (see Dear 2015; MacLeod 2017; de la Bellacasa 2017; Rhodes 2002). Work in the GeoHumanities typically emphasises the social and affective experience of place and social contexts (Dear 2015), in ways that offer much promise for novel accounts of living in the city. This stands in contrast to more conventional interest in populations and the epidemiology of disease conditions, common to public health and much social science research, that has typically dominated accounts of mental health, disadvantage and social exclusion in cities (see Duff 2014; Rhodes 2002).

Even so, to express a concern for practices of health-in-the-making in the face of housing and/or mental health problems in cities is not to reject the critical importance of systematic public responses to these problems. Nor is it to endorse the view that practices of health-in-the-making are somehow equal to the social, affective and material challenges associated with housing and/or mental health problems in cities. My interest in health-in-the-making – and the enabling or health-promoting aspects of urban life for people experiencing housing and/or mental health problems in cities – lies precisely in the extent to which such a focus may furnish grounds for novel interventions in urban life. This aim also warrants consideration of a broader sweep of conceptual and theoretical debates across the GeoHumanities in order to open up fresh insights into the ways health is lived in place, what it means to be healthy or to struggle with one's health in conditions of significant social and material disadvantage, and the ways health, in its individual and collective valences, is affected by diverse social, material and affective forces (see Duff 2014). Discussions across the humanities of the meaning and experience of 'place', 'health', 'inclusion' and 'disadvantage' have inspired significant innovations in peer-based and community-led health promotion and harm reduction initiatives in cities all over the world, reframing local problems of homelessness, mental illness and substance use (see Bourgois 1996; Duff 2017; Race 2017; Rhodes 2002; Southgate and Hopwood 2001; Lancione 2016). Sensitivity to, and an abiding interest in, the theoretical and conceptual contributions offered across the GeoHumanities has been a key feature of affective and performative studies of homelessness, to which this chapter seeks to contribute (see Dear 2015; Duff 2014; Lancione 2013; Cloke et al. 2008; Vasudevan 2015).

Affective and performative geographies of homelessness primarily emphasise the social, affective and material practices, events and relations by which the city is lived as it is encountered (Cloke et al. 2008; Duff 2017). They pay attention to the affective contours of urban life – the transitions in feeling-states, expressive capacities and embodied resonances (Thrift 2004) – by which a contingent, situated and lived city is realised or enacted. Such approaches reject the notion of an objective, material city that awaits human encounters within it, and emphasise instead the pulses, rhythms and relations of the city's making and doing (Lancione 2016). This chapter adds to this literature by offering conceptual insights into the affective and performative aspects of health-in-the-making, while offering novel empirical accounts of the lived experience of health and wellbeing in specific homeless communities in Melbourne. I address the first aim by developing the notion of 'niching' (Bister et al. 2016) in order to clarify what I mean by health-in-the-making and the ways it unfolds in homeless communities. I argue that select urban places support practices of health-in-the-making by opening up access to *enabling niches* with their specific capacities to affect and be affected. In developing this argument, I reference debates across the humanities for insights into the diversity of urban ecologies, contexts and dwelling practices. My goal is to establish a plausible conceptual account of the character of urban niches – and the varied mechanisms of their production – to inform my ethnographic analysis of health-in-the-making. I am especially interested in determining what the notion of niching may be able to teach us about the everyday practices of health-in-the-making by which homeless people make urban life more bearable. Turning to my ethnographic data, I indicate how niches support the expression of novel durational agencies and affective capacities in contexts of great adversity. These reports will help to elucidate the everyday tactics by which individuals share techniques for managing precarious life in the expression of an assemblage of health (Duff 2014). I will conclude by considering the implications of my analysis for efforts to articulate a novel politics of homelessness grounded in an affective right to the city.

The Niching of Care in Urban Life

Bister, Klausner and Niewöhner's (2016) discussion of 'niching practices' offers a useful means of elaborating my notion of health-in-the-making and the ways it unfolds in urban assemblages of varying kinds. I am particularly interested in the spatial/material, social and affective aspects of niching and how these aspects shape everyday experiences of health in the city. I should start though by reiterating Bister, Klausner and Niewöhner's (2016: 191) point that any analysis of niching practices should be treated metaphorically and not as a literal application of concepts derived from Niche Construction Theory (NCT) as this theory has emerged in evolutionary biology and anthropology (see Downey 2016 for a review). In the latter, the idea of the niche is taken to describe 'the relationship between environmental alteration and evolutionary dynamics' (Downey 2016: S53). It is intended to describe a measurable, observable process by which particular evolutionary, environmental, adaptive and/or selective pressures determine the *placement* of a population (or species) within a territory, where this placement serves selective or adaptive benefits, or both (see Fuentes 2015). In this respect, a niche is 'not simply an environmental location, but a behavioural-ecological conjunction' (Downey 2016:S54). Bister, Klausner and Niewöhner (2016: 190–191) seize on the relational underpinnings of this 'behavioural conjunction' to reimagine the varying 'practices of rendering urban assemblages habitable' by which more 'bearable' modes of urban dwelling are invented, adapted and lived in cities.

Interest in the 'everyday practicalities of dwelling' (McFarlane 2011: 951) that such a focus demands is intended to shed light on the practices, techniques and relations by which the city's inhabitants create 'socially, biologically and ecologically viable surroundings' (Bister et al. 2016: 191). It is to inquire after the ways 'bearable' atmospheres are crafted in urban spaces in ways that sustain varying modes of living-well in the city. These modes are mainly sustained through the specific social, material and affective resources, supports and relations that niching practices avail to sustain informal practices of health and wellbeing in cities (see also Andrews 2017). While Bister and colleagues go on to detail three unique examples of niching drawn from ethnographic data collected in Berlin among people living with a psychiatric diagnosis, their analysis does not elaborate the specific means by which urban niches are spatially, socially and affectively crafted in cities. I would like to begin therefore by offering a brief conceptual elaboration of niching drawing on recent affective and performative geographies of homelessness and the broader innovations in anthropology and philosophy that have inspired so much of this work (Cloke et al. 2008; Lancione 2017; Hénaff 2016). These literatures will help flesh out the conceptual and empirical diversity of urban niches and the principal mechanisms of their production. With recourse to this material, I argue that study of the spatial, social and affective aspects of niching offers the strongest clarification of the nature and production of urban niches and the ways they support, enable and extend diverse practices of health and wellbeing in cities.

A key feature of Bister, Klausner and Niewöhner's (2016) account of 'niching' is the importance of place and spatialisation. Niching is, first and foremost, a practice of place-making, of the creation of space and the emplacement of care in the city. Yet Bister and her colleagues offer few elaborations of the character of this place-making, suggesting the value of more careful conceptual consideration of the spatial aspects of niching. In my own work, I have found Judith Butler's (2015) notion of the 'right to appear' – the means by which urban collectives claim a right to civic engagement, a voice and a political stake in the city – to be a useful way of accounting for the spatial dimensions of caring practices in urban settings (Duff 2017). This work also highlights the value of the kinds of interdisciplinary dialogue that lie at the heart of the GeoHumanities (Dear 2015). Butler (2015: 71) argues that the right to appear presupposes a *space of appearance*, inasmuch as space is central to the body's participation in urban life, just as space is affectively transformed in the event of this participation (McFarlane 2011: 662–666; Lancione 2016: 149–151; Duff 2010: 883–885). For my own purposes, this suggests the urgent need to 'consider the bodily dimensions of action, what the body requires and what the body can do' (Butler 2015: 73) in any discussion of the ways bodies craft niches in urban space and what this crafting means, does or transforms in and of urban life. Butler's analysis provides useful indications of the spatialisation of niching practices in cities, of the ways niches are crafted in place, in the myriad spacings of urban life.

Such practices are inaugurated in the affective and material inhabitation of place, where this habitation makes use of the material affordances of urban space, even as it transforms space in accordance with the expression of a novel sociability, a desire for a more 'bearable' mode of inhabiting the city 'in defiance of the conditions of precarity' (Butler 2015: 70).

Niching practices are, in this respect, intimately concerned with the experience of place, and with the ways bodies interact and affect one another in space, in the spatialisation of sociality. Of course, this characterisation is not so far from the more ecological accounts typical of NCT and Bister and colleagues' (2016) useful discussion of recent anthropological treatments of this theory. What is different is a more overt insistence on the social and material forces by which spaces of appearance are crafted and maintained in cities. Butler's (2015) discussion of this space, and the varying practices of control, resistance and self-determination that shape it, offers a number of insights into the specific habits, practices, interactions and collective struggles by which 'urban assemblages' are rendered more 'bearable' or 'hospitable' in cities (Bister et al. 2016: 190). According to Butler's analysis, we should treat all niching practices as political struggles in and for place, in which individuals and groups seek to maintain a meaningful and enduring space of appearance. This struggle is itself part of the broader contest by which a *right to the city* – a right to participate in the full range of social, economic and political activities that characterise urban life (Lefebvre 1996; Harvey 2008) – is claimed in urban space. Lefebvre's (1996) enduring insight is that the legal and political warrant of this right is limited, insofar as it is entangled in claims to private property and the regulation of access to public space and the urban imaginaries that frame such space. This is particularly true of the city's marginalised communities for whom the right to the city has always been fraught (Marcuse 2009). Indeed, niching practices ought to be understood in these communities by way of the *spatialisation of the right to the city* (Iveson 2011), in which the cultivation of a safe space of appearance figures as the quintessential performative step by which the right to the city is claimed or asserted in place. Niching is, in these respects, indiscernible from the emplacement of the right to the city, from its performative expression in place in the expression of a space of appearance.

Characterising niching practices by way of recent discussions of the right to the city calls attention to the specific practices, interactions, techniques and modes of solidarity by which urban space is implicated in the expression of social, cultural, aesthetic and political struggles in urban settings. I mean by this analysis to conceptually position niching practices in terms of the social and material assertion and enactment of a right to the city, of a right to meaningful participation in urban life that is itself literally grounded in an embodied appeal to a specific space of appearance. Niching is intimately embedded in the creation and expression of diverse caring practices, as Bister and colleagues (2016) note, yet these caring practices should themselves be positioned in the wider context of specific social and political struggles by which a right to the city is asserted and embodied. The right to the city is the social, affective and material guarantee of the caring practices urban niches enable and sustain. In this respect, efforts to extend the enfranchisement of a right to the city among the city's marginalised communities may offer parallel means of promoting the niching practices documented in Bister and colleagues' (2016) work by exposing the social and political conditions of this care. The niching practices that sustain diverse urban geographies of care are themselves grounded in the enduring legitimacy of a universal right to the city for all its inhabitants.

The critical thing however, as I have tried to show elsewhere (Duff 2016b, 2017), is that this right must be lived and felt as a mode of affective emplacement in the city. I have argued so far that niching practices ought to be understood in terms of specific spatial, material and social processes by which a right to the city is enacted. Yet this right is meaningless in the absence of the kinds of affective resonances with which the right to the city is lived in a body's becoming city, becoming machinic, becoming emplaced, to return to some of Hénaff's (2016) analysis. The right to the city has to mean some thing to some body. It has to effect some kind of felt transition in a body's capacities or powers-of-acting (Deleuze 1988: 49) for it to be meaningful and for it to provide useful resources for extending urban struggles against exclusion and inequality (Cloke et al. 2008; Purcell 2002; Marcuse 2009). By foregrounding the *affective embodiment* of the right to the city, I have aimed to shift discussions of this right from a social or juridical register to a more affective one (Massumi 2002). Inasmuch as the right to the city has to be invented rather than conferred – to the extent that it describes an *oeuvre* rather than an *entitlement* (Lefebvre 1996: 149) – one must emphasise the ways this right affects bodies in place, where affect is understood by way of a transition in the (precarious) body's powers of acting, lived and felt as a change in its capacities and sensitivities (Deleuze 1988). These transitions mark the 'lived materialities' of struggle and survival by which precarious groups stake a claim to urban life. (Vasudevan 2015: 339). They are central, in this regard, to the lived expression of niching practices insofar as niching represents an attempt to render urban life more 'bearable', where bearable life must be understood affectively as a transition in a body's becoming-with the city and with the human and nonhuman assemblages that comprise this city (see McFarlane 2011; Cloke et al. 2008).

What I propose to do now is to turn to my ethnographic data for further insights into the spatial, social and affective aspects of niching practices in cities. My goal is to extend the argument that niching practices ought to be understood, affectively and performatively, as instantiations of a lived and felt assertion of the right to the city. I will close with some brief reflections on the implications of my argument for affective and performative geographies of homelessness and the caring practices these geographies often describe.

Ethnographies of Homelessness in Melbourne

The material presented below is drawn from two ethnographic studies of homelessness conducted in Melbourne. These studies investigated pathways into and out of homelessness and the social, affective and material conditions of living without secure housing (Duff 2014, 2016a, b). Forty-five adults aged between 22 and 49 years participated across the two studies. Most participants were recruited via specialist housing services in the Northern and Western suburbs of Melbourne. Data collection commenced in March 2013, with the last of the fieldwork completed in August 2017.

Alert to developments in affective methodologies (Knudsen and Carsten 2015), and discussions of non-representational methods (Vannini 2015), each study employed observational, interview and visual methods to capture the affective experience of homelessness and its social and material resonances in encounters between bodies, human and nonhuman (Duff 2014, 2016a). With interests in the affective and performative force of appearance, in the physical and embodied experience of homeless people's occupation of urban space (Vasudevan 2015: 321–324), and in the social, spatial and material contexts of this appearance, data analysis emphasised the affective, spatial and performative 'messy-ness' of the 'entangled worlds' that homeless bodies participate in (Dowling et al. 2017: 3). In presenting my data in the sections below, I hope to identify some of the varied social, affective and material practices, relationships and contexts by which niches are crafted in urban space in support of the health-in-the-making of homeless bodies. I will be further guided in these efforts by Dewsbury's (2010: 321–24) seven injunctions for the conduct of 'affect-based' qualitative research and Anderson and Ash's (2015) more recent account of 'atmospheric methods'.

Spatial, Social and Affective Aspects of the Niching of Care

Like almost all other affective and performative analyses of homelessness in urban settings, my research confirms the enduring centrality of *place* for homeless people (see Cloke et al. 2008; Lancione 2016). Equally salient are the social, affective and material struggles that follow from the absence of a secure place. This is why the study of niching practices is potentially so useful, for it helps to shed light on the character of spatial struggles in cities, and the consequences of living in the city without a secure place. Indeed, it may be argued that niching is the principal response to spatial insecurity for homeless people. Insofar as all life needs a place (Downey 2016), niching may be regarded as a kind of primal struggle to reclaim place, to stake a claim in the city. The niching practices I observed in my research were mostly unstable and highly tenuous attempts to reclaim a place in the city, a space of appearance for the assertion of the bare facts of insecure life in Melbourne. Mostly, these niches lasted for a few days at best, supporting some semblance of unreliable shelter from the elements, while enabling a kind of dramaturgy of public appearance by which homeless individuals performatively expressed the social, material and affective struggles of homelessness in the hope, usually, of eliciting support from passers-by in the form of loose change, comfort, material sustenance or simple conversation. This usually involved setting up a temporary shelter crafted from found objects in doorways or clear spaces on one or another of the main thoroughfares running through Melbourne's city centre. Some select examples follow (Figs. 10.1 and 10.2):

Fig. 10.1 Discarded belongings on Swanston Street (author photograph)

Fig. 10.2 An urban niche (author photograph used with subject permission)

The two examples illustrate a kind of bare niching by which a highly unstable and insecure space of appearance was crafted in the city. Following the conceptual analysis offered in the previous section, this niching may be said to entail spatial and material, social and affective elements. I will briefly review each aspect in order to flesh out the means by which niches are crafted in Melbourne in conditions of extreme disadvantage, but also to highlight how critical these niches are, no matter how unstable and insecure they inevitably remain. As I've noted, space is central to the formation of niches in urban settings, and my ethnography revealed several sites across Melbourne's city centre that are recurrently appealing settings for the staging of diverse niching practices. As others have noted (Cloke et al. 2008; Lancione 2016; Vasudevan 2015), high traffic areas throughout the city centre, with large numbers of pedestrians and good access to transport and social support services, are enduringly appealing sites for homeless individuals to linger in place. In Melbourne, local convenience stores, including franchises in the 7/11 chain, remain popular sites, and I observed diverse niching practices outside numerous 7/11 stores at one time or another during the fieldwork. There are varying reasons for the occupation of this kind of space, as Fred explained:

> Nah mate, it's just easier here with the tram nearby and lots of people walking past, so there's always someone who's going to stop for a chat and maybe give you a bit of money or something. Sometimes people will buy me a sandwich from the 7/11 or some sushi from the store round the corner too which is good. And it's never boring! There's always stuff going on and people to look at and someone to talk to. So I'll spend hours here before I go on to [service provider] for a meal and then try and sort out somewhere to crash for the night.

These niches sustain all kinds of social and material practices of support, sociality and sustenance. They are crafted from the bare facts of the precarious body's occupation of place, sustained through a kind of material marking of a border that grounds a claim to space, my space, for the duration of my occupation of it. Fred returned to the same 7/11 for several days at a time, leaving materials behind him to return to them again, to be rearranged on each occasion in a novel claim to space. This niching is spatial and material insofar as it relies on the specific practical affordances of the site: a space of high visibility affording some sense of safety and security with customers constantly exiting the convenience store, some with loose change handy from small purchases they might willingly deposit in Fred's nearby cap. Yet this niching is also social and affective insofar as this specific site is made meaningful for Fred in the social interactions it potentially avails. Fred spoke often of the boredom of long days 'with nothing to do while you wait to fight for a bed', and so the opportunities for casual conversation that this particular niche offered were highly valued. With high volumes of pedestrian traffic, and an adjacent major tram hub, Fred described a large number of casual interactions with passers-by, some of whom asked casually about his welfare, but most simply offered brief conversations regarding the football, weather or similar distractions. Of course, Fred also noted spending long periods of time with other homeless people, friends and acquaintances and with outreach staff from local homelessness and welfare services. All these interactions may be said to have affective dimensions to the extent that Fred alluded to the ways each supported different kinds of novel capacities;

capacities to endure boredom, renewed capacities to ask for help, assistance and comfort from others, along with a tenuous capacity to endure the struggle of living without secure housing.

Paramount above all else though was a feeling of the abeyance, for a moment, of struggle; the struggle to find security, comfort and support, sanctuary from fear. Most participants spoke of the near constant threat of conflict in the city ranging from petty disputes with peers through to fears of violent interactions with strangers, security officers and/or police. This threat engendered a unique atmosphere of vulnerability, fear and suspicion that was lived as a kind of heightened arousal, an expectation of imminent conflict. Some participants spoke, in this respect, of the value of alcohol and cannabis as tools for relieving or temporarily suspending this fear of threat and struggle. Certainly, I observed numerous interactions where a palpable atmosphere of struggle was ever present, sweeping bodies up in preparation for conflict (see also Cloke et al. 2008). The niching by which a highly visible space was crafted in a shop front or square of pavement outside a convenience store offered some affective respite from this atmosphere, where visibility afforded some protection from threats, and the milling crowds afforded the prospective comfort of sociality, of a conversation that might carry one away for a time.

A similar experience of retreat or withdrawal from the bustle of Melbourne's street life was observed in slightly more enduring form around a multipurpose sports facility on a major university campus in the city centre. Unfenced, the facility included two basketball courts, multiple all-weather table tennis tables, expanses of synthetic grass with raised platforms for seating and a covered BBQ area with multiple gas BBQ stations. The facility is located behind a large teaching building, one street back from a major north/south thoroughfare in the city centre. It is mostly used by students between classes or for improvised 'pick-up' games of basketball. During the fieldwork period, it was also observed to be popular among young skateboarders for the unique material affordances offered at the site, although university security periodically made efforts to move the board-riders on following noise complaints from staff in the adjacent building. Attention from security staff waxed and waned, however, with the activity of the riders, creating opportunities for local homeless individuals to make novel use of the space (Fig. 10.3).

The material affordances of the site, especially the availability of cooking facilities, were of primary importance throughout the observed interactions in this space. The space afforded a degree of privacy, given its remove from the city centre and the absence, in the main, of foot traffic from other parts of the city. While students frequented the site during the middle part of the day, particularly around lunch-time, it was relatively quiet for long stretches at a time, further explaining the noted absence of security guards. The space itself offered support for social interaction, for sitting, talking, sleeping and resting, cooking, refilling water bottles, reading and playing cards, all of which were observed at one time or another during the observational period. In many respects, the site functioned for homeless people in much the same way as it had, presumably, been designed for students to use, as a space to avail informal interactions around eating and the sharing of meals, resting and physical leisure activities. Noting the importance of games, Karen shared how:

Fig. 10.3 Lunchtime at the BBQ (author photograph)

> There's a $2 shop across the road, like all cheap stuff from China, wherever, and Bill and I decided the other day to buy some table tennis gear there. Some of the students here told us about it so we got it all cheap and then played table tennis all afternoon. We even sold the stuff back to some other kids the next day for a bit of a profit (laughs).

Tom and Brian emphasised the value of the gas BBQs and the numerous occasions on which they had enjoyed simple meals there with friends and acquaintances, particularly in the warmer months between October and April. Tom recounted how for weeks at a time his day would be organised around sitting outside convenience stores and on street corners for long enough to collect enough cash to afford 'a pack of cheap sausages, some sauce, a loaf of bread and maybe some beers' for a BBQ. He added that, 'it's a bugger because you need goldies (AUD $1 and/or $2 coins) to run the BBQ so you gotta save those, so it all takes a bit longer (to collect enough cash for a meal)'. Brian reminisced how 'during the summer time, when there's hardly anyone about, we could have the whole place to ourselves, and you'd never see a security guard or a cop, so we'd just have our own party'. Working in the adjacent building, I'd often come down to observe the same group, including principally Tom and Brian, grilling sausages on the BBQ and then sharing them round a large group as others drank beer or wine and laughed and shared stories. This activity was remarkably akin to the ways student groups used the space during the teaching semesters. The niching practices I observed here could be regarded as orders of conviviality, social interaction, relaxation and support. As I got talking to Brian and Tom over a period of several weeks, I learned more about how critical these kinds of spaces are for otherwise unremarkable moments of sociality, leisure and exchange. With the city's parks and gardens typically subject to pervasive security and surveillance efforts that work to exclude or 'move on' homeless people (see Cloke et al. 2008; Lancione 2017), spaces like this university square afforded some respite from the constant need to 'keep moving' as Brian put it.

This niche involved material, social and affective affordances as it supported material sustenance, social interaction and feelings of respite and repose. The practices of health-in-the-making it sustained typically included respite from the struggles of homeless life in the city, moments of subjective wellbeing and a practice of social inclusion. I'd like to close now with some final thoughts on the character of these practices of health-in-the-making and what they might reveal about a novel politics of homelessness in urban settings.

An Affective Politics of Homelessness

Urban spaces are mostly hostile spaces for homeless individuals. City life is lived for the homeless in an atmosphere of ever-present, pervasive risk, hostility, indifference and exclusion. Yet this does not mean that all homeless individuals are simply subject to this atmosphere, with numerous instances throughout my fieldwork of a kind of artful resistance to the struggles of urban life (see also Cloke et al. 2008). This is precisely how I would like to position the notion of health-in-the-making, as a way of thinking about how homeless individuals experience city life and the varied means by which they attempt to make this life 'more bearable' (Bister et al. 2016: 191). In this respect, my notion of health-in-the-making is akin to recent feminist accounts of the ethics of care (see Puig de la Bellacasa 2017 for a review) and their descriptions of the diverse efforts by which individuals and groups struggle to 'maintain, continue and repair' a means of living 'as well as possible' (Tronto 1993: 103, cited in Puig de la Bellacasa 2017: 3). For Puig de la Bellacasa (2017: 11–13), the ethics of care that these practices describe necessarily evokes a politics, given the ways power and discourse differentially shape the capacities of individuals and groups to sustain practices of care. This politics points, in turn, to the 'situated material conditions' (Puig de la Bellacasa 2017: 11) that ground expressions of care, either mediating and facilitating the capacity to care or constraining and attenuating this care, depending on the character of these conditions (see Tan and Atkinson, this volume, for discussions of care). Attending to these conditions also points once again to the value of the GeoHumanities for discussions of homelessness in the city, drawing together the spatial insights of human geography with a sensitivity to the social and affective dimensions of care, place and support common to the humanities.

It is with this kind of cross-pollination in mind that I have sought to emphasise the material and spatial conditions of niching practices within homeless communities in Melbourne. I have focused my analysis on the ways niches enable practices of caring, of health-in-the-making, among homeless groups according to the situated material conditions of this niching. Health-in-the-making, in the ways I have described it throughout this chapter, may be understood in terms of recent discussions of the ethics of care to the extent that it captures the artful, spontaneous, planned and accidental *practical doing* by which care is effected in the city in conditions of profound disadvantage. Like care, and caring practices, health-in-the-making captures the fragility, the always-uncertain tensions and struggles, by which homeless people seek to maintain a more bearable space for living in the city. This living can scarcely be said to necessarily equal 'living well' (see Puig de la Bellacasa 2017), just as it is not always bearable, let along hospitable (see Bister et al. 2016). Yet this is precisely the point of emphasising the muddling-through of health-in-the-making, the always-unfinished struggle by which the precarious body, to return to Butler's (2015) analysis, works to fashion or assemble means that are 'good enough' for now to support some measure of health and wellbeing. This muddling-through may be partially understood at the level of practice, yet it should also be understood affectively insofar as the lived experience of 'living well' is guided in the first instance by a kind of affective resonance by which the body is guided towards its own *becoming-well* (see also Duff 2014: 79–84).

It follows that regardless of their practical and discursive resonances, neither niching, health-in-the-making nor care itself should be understood as primarily deliberative or intentional. I would argue that they are affective before they are intentional, guided by a line of becoming well by which the body is insinuated into an assemblage of forces. Puig de la Bellacasa (2017: 20) argues that care should be regarded as a 'force distributed across a multiplicity of agencies and materials [that] supports our worlds as a thick mesh of relational obligations'. The key to an affective politics of homelessness lies in the identification of novel means of supporting this mesh, of strengthening these relational obligations, and opening up new forces of becoming in and for homeless communities. My notion of health-in-the-making may indeed be regarded this way, as an attempt to understand the forces, relations and materials that may sustain a 'thicker mesh' of supports for homeless bodies. Of course Bister and colleagues' (2016) work on niching practices can also be understood this way, along with recent feminist discussions of the ethics of care. The conceptual, empirical and analytical goal that is shared across these scholarly projects is the task of identifying how 'worlds', 'assemblages', 'niches' or 'places' are supported and threatened in disadvantaged communities, the extent of the caring modalities that exist within these worlds and how these modalities may be enhanced, 'thickened' or promoted. Care happens in homeless communities, as it is distributed across diverse assemblages of bodies, forces, materials, affects and relations. It is a mistake in this respect to think that the homeless need simply to be cared for.

Certainly all homeless individuals need greater assistance from formal welfare services. They need help accessing income support and secure housing. What we learn, though, by focusing on practices of health-in-the-making, on niching and care, is that the central goal ought to be identifying means of sustaining and strengthening the relational meshes that already entangle homeless individuals in complicated networks of care and support. The point of this chapter, though, is that we currently aren't very good at recognising these forces. We haven't been in the habit of attempting to understand the relations of care, formal and informal, that homeless communities themselves struggle to invent and sustain. For fear of being seen to overlook the force of structural inequalities, of poverty, power and politics, scholars have typically treated the homeless as marginalised subjects in need of formal care. This is care from outside, above and beyond homeless communities, the care that cares for those that can't care for themselves. This is the reason why Puig de la Bellacasa (2017) emphasises the complicated politics of care. Care is always political because it inevitably instantiates and then reifies asymmetrical relations between carers and the cared-for. Focusing on the situated character of care and caring, of niching, or health-in-the-making, is an attempt to disrupt these asymmetrical power relations by returning to the problem of care as a situated, relational doing, to the force of care as an affective bond between bodies. Of course this project may yet fail, and the effort to understand the affective character of health-in-the-making may not deliver novel strategies for supporting homeless people to assert their right to the city. Yet we should never forget that the conventional strategy of treating the homeless as a group in need of care has not always secured this right either. The promise of a *novel politics of care* lies in uncovering *novel practices of care*, particularly in places where we had not expected to find them.

Acknowledgements The author extends his sincere thanks to the individuals who participated in the fieldwork from which this chapter is drawn and the agencies that supported it. Formal approvals to conduct this research were received from Monash University's Human Research Ethics Committee. The completion of this research and the preparation of this chapter were supported by the award of a Vice Chancellor's Senior Research Fellowship from RMIT University.

References

Anderson, B., & Ash, J. (2015). Atmospheric methods. In P. Vannini (Ed.), *Non-representational methodologies: Re-envisioning research* (pp. 34–51). London: Routledge.
Andrews, G. (2017). 'Running hot': Placing health in the life and course of the vital city. *Social Science & Medicine, 175*, 209–214.
Bister, M., Klausner, M., & Niewöhner, J. (2016). The cosmopolitics of 'niching'. Rendering the city habitable along infrastructures of mental health care. In A. Blok & I. Farias (Eds.), *Urban cosmopolitics. Agencements, assemblies, atmospheres* (pp. 187–206). London: Routledge.
Bourgois, P. (1996). *In search of respect: Selling crack in El Barrio*. New York: Cambridge University Press.
Butler, J. (2015). *Notes towards a performative theory of assembly*. Cambridge: Harvard University Press.
Cloke, P., May, J., & Johnsen, S. (2008). Performativity and affect in the homeless city. *Environment and Planning D, 26*, 241–263.
De Certeau, M. (1984). *The practice of everyday life* (S. Rendell, Trans.). Berkeley: University of California Press.
de la Bellacasa, M. (2017). *Matters of care: Speculative ethics in more than human worlds*. Minneapolis: University of Minnesota Press.
Dear, M. (2015). Practicing geohumanities. *GeoHumanities, 1*, 20–35.
Deleuze, G. (1988). *Spinoza: Practical philosophy*. San Francisco: City Lights.

Dewsbury, J.-D. (2010). Performative, non-representational, and affect-based research: Seven injunctions. In *The SAGE handbook of qualitative geography* (pp. 321–334). London: SAGE Publications.
Dowling, R., Lloyd, K., & Suchet-Pearson, S. (2017). Qualitative methods ii: More-then-human methodologies and/in praxis. *Progress in Human Geography, 41*, 823–831.
Downey, G. (2016). Being human in cities: Phenotypic bias from urban niche construction. *Current Anthropology, 57*, S52–S64.
Duff, C. (2010). On the role of affect and practice in the production of place. *Environment and Planning D, 28*, 881–895.
Duff, C. (2014). *Assemblages of health: Deleuze's empiricism and the ethology of life*. Rotterdam: Springer International.
Duff, C. (2016a). G major to D major to A minor 7 (a progression to recovery). In J. Coffey, S. Budgeon, & H. Cahill (Eds.), *Learning bodies* (pp. 173–189). London: Springer.
Duff, C. (2016b). Atmospheres of recovery: Assemblages of health. *Environment and Planning A, 48*, 58–74.
Duff, C. (2017). The affective right to the city. *Transactions of the Institute of British Geographers, 42*, 516–529.
Fuentes, A. (2015). Integrative anthropology and the human niche: Toward a contemporary approach to human evolution. *American Anthropologist, 117*, 302–315.
Galea, S., & Vlahov, D. (Eds.). (2005). *Handbook of urban health: Populations, methods, and practice*. Dordrecht: Springer Science & Business Media.
Harvey, D. (2008). The right to the city. *New Left Review, 53*, 23–40.
Hénaff, M. (2016). *The city in the making* (A-M. Feenberg-Dibon, Trans.). London: Rowman and Littlefield.
Iveson, K. (2011). Social or spatial justice? Marcuse and Soja on the right to the city. *City, 15*, 250–259.
Jacobs, K., Kemeny, J., & Manzi, T. (2003). Power, discursive space and institutional practices in the construction of housing problems. *Housing Studies, 18*, 429–446.
Knudsen, B., & Carsten, S. (Eds.). (2015). *Affective methodologies: Developing cultural research strategies for the study of affect*. London: Palgrave.
Lancione, M. (2013). Homeless people and the city of abstract machines: Assemblage thinking and the performative approach to homelessness. *Area, 45*, 358–364.
Lancione, M. (2016). Racialised dissatisfaction: Homeless management and the everyday assemblage of difference. *Transactions of the Institute of British Geographers, 41*, 363–375.
Lancione, M. (2017). Micropolitical entanglements: Positioning and matter. *Environment and Planning D, 35*, 574–578.
Lefebvre, H. (1996). *Writings on cities*. Oxford: Blackwell.
MacLeod, K. (2017). *Wellbeing machine: How health emerges from the assemblages of everyday life*. Durham: Carolina Academic Press.
Marcuse, P. (2009). From critical urban theory to the right to the city. *City, 13*, 185–197.
Massumi, B. (2002). *Parables for the virtual: Movement, affect, sensation*. Durham: Duke University Press.
McFarlane, C. (2011). The city as assemblage: Dwelling and urban space. *Environment and Planning D, 29*, 649–671.
Purcell, M. (2002). Excavating Lefebvre: The right to the city and its urban politics of the inhabitant. *GeoJournal, 58*, 99–108.
Race, K. (2017). *The gay science: Intimate experiments with the problem of HIV*. London: Routledge.
Rhodes, T. (2002). The 'risk environment': A framework for understanding and reducing drug-related harm. *International Journal of Drug Policy, 13*, 85–94.
Rydin, Y., Bleahu, A., Davies, M., Dávila, J. D., Friel, S., De Grandis, G., et al. (2012). Shaping cities for health: Complexity and the planning of urban environments in the 21st century. *Lancet, 379*(9831), 2079–2108.

Southgate, E., & Hopwood, M. (2001). The role of folk pharmacology and lay experts in harm reduction: Sydney gay drug using networks. *International Journal of Drug Policy, 12*, 321–335.
Thrift, N. (2004). Intensities of feeling: Towards a spatial politics of affect. *Geografiska Annaler, Series B, 86*, 57–78.
Vannini, P. (Ed.). (2015). *Non-representational methodologies: Re-envisioning research.* London: Routledge.
Vasudevan, A. (2015). The autonomous city: Towards a critical geography of occupation. *Progress in Human Geography, 39*, 316–337.

Chapter 11
Truth or Dare: Women, Politics, and the Symphysiotomy Scandal

Oonagh Walsh

Abstract In 2016, a Redress Scheme was launched in Ireland to compensate women who had undergone the procedure of symphysiotomy in childbirth in the decades between 1940 and 1990. The scheme was the culmination of an exhaustive process that had seen two failed attempts at an independent report, the fracture of the main patient representative group, constant lobbying of Dáil Éireann, and a widespread media campaign that crafted a David-and-Goliath narrative for public consumption. Inevitably, the reality was somewhat different. Throughout the process, emphasis was placed upon the importance of women's voices and the primary testimony of those who had undergone the procedure. Yet, for one of the survivor groups, only certain kinds of voices were permitted, those who adapted their histories to the narrative that refused complexity or subtlety. This essay examines the manner in which a variety of women's voices were crafted, amplified, and silenced in the service of a specific agenda. It also interrogates the role of the media in driving that narrative and raises the vexed question of how voices are heard, or silenced, in the age of social media.

There are several anecdotes I tell my students when we discuss changes in the patient/practitioner power relationship over the years: here are two. The first is from 1945 and concerns the then county surgeon at Kilkenny General Hospital. Accompanied by a flotilla of staff including the matron, ward sisters, and junior doctors, the surgeon conducted his rounds, pausing at each bed to grill the nervous white-uniformed acolytes about the occupant's malady, treatment, and progress. The exclusive clinical jargon passed back and forth over the unfortunate patient's body without acknowledgement of the person lying anxiously beneath. One man, whether braver than the rest, or perhaps more worried about his condition, waited until the surgeon was ready to sweep away with his entourage, plucked up his courage and asked: 'Excuse me doctor: can you tell me what's wrong with me?' Looking

O. Walsh (✉)
Glasgow Caledonian University, Glasgow, UK
e-mail: oonagh.walsh@gcu.ac.uk

S. Atkinson, R. Hunt (eds.), *GeoHumanities and Health*, Global Perspectives on Health Geography, https://doi.org/10.1007/978-3-030-21406-7_11

directly at the patient for the first time, the surgeon drew himself up to his full height and drawled: 'The only thing that's wrong with you is that you talk too much'.[1] The second incident occurred in 1992, at St. James' (private) Hospital in Dublin. The physician conducting his rounds stopped at the end of one of his female patient's bed. Without drawing the curtains or lowering his voice, he outlined the woman's condition and treatment – surgery for Crohn's disease – telling her to expect a visit from the surgeon to schedule her operation. Stepping to the next bed, the physician said to the bewildered occupant: 'What applies to her, also applies to you'. As he turned on his heel, the second woman (who did not know that she had Crohn's disease nor that she might need surgery) dissolved into tears, to be comforted by her neighbour who said: 'Next time, maybe we'll move into the same bed, and then he won't even have to walk!'[2]

Although both incidents were shocking enough to be remembered, they were not entirely unusual nor regarded by staff and patients as anything more than the rather insensitive actions of medical men who were regarded as gods. It was after all the role of specialists to cure and the role of the patient to accept the skilled diagnosis and treatment of medical professionals. In most medical engagements, the patient was the relatively passive recipient of whatever treatment was deemed appropriate, and the notion that an ill patient had any role in determining that treatment would have been met with puzzlement.[3]

Times have changed and largely for the better. The training of medical students places substantive emphasis upon the need to develop an empathetic approach to patients, and to recognise the person, as opposed to only the ailment that is presented to them. Although the new approach has met with mixed success – it seems that the sheer volume of information and clinical skills development required from students makes the more leisurely and reflective process of empathetic growth very difficult[4] – it has nonetheless become a core part of western medical student life. Few would argue that the training of doctors who make an effort to engage with the whole patient is a bad thing, but the move towards a patient-centred approach has had significant and unexpected implications for medical practice today. What began as a necessary recalibration of power differentials has moved in some areas into a reversal of hierarchies, one that moreover dismisses empirical clinical expertise in favour of often unsubstantiated lived experiences. This has produced an allied and alarming loss of that contested entity – truth – to the point that the facts of a case are accorded less significance than the emotional impact of clashes framed within a David-and-Goliath narrative. This chapter examines the manner in which the deliv-

[1] Recounted by Evelin O'Hea to Dr. Niall Walsh in the 1960s.

[2] Recounted in 1992 by Carina O'Reilly, a patient on the ward.

[3] There are many complex variations of the patient/practitioner relationship, of course, and research in psychiatric history, for example, has shown far more nuanced engagements between the two, with patients successfully resisting treatments and negotiating degrees of authority with doctors and nursing staff.

[4] Interestingly, there is some evidence to suggest that students become less empathetic as their training advances. See Neumann et al. (2011).

ery of medical services past and present has become part of a discourse that not merely prioritises the patient over the practitioner voice but seeks unhelpfully to dismiss formal medical expertise. It looks at a number of politico-medical scandals that have shaken established certainties and seeks to position these cases within GeoHumanities approaches that allow 'sedimented patterns of culture, economics, [and] politics' to be both contextualised and challenged.[5] I hasten to say that my concern lies not with the vital process of challenge that ensures an effective co-ownership of healthcare delivery, one in which patients work with medical staff to agree their therapeutic regime. Rather, it is the reckless disregard for that contested entity, truth, that is worrying.

The Symphysiotomy and Other Scandals

In Ireland, there have been several horrific revelations in recent decades that have profoundly shocked the country, and each has revolved around questions of the abuse of power by the State, the Church, or the medical profession. The first was the discovery of appalling levels of abuse of children in Church-run orphanages, Houses of Industry, and reformatories throughout the twentieth century. The Ryan Report, published in 2009, catalogued in truly shocking detail the torture, rape, and abuse of children and caused a fundamental breach between the Church and its congregations.[6] The second scandal related to the investigation by Judge Maureen Harding Clark into peripartum hysterectomy in Our Lady of Lourdes Hospital in Drogheda, Co. Louth.[7] Her report exposed a dysfunctional healthcare system, unique to this one hospital, in which the founding obstetrician and the Medical Missionaries of Mary offered maternity care that conformed to the strictest Catholic doctrine.[8] This hospital was also implicated in the symphysiotomy scandal, continuing to use the procedure many years after it had been abandoned everywhere else in Ireland. The reports associated with both investigations amassed enormous amounts of data, trawling through archives held by public and private bodies: there was evidence in the hysterectomy case that records had been removed illegally, meaning that the investigation depended heavily upon patient testimony.[9] The 'Commission to Enquire into Child Abuse' interviewed all traceable former inmates of the Church

[5] Stanley Fish, 'The Triumph of the Humanities', *The New York Times*, June 13th, 2011.

[6] http://www.childabusecommission.ie/rpt/

[7] Judge Maureen Harding Clark, S.C., *The Lourdes Hospital Inquiry: an inquiry into peripartum hysterectomy at Our Lady of Lourdes Hospital, Drogheda* Dublin: 2006.

[8] This meant no prescribing of artificial contraception, even when it was legally available in Ireland, no sterilisation for contraceptive purposes, and no advice on terminations in cases of fatal foetal abnormality.

[9] Patients played a key role in the enquiry, working with the investigating team to shape the direction and the report. 'For the most part, this group presented as sound women accompanied by supportive spouses. The vast majority but not all were former patients of Dr. Neary. They were able to present their history without either high emotion or rancour.' Harding Clark, p. 49.

homes and a Redress Scheme established to compensate victims, which is likely to cost some €200 million by the time all claims are addressed. So, the model in Ireland was one that depended heavily upon primary testimony, supported by archival data, and that favoured financial compensation for victims as a just conclusion.

Both cases caused shock in Irish society and forced citizens to question previously unassailable institutions such as the Catholic Church. Ongoing allegations of abuse by Catholic priests outside as well as within Church-run institutions have consolidated a sense that such corruption was endemic, and the revelations have led to a presumption that there is no charge against professed religious that can be defended. Perhaps this is a natural overreaction to not just decades but a century and a half of Catholic Church dominance in almost every aspect of Irish life. However, a key consequence is that a narrative of Church guilt is established before any investigation takes place, and it becomes extremely difficult to raise a counter-narrative or to engage with the possibility of an alternative account of any engagement. This has significant implications for contested events, especially those that require financial resolution by the State. Indeed, the State itself has been implicated in these horrendous cases, for failing to exercise oversight and supervision of Church-run institutions (which legally were supposed to be inspected annually) and permitting an unchallenged exercise of power over vulnerable and disenfranchised subjects. The presumption of guilt, and a continually reiterated account of Church and State perfidy, has created an environment in which it is difficult to suggest that the actual facts of a case might be more complex than it initially appears.

This tendency touches upon a wide range of issues relating to Ireland's attitude towards the areas of sexuality and historic social mores in particular. This may be seen most vividly in the current investigation into exceptionally high infant mortality rates in Mother and Baby Homes throughout Ireland. The discovery of a mass grave of infant remains has revealed a shockingly high death rate amongst the children born and raised in one of these institutions in Tuam, Co. Galway. Initial investigations have revealed that up to 800 bodies with an age range of between 35 foetal weeks and 2–3 years were buried in a large underground chamber on the grounds of a Mother and Baby Home run by the Bon Secours nuns between 1925 and 1961. The causes of death appear to be congenital debilities, infectious diseases, and, most dreadful of all in its implications, malnutrition.[10] A comprehensive Commission of Enquiry has been established, which has released three interim reports, with a final report due in 2019. Even as the report is being compiled, it is clear that at the very least, appalling neglect of the children took place and, at the worst, active abuse. There is thus a painful process ahead of the Commission team, and the broader public, as inevitably horrific cases are brought to light.

Before this process has properly begun, however, the media has already positioned the Homes in an unproblematic and simplistic Church and State narrative that elides all of the complexities of what really happened. What is more alarming is the willingness of government to comply with this narrative, one that is likely to

[10] The neglect of the children in the Tuam home was known by contemporaries: the 1947 Inspector's Report noted that many were emaciated, pot-bellied, and apathetic.

sit well with the public. One example from the Mother and Baby Homes that illustrates this issue is the burial place of the unfortunate infants. The chamber in which they were discovered appears to have been originally constructed as a septic tank, dating back to when the building was a workhouse (originally constructed in 1839). It is as yet unknown whether the tank was ever used – it appears not to have been – and there is no evidence to suggest that it was a tank for waste by 1925 when the Mother and Baby Home opened. It seems that these empty chambers were used as a burial space for the infants, a fact which carries its own emotional burden. However, media coverage has consistently stated that the bodies were found in a septic tank, leading to a truly monstrous presumption that the children's bodies were flushed like excrement into the waste system. The effect of this irresponsibility is felt at the highest levels. The Minister for Children and Youth Affairs, Catherine Zappone, had a brief meeting with the Pope on August 25 2018, and raised the issue of the Church's responsibility for the inmates of the Tuam Home. In a radio interview on the 27th, she related what she described as the precise words she used to the Pontiff, which included the statement: 'Children's remains were found in a sewage system there'.[11] Despite the efforts of the woman who discovered the mass grave to clarify the nature of the burial ground,[12] this compelling inaccuracy has taken an unshakable hold, with international as well as national press coverage repeating it as fact.[13] Zappone is undoubtedly reflecting a shared horror at what the country believed an unspeakable act, but as the Minister responsible for the Commission to investigate the operation of the Home, she has full access to the facts surrounding the children's remains. When she perpetuates an untruth with such serious implications for the public's engagement with medicine, history, and truth, she contributes to a dangerously simplistic narrative that further undermines the social compact.

There is another important aspect to this wilful neglect of fact. The Commission has emphasised the importance of researching and writing a comprehensive social history as part of the investigation, in order to ensure that the lives of those who suffered, and died, in these institutions are placed in the appropriate historical context.[14] Without this, grasping the significance of behaviours and processes is

[11] Catherine Zappone in an interview with Miriam O'Callaghan on the 'Today' programme, 27/8/18. https://www.rte.ie/news/pope-francis-in-ireland/2018/0827/987741-pope-francis-katherine-zappone/

[12] Rosita Boland, 'Tuam mother and baby home: the trouble with the septic tank story', Irish Times, June 7th, 2014.

[13] *Press and Journal,* August 27, 2018, 'Between 1925 and 1961 around 800 infants were buried in a septic tank at the home for unmarried mothers.' *Belfast Telegraph,* 'The Sisters of Bon Secours nuns in charge of Mother and Baby Home in Tuam, Co Galway, where the burial of hundreds of babies in a septic tank mass grave has been uncovered have said they no longer hold the burial records.' June 6, 2014. An *Irish Times* interview with Catherine Corless on June 7, 2014, quotes from other newspapers: 'Bodies of 800 babies, long-dead, found in septic tank at former Irish home for unwed mothers' – *The Washington Post*; 'Nearly 800 dead babies found in septic tank in Ireland' – Al Jazeera; '800 skeletons of babies found inside tank at former Irish home for unwed mothers' – *New York Daily News*; 'Almost 800 'forgotten' Irish children dumped in septic tank mass grave at Catholic home' – ABC News, Australia.

[14] Mother and Baby Homes Commission of Investigation Interim Report, September 16, 2016, p. 1.

impossible, and the determination of responsibility becomes extremely difficult. When the then Taoiseach, Enda Kenny, made a statement in Dáil Eireann in 2017, he mentioned the importance of a collective acknowledgement of responsibility for the treatment of unmarried mothers and their babies. Stating 'no nuns broke into our homes to kidnap our children. We gave them up to what we convinced ourselves was the nun's care',[15] he faced an immediate and personalised backlash that effectively closed down the vital acknowledgement and discussion of this aspect of Ireland's historically problematic relationship with sexuality and social mores. In their haste to assure the public of their shared revulsion at the deaths, TDs unwittingly simplified the narrative – the Church, with the benefit of the State's blind eye, was solely responsible – and the active collusion on the part of families in sending their unmarried mothers and heartbreakingly vulnerable children to these homes is simply glossed over. Yet, this is a vital component of the story and of Ireland's history. But it was in the following day's debates that an even more pressing issue emerged. The Taoiseach was responding to questions from opposition members on the Commission's Interim Report. In pressing for its release, criticising its parameters, and questioning its terms of reference, the deputies were seeking to direct what is supposed to be an Independent Commission and stating what the conclusions should be in advance of an investigation.[16] It is very unlikely that those who spoke so passionately about the appalling circumstances in the homes – there are 18 homes currently under investigation – intended their words to impact negatively on a highly sensitive and complex process, yet that is precisely the effect. In hastening to reflect a public presumption that the Catholic Church was exclusively responsible, and to be seen to lead the condemnation, they are preventing the actual complex, problematic, and uncomfortable truths about a broader complicity in these tragedies.[17]

This is reflective of a presumption that it is the government's intention to deny justice, which should be through the actual judicial system as opposed to a community-based, natural justice framework. This may partly be a result of the fact that Ireland has traditionally offered court-ordained compensation payments that are far higher than those of other countries: between 2012 and 2017, the Health Service Executive paid half a billion euro in out-of-court settlements, with an average payout of over €172,000 per claim.[18] Ireland's libel case awards dwarf those in Europe, with an average award of €90,000 in comparison with the Dutch average of €2,500.[19] Personal injuries awards are 4.4 times higher than in the UK, with a lower

[15] Leader's Questions, Dáil Éireann Debates, March 7, 2017.

[16] In response to Deputy Catherine Connolly, the Taoiseach stated: 'The Deputy is asking me to direct an independent commission to do certain things….I expect Deputy Connolly will allow the commission of investigation to report its official findings. That is its duty, its responsibility, and why it was set up.' Ibid, March 8, 2017.

[17] Mother and Baby Homes Commission of Investigation, 'Terms of Reference' http://www.mbhcoi.ie/MBH.nsf/page/Terms%20of%20Reference-en

[18] *Irish Times,* August 20, 2018, 'Nearly €500 m paid out in settlements against HSE in five years'. The average award has doubled in recent years.

[19] *Irish Times,* February 15, 2017, 'Irish system is unusually generous with libel damages'. The record for a libel case in Ireland is an extraordinary €10 million.

level of proof of injury required.[20] The cry of 'Give them their day in court' is a fundamental (and wilful, given it is being robustly supported by lawyers) misrepresentation of what the courts are for and how they operate. It also means that litigation, not mediation, is the first response to a group grievance. But there is little reflection on how courts actually operate, and the fact that litigation is expensive, labourious, and stressful in the extreme is largely ignored. The judicial system is inherently adversarial, and engagement with the process, whether as a claimant or a respondent, automatically places both in a combative situation. Moreover, most people fail to grasp that the law depends upon precedent, statutory sentencing and is above all evidential. But in Ireland, not only is it practically the first recourse but is frequently a process that depends heavily upon a concomitant public trial by media.

In 2016, a Redress Scheme was launched in Ireland to compensate women who had undergone the procedure of symphysiotomy in childbirth in the decades between 1940 and 1990. The scheme was the culmination of an exhaustive process that had seen two failed attempts at an independent report, the fracture of the main patient representative group, constant lobbying of Dáil Éireann, and a widespread media campaign that crafted a David-and-Goliath narrative for public consumption. To the non-medical ear, the operation sounds impossibly barbaric. Symphysiotomy is a procedure in which the cartilage that holds the front of the pelvic girdle at the pubic joint (pubic symphysis) is partially severed during labour, allowing the pelvis to enlarge slightly, but sufficiently, to permit the baby to be born vaginally. The pelvis is not sundered: if it parts, the procedure has not been performed properly. After delivery, the pelvis requires tight bandaging to allow the joint to heal and to prevent long-term injury to the mother. In most cases, the scarring of the joint creates a fractional permanent enlargement of the pelvic circumference: this was historically believed to offer small-framed women a greater chance of vaginal deliveries in the future. Symphysiotomy can also occur spontaneously and is a rare but recognised event in late pregnancy. It is seldom used in the developed world, with the exception of some breech births or in cases of severe shoulder dystocia, where the baby's head has been delivered, but the leading shoulder becomes trapped behind the pubic bone. In both of these cases, caesarean section is not normally an option as labour is too far advanced and the baby is likely to become distressed if not delivered swiftly. Although western medicine would regard the intervention as very much a last resort, it can be a life-saving procedure in the developing world. In areas where very young women – barely out of childhood themselves – experience long, difficult labours, and their underdeveloped bodies struggle to deliver their full-grown, full-term babies, a symphysiotomy can be a life-saving operation. It can be safely performed by a trained midwife with minimal equipment – local anaesthetic, sterile scalpel, and a surgical girdle for post-operative care – and can help to prevent the long labours that routinely result in devastating consequences such as obstetric fis-

[20] *Irish Independent,* July 22, 2018, 'Revealed: personal injury payouts four times higher in Ireland than in UK'.

tula.[21] Outside of western countries, access to the sophisticated level of medical care represented by caesarean section is impossible for many, so symphysiotomy can offer a means of preventing trauma in delivery and preserving the life of mother and child.[22]

Nonetheless, and perhaps partly because its nature makes both men and women squirm, the operation seems a relic from the dark ages. The first recorded use of the procedure was in eighteenth-century France, in the case of a woman with dwarfism whose previous babies had died at birth.[23] Although never a common intervention, it was used throughout Europe from the mid-nineteenth century onwards in cases of obstructed labour, where there was a clear disproportion between the baby's head and the mother's pelvis. The first reports relating to the use of symphysiotomy in Ireland began to emerge in around 2000, with a survivors group named Survivors of Symphysiotomy (SOS, later SOS Ltd) linked with an advocacy group named Patient Focus to draw attention to the little-known procedure.

The issue came to national attention as a result of a broadcast on RTE by investigative journalists associated with the 'Prime Time' programme.[24] Following the programme, 'Brutal Procedure: the symphysiotomy scandal' (February 2010), women began to come forward, providing often extremely explicit and harrowing testimony of the manner in which the operation had been performed and the lifelong pain and damage it had caused. Public perceptions of symphysiotomy were shaped by a number of emotionally compelling factors. In the first place, the women claiming to be affected were almost all elderly and many very frail, and their appearances on television provoked a natural wave of sympathy. Secondly, the highly specific accounts, many identical to each other in depth of detail and brutality, seemed highly persuasive.[25] Thirdly, the Irish public were still reeling from the two significant scandals above involving the Catholic Church. Three separate groups represented the women who had undergone the procedure. The first was 'Survivors of Symphysiotomy Ltd', founded in 2000. The second was 'Patient Focus', an advocacy group that had successfully represented victims of the caesarean-hysterectomy

[21] This happens to between 50,000 and 100,000 women worldwide each year, the vast majority in the developing world. Most are shunned and may be turned out of their communities, and the condition is only curable by surgery. Obstetric fistula is caused exclusively by obstructed labour and in many cases is the result of pregnancy and labour in very young women and indeed children. The WHO estimates that 2.5 million girls under the age of 16 give birth each year, with an additional 16 million aged between 15 and 19. These are the group most likely to die in childbirth or shortly afterwards, and the health of those who survive is frequently severely compromised. Ganchimeg et al. (2014).

[22] Obstetric fistula needs assessment report: Findings from nine African countries. [Report]. New York, United Nations Population Fund and EngenderHealth, 2003. 95 p. www.unfpa.org/fistula/docs

[23] Oonagh Walsh, *Report on Symphysiotomy in Ireland, 1944-1984* Dublin: Department of Health, 2014. pp. 16–17.

[24] https://www.rte.ie/news/player/2010/0218/2704182-brutal-procedure-the-symphysiotomy-scandal/

[25] So similar were the accounts that the final Redress Scheme noted a high likelihood of unconscious or deliberate imitation.

scandal. The third was 'SOS', a breakaway group from the original 'Survivors of Symphysiotomy' and the highest profile of the three. SOS implemented a media campaign that ensured a high level of visibility for the women who claimed they had undergone the procedure, including a fascinating performative element that was quite new in Irish public discourse. Members of SOS travelled to Dáil Éireann at regular intervals; vigils were held outside Parliament buildings, with women holding baby bootees and jackets; and politicians tumbled over each other to be seen to present Dáil addresses that unquestioningly accepted the narrative of deliberate harm inflicted by medics who were allegedly in thrall to the Catholic Church. Teachta Dála members wept as they repeated the stories told to them of women being held down, surrounded by medical students, while a surgeon used a long-bladed saw on their pubis. There was widespread media coverage, with the procedure and the campaign featured extensively on the major current affairs programmes including 'The Frontline' and 'Tonight with Vincent Browne', as well as a constant stream of print pieces that repeated this narrative of wilful betrayal.[26] In response, the government attempted to commission a series of reports to establish the extent of use of symphysiotomy in Ireland and place its use in Ireland in an international context. It is not an exaggeration to say that the whole country became involved in a collective ritual of self-abasement, with journalists outdoing each other in their efforts to register hyperbolic flights of self-flagellation and remorseful apology.

This ran for the best part of 10 years, and over the course of that decade, both the first-hand accounts of the procedure and the demands for compensation increased in ferocity and scale. The original campaigners had sought an explanation and an apology, with entitlement to a medical card so that they could access services to address their medical needs. But as the media campaign intensified, and politicians sought to outdo each other in terms of suggested redress, the demands for financial compensation increased exponentially. By the time Judge Harding Clark interviewed women involved in the Redress Scheme, members of SOS had stated that that what would bring closure to them was an apology and substantial compensation. They wanted '€250,000 at the lower end of the spectrum of injuries, pain suffering and disruption of normal life to a maximum of €450,000 at the upper end'.[27] When the Redress Scheme completed its work, some 399 women received compensation (this figure included several cases that Harding Clark thought on balance of probability had not had the procedure, but she made a deliberate decision to find in the applicants' favour wherever possible).[28] The final scheme offered three levels of compensation: €50,000 for having undergone the procedure, €100,000 for proven long-term

[26] Marie O'Connor, spokesperson for SOS, and Colm McGeehan, lawyer for the group and O'Connor's ex-husband, had a national platform to reject the Report and its recommendation of a Redress Scheme.

[27] Pro-forma letter produced by SOS for Judge Yvonne Murphy in response to her consultation process. Appendix 4, Murphy (2014).

[28] 'As independent Assessor, I determined that once symphysiotomy or pubiotomy was established, a compassionate and generous view would be applied to the assessment of each claim. All objective evidence received or obtained was carefully considered and any doubt was applied in an applicant's favour.' Harding Clark (2016, p. 10).

impact, and €150,000 for lifelong impairment. Working on a calculation of an average of 133 women (out of the 399 successful applicants) per category as suggested by SOS above, this amounts to the sum of just under €140,000,000, in addition to the ongoing costs of medical cards and home modifications already agreed to by the government. A justifiable expense, perhaps, for lifelong suffering.

And yet, not all was as it appeared. Perhaps the most shocking element of the whole symphysiotomy case is the fact that a full one-third of the women who applied to the scheme, and furnished excruciating and compelling testimony as to their experiences, never underwent the procedure at all. This included some of the most vociferous and public figures, who appeared in Dáil Éireann to standing ovations, and spoke frequently on television and radio of their experiences. They furnished Judge Harding Clark with vivid first-hand accounts of the operation and its aftermath, stating that 'I will never forget the sound of the saw cutting into me'; 'the negative impact of having been abused and assaulted while in a state of unconsciousness changed my whole course of life'; 'I could not leave the house as I was incontinent'; and 'The symphysiotomy damaged me for the rest of my life'.[29] It would not be surprising to find a small number of people in any large group who deliberately or unwittingly lie about an experience. But for a full one-third of applicants to lay claim to an exceptional procedure, whose alleged after-effects were life-changing and provable, is extraordinary. This is not merely an issue of misappropriated public funds. It is the complete destruction of an identity based upon violation and victimhood, which must be entirely devastating for those women who were wrongly convinced that they had been deliberately brutalised by their obstetricians, in cahoots with the Catholic Church. How does one go on when faced with the proof that what they believed to be the formative event of their life never happened?

In pondering the dramatic misrepresentations that were commonplace throughout the process, Harding Clark foregrounded the role of the media and its reckless use of language:

> 'There can hardly be a person in Ireland who has not been exposed to reports of the procedure described as butchery akin to Nazi medical experimentation; aggravated sexual assault; and a form of female genital mutilation causing lifelong disability, chronic pain, mental suffering and family breakdown, and much more. It is viewed as a procedure synonymous with barbarism and pain, but not just barbaric and painful but unnecessary and unwarranted, creating a legacy of countless victims whose lives were permanently ruined.'[30]

Less remarked upon by Harding Clark was the role played by the legal profession itself and especially by SOS's legal advisers. In a remarkable conflict of interest, SOS's spokesperson was the ex-wife of the group's key lawyer, and the junior counsel was their son.[31] This was the group that opposed the first two governmental

[29] Harding Clark (2016, pp. 100–01).

[30] Harding Clark (2006, p. 96).

[31] 'The SOS group is chaired by Marie O'Connor and its legal adviser is Mr. MacGeehin [sic – McGeehan], her former husband. Their son, Ruadhán Mac Aodháin, is also the group's legal adviser and has worked as junior counsel on the court cases.' Paul Cullen, 'More symphysiotomy

efforts to produce an empirical report on the procedure and unilaterally dismissed the third 'Report on Symphysiotomy in Ireland, 1944–1984' (written by this author), as well as the 'Independent Review of Issues Relating to Symphysiotomy', and the 'Surgical Symphysiotomy *ex gratia* Payment Scheme' as 'whitewashes'. They insisted that litigation, or an agreed settlement based upon an exceptional case – that of Olivia Kearney, who was awarded €450,000 for a symphysiotomy performed *after* a caesarean section – was the only acceptable option.[32] Of course, the concomitant reward for legal representatives was huge, if every woman secured such an award, but a Redress Scheme that capped legal fees and avoided the adversarial court system was far less lucrative.[33] Once the scheme was announced, law firms leaped with enthusiasm into the roles of legal champions, as well as ambulance chasers.[34] It was an unedifying but arguably inevitable outcome of a process that had positioned the government as seeking to 'deny justice'.

The symphysiotomy case exhibits the ease with which complex issues of accountability, access to political power, and the responsibilities of the media to not merely report accurately but to act as a bridge between specialist medico-legal knowledge and the general population can fail. The consequences of this failure are many and long-reaching. This is not an abstract, academic reflection on falling media standards or politicians' fears of alienating voters. Rather, the implications of allowing, or indeed facilitating, dangerous untruths include the destruction of self-identity, reckless financial commitments in the hundreds of millions, and a permanent breach of trust in institutions – medicine and the state – that are supposed to be the first lines of defence for the defenceless.

So what has been the result of this lengthy, fraught, moment in Irish medical history? On the positive side, 399 women have received a substantial financial settlement and apologies from the Irish government, the Institute of Obstetricians and Gynaecologists, and the Medical Missionaries of Mary on behalf of those who undertook the operations. One also hopes that they now have a degree of relief that the campaign has come to a positive conclusion. But for a substantial number of women (close to 200), their entire sense of self, and their *raison d'être*, has suddenly been destroyed. Having constructed not merely an identity as a 'survivor', or a 'victim', or an unwitting and unwilling subject of 'medical experimentation' – all distressing designations in their own right – but one of campaigners for justice against

---

cases still in legal pipeline, says solicitor' *Irish Times* Nov. 24th, 2016.

[32] This was reduced to €325,000 on appeal to the Supreme Court. The original award was extremely high, as there was no medical justification for performing the symphysiotomy after the baby had been delivered. Kearney –v- McQuillan, 11/7/2012, http://www.supremecourt.ie/Judgments.nsf/1b0757edc371032e802572ea0061450e/eb1050c58b4422b080257a3f0042d60d?OpenDocument

[33] McGeehan's firm nonetheless secured €1.1 million in fees from the Scheme, a not inconsiderable sum.

[34] Once the Redress Scheme was confirmed, legal firms began offering their services as agents to claimants. Several attempted to fold-in costs from previous litigation in addition to legitimate costs, confirming the sense that Redress Schemes were a 'victimless' source of funds. *Irish Times*, December 31 2015.

a corrupt medico-religious cabal, they are left with nothing save a public confirmation that at best they were deluded and at worst liars:

> 'I must confess to a feeling of great sadness for those applicants who somehow believed that they were victims who had survived symphysiotomy only to learn that they were mistaken and that their recollections were based on error. Elderly women should not have to face such emotional turmoil. I have no doubt that some of them have spent good money presenting their claims which cannot be recouped. Those who have been active in representing themselves as victims to the media must now retrace their lives and must be understandably upset.'[35]

The symphysiotomy case has had other repercussions. In Ireland and beyond, owing to the loaded and inaccurate representations of the procedure and its purpose, it is damned by association. In the developing world, however, it has the potential to save the lives of mothers and babies and to prevent at least some of the horrific lifelong afflictions of vaginal fistula. Worldwide, some 303,000 women die in childbirth each and every year, and in parts of Asia, over 12% of those women die as a result of obstructed labour, the circumstance that symphysiotomy is exclusively indicated for (Lale Say et al).[36] Of those young, often undeveloped mothers who survive, there are an estimated 50,000 to 100,000 who suffer the physical and mental repercussions of fistula.[37] Permanently doubly incontinent, many are shunned by their communities and rejected by their husbands, leading vulnerable and marginalised lives. In parts of Africa and Asia in particular (though by no means exclusively), girls as young as 12 years of age endure labours lasting several days that often end with the permanent incontinence of the mother and in many cases with the death of both mother and baby.[38] Women who live in remote areas with no timely access to hospitals with modern operating theatres that can offer safe caesarean sections suffer painful and debilitating obstructed labours that can in many cases be alleviated by symphysiotomy. All that is required is a trained midwife with local anaesthetic and a sterile scalpel and gloves, and a life-saving and health-preserving intervention can be performed. Symphysiotomy was described as a 'third-world operation' by some SOS campaigners, with no reflection on the fact that it can be a life-saving operation in the third world.[39] The Eurocentric view that drove the campaign had the perhaps unintended consequence of ensuring that research into its use and utility by western medical professionals was largely abandoned. It is ironic that

[35] Harding Clark (2016, pp. 102–3).

[36] Lale Say et al, 'Global Causes of Maternal Death: a systematic review', The Lancet, Vol 2, Issue 5, 2014, pp. 323–33.

[37] Tunçalp et al, 'Measuring the incidence and prevalence of obstetric fistula: approaches, needs and recommendations', Bulletin of the World Health Organisation 2015; 93: pp. 60–2.

[38] '…girls 10–14 years of age are 5–7 times more likely to die of childbirth [than women over the age of 20]… Many times, obstructed labor leads to fistulas; the pressure of the foetal head on the vaginal wall causes tissue necrosis, and fistulas develop between the vagina and the bladder or rectum after the necrotic tissue sloughs. More than two million adolescents are living with fistulas… Mortality rates are 73% higher for infants born to mothers <20 years of age than for those born to older mothers.' Nawal M. Noor (2006).

[39] Harding Clark (2016, p. 34).

a campaign that claimed to fearlessly stand for women's rights against an oppressive male medical profession actually did so at the expense of the most vulnerable and marginalised women in the developing world.

This is, to an extent, symptomatic of broader trends in the western world in relation to rights, responsibilities, and a new perception of the control and direction of resources. Allied to a reluctance to engage with often difficult truths – sometimes patients will suffer and die – is an underlying hostility to the medical profession. Press reports of near-miraculous tales of survival are invariably couched in adversarial terms, even while a narrative of triumphal success is presented. The formulaic tale is as unvarying as a fairy tale: the patient 'defied the doctors' who had 'given them [a specific number of] months to live'. This scenario is so common that there is no reflection on the actual meaning of the words, although the inference is clear: instead of two sides working together towards the shared goal of recovery, there is an autocratic medic like Atropos, the eldest of the three fates who wielded the 'abhorred shears' that cut a human life. Ironically, one of the key elements in this newly toxic relationship between some patients and practitioners is a movement that began as a means of drawing the two together in a shared commitment to human wellbeing. Alternative or Complementary Medicine was intended to be just that: an additional weapon in the disease armoury and one that helped to address the vital role of the spirit in support of the body. Approaches have broadened well beyond the originary field of homoeopathy, which although dismissed by many physicians because of its lack of an empirical research base was nonetheless accepted as a practice that empowered many patients and assisted in dealing with illness and aiding recovery. Now there is a multi-million-pound industry that embraces an enormous range of therapies, and a considerable portion of the NHS's limited budget was until 2018 assigned to complementary medicine.[40]

It is a fascinating field: despite no evidence base, it nonetheless makes radical claims for cures where conventional medicine has not only failed, but has, it is sometimes claimed, actively damaged health. As in any instance where there is a contest over resources, and a great deal of money to be made, clashes have become heated, with each side dismissing the other in the most emphatic of terms.[41] What is interesting to note however is how alternative practitioners have adopted the language of medicine, and a pseudo-scientific discourse, to convince patients to dismiss mainstream medical practice. Of course, many non-medical fields have long utilised pseudo-science in the service of marketing: without exception, the cosmetic industry depends upon what Ben Goldacre terms 'Bad Science' to sell products. Beautiful people in white coats and wearing glasses gaze thoughtfully into test

[40] National Health Executive, 'NHSE to stop prescribing homeopathic treatments in £190 m savings drive' (24/7/17), http://www.nationalhealthexecutive.com/Health-Care-News/nhse-to-stop-prescribing-homeopathic-treatments-in-190m-savings-drive

[41] Emma Weinbren, 'England chief medical officer sparks fresh homeopathy row', *Chemist and Druggist*, 29/1/2013. England Chief Medical Officer Professor Sally Davies declared in response to Parliamentary questions that homeopathy is 'rubbish'. https://www.chemistanddruggist.co.uk/news/england-chief-medical-officer-sparks-fresh-homeopathy-row

tubes, selling not merely the hope of beauty, but a scientific guarantee of it.[42] Similarly, the most enthusiastic critics of mainstream medicine have co-opted a highly medicalised discourse, using the language of the laboratory to offer non-laboratory alternatives. It becomes increasingly difficult for the public to identify what claims have merit and what is likely to cause active harm. In the age of internet diagnosis, and the huge power of the deep web to source any unregulated drug, this is a difficult position.

The hostile positioning of both practices – conventional and alternative – has other repercussions. From a position where both sides viewed each other with a degree of scepticism, there is in some areas now an open antagonism and a suspicion that treatments are withheld or modified unnecessarily in order to preserve the authority of the medical profession. Some tensions may be explained as a 'public disdain for, or mistrust of, orthodox science or concerns about the way in which medical care is delivered', making encounters with medical professionals potentially fraught.[43] Recent years have seen an explicit rejection of perceived 'elites' – Donald Trump's electoral rallying cry – who are presumed to act against the 'little man'. Brexit has allegedly been fuelled by those 'left behind', building upon a sense that those in power have served only their own interests but will now be held accountable. This sense that authorities are now answerable to those they serve is a largely positive development, not least in forcing previously unassailable bodies, professions, and systems to consider their relationships with the 'end user'. In medicine, the increasing accountability of doctors (vital to maintain standards), and the growing emphasis upon patients as 'active participants in their care', has transformed medical treatments in many positive ways.[44] Patients have also been encouraged to question their doctors and nurses and medical staff trained in the use of 'plain English' to ensure that treatment options are understood by the recipient of care.[45] This has had many beneficial effects, not least the proven health benefits of enhanced recovery, as many feel more in control of their disease (see NICE 2012). It has however also led to considerable suffering for individuals as well as damage to society's relationship with medicine, as engagements increasingly take place within an antagonistic framework, with a presumption that high-ranking professionals operate with ulterior motives and with a specific hostility towards those whom they serve.

---

[42] The actress Jennifer Aniston seemed to ironically reference this peculiar blend in her campaign for L'Oreal shampoo, when having extolled the beauty element of the brand, would turn to camera and say 'And now for the science bit.'

[43] Angela Coulter, 'Killing the Goose that Laid the golden Egg?', British Medical Journal, 2003: Jun 14; 326 (7402), p. 1280.

[44] NHS England, 'Enabling people to make informed health decisions' https://www.england.nhs.uk/ourwork/patient-participation/health-decisions/

[45] The NHS has a range of resources for health professionals that suggest ways of communicating health advice in terms that the layperson can understand. http://www.healthliteracyplace.org.uk/tools-and-techniques/techniques/use-simple-language/

A heart-breaking example is the case of Charlie Gard, the child who became a rallying cry for a series of vested interests. Charlie was born with MDDS, a rare and terminal condition which causes progressive muscle wastage, brain damage, and eventually a failure to breathe unaided.[46] He was under the care of a medical team at Great Ormond Street Hospital, and as he approached his first birthday, his condition had deteriorated to the point there was no prospect of maintaining his quality of life. His distraught parents believed that there was still a hope of maintaining life, if not a recovery, and turned to social media for support. It was truly awful to see how quickly a well-intentioned campaign deteriorated into accusations that medical staff wanted to 'kill' the child and were 'bored' with his parents. Death threats were sent to staff, not by a small number of people, but by thousands, and doctors and nurses were accosted in the streets.[47] The public response to the Gard case indicates a worrying trend towards not merely a resentment of traditional medicine but an outright rejection of expertise. As Hadley Freeman noted, 'This mood [reaction to the Gard case] deifies feelings over facts, and insists that any opinion is worth as much as that of an expert – more, in fact, because as we learned last summer in the lead-up to the EU referendum, experts are now a dark elite cabal and not to be trusted'.[48]

Conclusion

Medical practice has changed dramatically in the past 50 years, and the pace of change is accelerating at a frightening speed. We are relentlessly bombarded with press reports of scientific breakthroughs that promise longer, better lives while also giving us contradictory advice over the dire consequences of over- and undereating, indulging in or abstaining from alcohol, or taking too much or too little exercise. But we can perhaps take some comfort from the fact that what we really depend upon to protect our health is faith in the people who treat us. Despite the controversies raised above, nurses and doctors remain overwhelmingly the people we trust most. The 2017 Veracity Index shows that 94% of people believed that nurses can be trusted to tell the truth, with doctors scoring second best out of 24 professions at 91% (the 5 least trusted professions are, in descending order, estate agents, journalists, professional footballers, government ministers, and at the bottom of the food chain, politicians in general).[49] Perhaps we would do well to consider how we might

---

[46] Infantile onset encephalomyopathic mitochondrial DNA depletion syndrome.

[47] 'Thousands of abusive messages, including death threats, have been sent to doctors and nurses at the children's hospital in London, the chair of the hospital revealed. She [a Scotland Yard spokesperson] said that staff were also facing abuse in the street. Families visiting their sick children had also been 'harassed and discomfited'. *The Guardian,* July 22, 2017.

[48] Hadley Freeman, 'My sister, the doctor, has graduated – just as the world has turned on medical experts', ibid.

[49] Ipos MORI Veracity Index, November 2017, Social Research Institute, p. 2.

draw upon the best that the 'old-fashioned' human relationship of trust between patients and practitioners can offer, tempered with a healthy and historically well-informed scepticism regarding the reality of medical fallibility. It is as partners, not combatants, that we can secure the best of health.

References

Coulter, A. (2003). Killing the goose that laid the golden egg? *British Medical Journal, 326*(7402), 1280.

Fish, S. (2011, June 13). The Triumph of the humanities. *The New York Times.*

Fuller, L. (2005). Religion, politics and socio-cultural change in twentieth-century Ireland. *The European Legacy, 10*, 41–54.

Ganchimeg, T., Ota, E., Morisaki, N., Laopaiboon, M., Lumbiganon, P., Zhang, J., et al. (2014). Pregnancy and childbirth outcomes among adolescent mothers: A World Health Organization multicountry study. *British Journal of Obstetrics and Gynaecology, 121*(S1), 40–48.

Goldacre, B. (2008). *Bad science.* New York: HarperCollins.

Harding Clark, M. (2006). *The Lourdes hospital inquiry: An inquiry into peripartum hysterectomy at our lady of Lourdes hospital, Drogheda.* Dublin: Dublin Department of Health.

Harding Clark, M. (2016). *The surgical symphysiotomy ex gratia payment scheme.* Dublin: Report to Minister for Health, Simon Harris, TD, 19th October.

Igde, F. A., & Sahin, M. K. (2017). Changes in empathy during medical education: An example from Turkey. *Pakistan Journal of Medical Sciences, 33*, 1177–1181.

Inglis, T. (1998). *Moral monopoly: The rise and fall of the Catholic Church in modern Ireland.* Dublin: University College Dublin Press.

Inglis, T. (2007). Catholic identity in contemporary Ireland: Belief and belonging to tradition. *Journal of Contemporary Religion, 22*, 205–220.

Murphy, Y. (2014). *Independent review of issues relating to symphysiotomy.* Dublin: Department of Health.

Neumann, M., Edelhauser, F., Tauschel, D., Fischer, M., Wirtz, M., Woopen, C., Haramati, A., & Scheffer, C. (2011). Empathy decline and its reasons: A systematic review of studies with medical students and residents. *Academic Medicine, 86*, 996–1009.

NICE. (2012). *Patient experience in adult NHS services: Improving the experience of care for people using adult NHS services* (NICE Clinical Guidelines, No. 138). London: The National Institute for Clinical Excellence.

Noor, N. M. (2006). Health consequences of child marriage in Africa. *Emerging Infectious Diseases, 12*, 1644–1649.

Rosita Boland, B. (2014, June 7). Tuam mother and baby home: The trouble with the septic tank story. *Irish Times.*

Say, L., Chou, D., Gemmill, A., Tunçalp, O., Moller, A.-B., Daniels, J., Gülmezoglu, A. M., Temmerman, M., & Alkema, L. (2014). Global causes of maternal death: A WHO systematic review. *The Lancet, 2*(6), 323–333.

Tunçalp, O., Tripathi, V., Landry, E., Stanton, C. K., & Ahmed, S. (2015). Measuring the incidence and prevalence of obstetric fistula: Approaches, needs and recommendations. *Bulletin of the World Health Organization, 93*, 60–62.

Walsh, O. (2014). *Report on symphysiotomy in Ireland, 1944–1984.* Dublin: Department of Health.

Part III
Practice

Central Interior

Sarah de Leeuw

It's summer in British Columbia
except not quite way up here,
even though the rivers are flooding
their banks, and the ground is loosening,
and my sister is living in Munich
right now, going to volleyball practices
in the evenings after work, working
out in purple spandex with her ropey
arms and platinum hair, a bombshell
who misses London, the city, the pace.

So when I'm driving early in the morning
to catch a flight going south, sky as light
as late last night, crusts of snow still
in the ditches, I catch my breath
at the small-townness of it all, at the hard-
drinking men hauling themselves into cabs
of logging trucks, mine and mill dust
floating in the air, a blur, inside

Dad are cells so sick it takes everything
to walk once around his garden
of new, sharp, tart rhubarb stems
sticking up almost like they hurt, so red
from composted earth, waxy leaves
opening, slackening fists.

From: de Leeuw, S. (2019). *Outside America* (p. 51). Gibsons, BC: Nightwood Editions.

Chapter 12
GARTNAVEL: An Experiment in Teaching 'Asylum Week'

Cheryl McGeachan and Hester Parr

Abstract We approach Gartnavel Royal Hospital in Glasgow, Scotland, as a site of pedagogy via foot, archival record, and film. We animate the mental health asylum as a contemporary space of learning and geohumanities education. We share elements of lectures, a field visit and workshops that bring undergraduate students up-close to asylum geographies past and present. We creatively render these materials on the page as means to explore educational and experimental encounters with shifting forms of medical intervention on the human mind-body.

This chapter discusses the development of creative approaches for teaching within the geohumanities by drawing attention to newly emerging 'asylum pedagogies' (de Leeuw et al. 2018). A key tenet of the geohumanities enterprise is to expose creative dimensions to site and practice, and yet, little focus has yet been given to how these elements can be developed into critical pedagogies (Dear 2015). Through one example of experimental teaching practice, that of 'Asylum Week', run several times as part of a senior honours degree course in the University of Glasgow, Scotland, this chapter showcases how creative connections to site, practice and embodiment enable the development of critical and creative pedagogies attuned to the geohumanities. In doing so, we wish to animate the mental health asylum as a contemporary space of learning and geohumanities education that develops 'hopeful pedagogies' whereby previously stigmatised people and places have the potential to be 'rescripted' (Parr 2008). We hence write of pedagogy in this innovative volume, bringing geohumanities and mental health together in new conversations and for possible transformation.

C. McGeachan (✉) · H. Parr
Geographical and Earth Sciences, Glasgow, UK
e-mail: cheryl.mcgeachan@glasgow.ac.uk; hester.parr@glasgow.ac.uk

S. Atkinson, R. Hunt (eds.), *GeoHumanities and Health*, Global Perspectives on Health Geography, https://doi.org/10.1007/978-3-030-21406-7_12

This chapter is written experimentally in a version of an acrostic built through G-A-R-T-N-A-V-E-L, the name of a psychiatric hospital local to our university, as an organising device:

Gartnavel
Asylum Week
Ruins/remains
Teaching
Nature
Archives
Violence
Experiment
Learning

In part, this is a response to debates in the geohumanities that seek to free the narrative chains of conventional writing in order to foster new, more experimental ways of seeing and knowing the subject-matters that we encounter (Last 2018). However, it also relates especially to this chapter's interest in new understandings of site and practice in geohumanities teaching. Stylistically, then, we employ a range of techniques that fuse together the stories of Asylum Week's teaching practice, the site of Gartnavel and its varied inhabitants across time and space. In using a range of juxtaposed materials, we wish to unseat the reader's standard assumptions about mental health while striving to 'undo' the mental hospital in our teaching, taking apart critical aspects of how it responds to this human condition we presently call 'mental illness'. Our adoption of an acrostic offers a means through which playfully to narrate this undoing.

Gartnavel Royal Hospital in Glasgow houses both a ruined Victorian asylum and a newly built mental health treatment facility, and, for us, it becomes a site of pedagogy via classroom, footfall, archival record and film. This chapter allows the reader access to the site – through staff and student experience – and to the pedagogic practices employed to bring the embodied geographies of mental health care to life. We invite the reader to 'walk' with the group through 'Asylum Week', a walk deliberately shaped not as a smooth path from start to finish but, rather, as a series of jolts, wrong-turns and stops (Wylie 2005). Such a tactic means that the chapter can be 'dipped into' by others seeking to develop their own set of geohumanities-inspired pedagogies, notably in relation to developing new ways of teaching mental health and its geographies.

This piece of writing is thus 'betwixt and between' the usual modes of academic and creative presentation, and, as such, this might suggest a sort of experiment (Last 2012). Here, we acknowledge the debates about what geohumanities experiments might be (see https://mutablematter.wordpress.com/ and Hawkins et al. 2015) and place them as relating to:

> Expressly spatialized work that pushes against more standardized dis/un/placed or space-neutralized biomedical understandings of health, illness, and human well-being (including where each of those are constituted). (de Leeuw et al. 2018)

Our concern is in communicating evocative suggestions for teaching practice in one course and how this might offer an exemplar for connecting health and geohumanities via progressive educational ambition. Reflections on the potential development of critical educational resources through creative practice have been an underlying component of the geohumanities enterprise, particularly in relation to transdisciplinary approaches and the incorporation of the 'digital' into research and teaching (Dear 2015). Yet, despite this, there is surprisingly little written about teaching geohumanities explicitly. Angela Last (2018 via https://mutablematter.wordpress.com/) argues that geohumanities in general risk being a rebranding exercise in cultural geography or environmental humanities. She insists on its value, however, as an interdisciplinary vehicle for particular kinds of challenging ideas. In her 'fantasy' geohumanities courses, Last suggests contemporary global crisis as one structuring logic for *teaching* geohumanities, with creative interventions between fields of knowledge and empirical materials as part of the unsettling possibility. This, Last suggests, provides a way of productively disrupting the discipline through a rewriting of both the past and future.

Following Last's provocations, our intention through Asylum Week, with its different temporalities and spatialities (see Fig. 12.1), is to demonstrate how the geohumanities-led practices which assemble across Asylum Week unsettle student assumptions about 'madness', mental health and illness by creatively illuminating shifting forms of medical intervention on the human mind-body. Our methods for achieving these aims chime with Hawkins et al.'s (2015) configuration of broad geohumanities approaches:

> geohumanities practices might be as much ethnographic methods, visual and textual analysis, or archival work as they are the shared work of exploration and mapping, or the processes of weaving words or fibres into poems or fabrics, making marks on page, photographic plate, canvas, or landscape. (Hawkins et al. 2015: 213–214)

Asylum Week's ambitions to entangle the ethnographic with the archival, and to consider both textual and visual analysis in a set of shared relations with site, place and space signals asylum pedagogies that subvert widely accepted understandings of embodied mental health practice and experience. If, as Hawkins et al. (2015)

ASYLUM WEEK

LECTURE: *Mind-body: historical institutional geographies of mental health care 1850s-1900s*

GARTNAVAL FIELD CLASS: *new and ruined asylum building - walking tour, psychiatrist talk and Q&A*

ARCHIVES WORKSHOP: *Encountering archival case notes and stories of experiments*

FILM WORKSHOP: *Hurry Tomorrow* (richardcohenfilms.com): power and medical control in a closed mental health ward in the 1970s USA.

LECTURE: *Mind-body: contemporary mental illness in a 20th-21st century somatic society*

Fig. 12.1 Asylum Week

suggest, the geohumanities should inspire transformative thinking, then, through the range of methods, practices and themes discussed below, we hope to demonstrate *in what ways* this transformative thinking can be developed around mental health.

In what follows, we share elements of lectures, a field visit, research materials and workshops that are intended as pedagogical vehicles to bring undergraduate students up-close to asylum geographies of past and present. Creatively presenting extracts from visual images and teaching notes relating to the walking tour of the old ruined asylum, patient case notes, staff diaries and film footage from other psychiatric places reveal how we encounter Gartnavel as a site-specific geohistory *and* as an iconic example of an institutional response to mental illness. While some sections simply gesture to our Asylum Week approaches, others give more specific details regarding the sessions undertaken and the materials, experiences and themes that students undertake during the project. In giving different exposures to Asylum Week's dimensions, we aim not only to highlight the significance of a geohumanities-centred approach to asylum pedagogies but also to aid in the nurturing of a range of experimental teaching methods through engagement with the exercises charted below.

G – Gartnavel

Asylum Week introduces Gartnavel via '*Let there be Light Again*', a textual collection of historical stories and records (Andrews and Smith 1993). *In practice*, though, the weathered sandstone comes into view gradually as our feet follow the undulating contours of the ground below. The remnants of the original asylum building of what was then the Gartnavel Royal Asylum, opened on this site in 1843, are the starting point for walking the site (Fig. 12.2). Gartnavel, built in the suburbs of Glasgow, began life as a large, classic grand Victorian complex, bearing all the traditional features of an asylum. Built to replace William Stark's original, loosely panoptic, city asylum, Gartnavel – meaning 'the place of the orchard' – became a beacon for modern mental health care in Scotland (Andrews and Smith 1993). Designed by architect Charles Wilson, the Tudor Gothic-style buildings had ideas of segregation at their core. The East Wing housed non-fee-paying or pauper patients, while the West was kept for fee-paying, wealthier patients. As Chris Philo (2014 n.p.) notes:

> Encountering old 'lunatic asylums' can send a shiver down the spine. Perhaps as eerie ruins reclaimed by forest, maybe as converted housing expunging all traces of past use or possibly as still-functioning sections of present-day psychiatric facilities, these massive relics of Victorian Britain (with earlier and later incarnations) are disquieting, thought-provoking phenomena.

They are also, he argues, 'remarkable spaces' where changing understandings about mental health and illness can be explored. Decades of scholarship is slowly

Fig. 12.2 Gartnavel Royal Mental Hospital, Old and New Buildings. (Photograph reproduced with permission of Ebba Högström)

peeling away the layers of asylum histories, revealing the intricate networks of human and non-human actors involved in creating, inhabiting and deconstructing such places (Philo and Andrews 2017; Andrews et al. 2007). As sites of continuous experimentation, scholars have begun to describe the asylum site as one of a laboratory for mental health; a place of testing, analysing, dissecting and observing (see Wallis 2017). In our own creative attempt to develop asylum pedagogies, we draw upon Gartnavel as both a set of remarkable spaces – full of material and ephemeral phenomena – and a site where students' conceptions regarding mental health/illness can be transformed, distilled and reimagined.

> Staff: 'Imagine yourselves as patients, walking here along this path 100 years ago, going from the chapel to the ward'

This instruction above, given on the walking tour, relates one purpose of the visit to Gartnavel, as students are invited to exercise their geographical imaginations through their immersed bodily experiences, engaging the lives of past and present patients.

A – Asylum Week

In inviting imaginative geographies via bodies, we connect to the core purpose of Asylum Week, a segment of an honours undergraduate course on *'Geographies of the Body'*. The course examines the body as a conjoint geographical space and lived place in 'Western societies'. Thinking through social and cultural geography, and also wider social theory, it examines the lives of different human groups whose body spaces have been/are considered problematic by those in positions of legal, social and medical control. This course thus tells a story of overt power and latent resistance, a story shaped around the space of the human body.

Students develop historical and contemporary perspectives on how human bodies have been managed and regulated through institutions, state legislation and civil-moral norms in Western society specifically. Critical perspectives on how people manage their own bodies in contemporary times ask about how individuals and collectives creatively respond to cultures of risk and modification that are productive of normative aesthetic and healthy physical forms. In this context the co-production of 'different' or 'stigmatised' (e.g. mad, deviant and medicated) body spaces is interrogated. The course examines how such bodies are imagined, embodied and regulated, but also how people might resist their regulation, and, in this way, the course can be understood as all about *contested body spaces*. It is within this frame that Asylum Week lectures, workshops and encounters seek to animate the contained mad/ill body. Asylum Week also traverses time and space, prompting students to dwell in the past as well as the present. Mad/ill bodies are presented as geographies of *becoming,* in that they are contextualised and constituted by *changing* regimes of power, care and control. These mind-bodies are not just, however, 'other', and students are encouraged to think of *our* bodies and *our* minds as the ones also irreducibly 'at stake' in these shifts. Such ambition is subject to the charge from Agnew (2012: 516), that the reach to the humanities in geography is still coloured by its traditional 'instrumentalist' understanding of the purpose of geographical knowledge: it must be *for* something else, another purpose. In Asylum Week, this 'instrumentalism' is not prescriptive, but simply hopeful, attempting to open up and scramble ideas about who is mad and ill and where they should be located. Below, we animate the centre of GARTNAVEL via pedagogical scrutiny.

R – Remains/Ruins

Throughout the site visit to Gartnavel, students are prompted to think through whether asylum is a *ruined concept* as well as a decaying place. Ruins and ruination remain seductive forms and processes for a range of geohumanities scholars (Murray and Lorimer 2015; Desilvey and Edensor 2013). As remnants of the past and beacons of alternative futures (Galviz et al. 2017), ruins provide a rich understanding of places as simultaneously 'transformed and stabilised' (Edensor 2011) and can become important sites of reflection for considering the meshing of past and present worlds and practices (Olden 2017). Asylums, as ruined places, evoke ideas of abandonment and liminality (Moon et al. 2015), ones circulating as we walk the boundaries of decay, listening to tales of past times. Gartnavel's liminality is bound up with the spatial proximity between the historic, crumbling old asylum building and the newly built psychiatric facility. Students are confronted with transformed spaces and stories of stigma and mental health reform. By positioning them in ruined situ, students become 'positioned on the edge of current experience where the asylum past is receding into the distance but where, for the moment it can be encountered in the present' (Moon et al. 2015). But, if 'asylum' is seen as obsolete, with its built spaces collapsing alongside its purpose as a 'set apart space' for those with mental health differences, then what is its resonance with the adjacent new build psychiatric facility?

> Staff: What does the concept of asylum mean here?

Encountering ruins in Asylum Week goes beyond site, space and stone. The asylum ruin can become a window into lives and practices framed in older medicalised models of care (Wallis 2017). Asylum Week thereby asks students to explore some of the spatial histories of Gartnavel through its psychiatric histories that open up embodied encounters of people within this now ruined place. Countering the claims that 'the geography of the body has been neglected when considering mental health' (Parr 1999: 182), patient case notes are enrolled, reanimating the ruin, allowing students to see, feel and hear the hidden bodies of the asylum.

Back in the classroom, Hazel Morrison's (2014, 2016) research on archival remains of clinical encounters in Gartnavel provides the basis of a workshop (Fig. 12.3). Focussing upon one of the hospital's most eminent superintendents, Dr. David Kennedy Henderson (1884–1965) and his development of a 'dynamic' approach to psychiatry, Morrison excavates patient stories from the material remnants in the NHS Lothian archives. Henderson's handwritten case notes take students into the heart of the clinical encounter in Gartnavel, drawing them into the everyday realities of medicalised bodies and minds. Through Morrison's work, students meet individuals at the start of their time in the asylum. In the waiting room, they momentarily 'hear' the sobs of Harriet Paterson as she protests against her 'forced' admission and sense the pain of Annabel Brown at her and her father's separation:

> Clinging to her father, very pale faced and distressed, not demanding to be taken away, but making no move to go to the ward with sister, and expressing fear and doubt – imploring her father to give her just another minute, kissing and embracing him in a manner painful to observe. He, himself, showed much indecision of mind, encouraging her in her embraces, asking [Medical Officer] if he were sure this was the best place for her, enquiring how she would be treated etc. (Case notes, quoted in Morrison 2016: 75)

Following admission, patients underwent a 'physical examination' behind drawn curtains in the ward. Students glimpse through the curtains via Henderson's case notes:

> She was extremely unwilling to allow a physical examination to be done. She refused to allow the chest to be examined saying that she understood this to be a mental hospital and that she did not see what examination of a patient's chest had to do with mental disease .... The necessity of making a physical examination was explained to her but she still persisted in her unreasonable attitude. After a great deal of delay and persuasion she allowed the examination asking in a sarcastic tone as it proceeded if the MO [Medical Officer] could find any evidence of mental disease from his examination. (Case notes, quoted in Morrison 2016: 76).

Group Exercise: The Clinical Encounter

Aim

To think critically about the different ways in which bodies are measured, categorised and 'seen' in mental institutions.

Task

Analyse case note materials relating to Dr. David Henderson's patients at Gartnavel during the 1920s. Using discourse analysis te chniques to highlight the different ways in which bodies are being measured and categorised in the various materials.

Compare and contrast medical materials with patient's own words. Discuss how they differ and why this matters in our analysis of embodied geographies.

Materials

Case note records and personal patient materials (including patients own writing) from Dr. Hazel Morrison's research into the 'clinical encounter' at Morrison (2016, 2014).

Fig. 12.3 Example of a teaching exercise used during Asylum Week

Once the processes of bodies being prodded, poked, measured and examined are highlighted, attention begins to focus on the mind through the lens of the 'mental examination'. Medical Officers studied and assessed the patient's 'stream of mental activity', edging inwards to the patient's inner worlds. Students become acquainted with patient Robert Norton's fear and frustration:

> He . . . discussed the way in which the questions were put, was suspicious of my motives in questioning him, and he refused to allow me to take any notes. He said that if I took notes he would refuse to speak to me at all . . . He constantly repeated the questions I asked him, he found fault with it in various ways, he complained about the manner in which it was asked, he took exception to the way I sat on a chair besides his bed, and eventually he refused to co-operate any further, finishing up by saying that he considered I was just as insane as he was. (Henderson as quoted in Morrison 2016: 78)

These snippets reveal the embodied human consequences of psychiatric intervention. People are turned into patients under the psychiatric-medical gaze, and lived experiences become clinical case material.

In the classroom, student groups consider the voices of people experiencing mental illness (Wynter and Smith 2017) by examining a selection of prose produced by former female Gartnavel patients from the 1920s and their letters home. Morrison (2014: 310) argues that, by engaging with the fluctuating perspectives of patients in their writing and by remaining open to the ambiguity of asylum histories, their interior landscape is better understood. Students begin to unpick deeper layers of narrative encounter in the asylum through patient letters:

> My Dear Mother … I implore and beseech you take me out of this – you couldn't know what you did when you put me among jibbering, moaning, screaming idiots. Come for me today and take me home, do not let me be another night enduring this living death. (Miss M.B. to her mother quoted in Morrison, 2014: 310).

Site-specific experiences of the ruined buildings of Gartnavel become populated with the raw extracts of human experience. The sounds of the site visit are replaced by the haunting voices of institutional life. Through combining the experiences of viewing the ruined site and reanimating the ruin through different narrative forms, students begin to recognise the embodied realities of past lives lived in asylum spaces.

T – Teaching

Teaching spaces in Asylum Week change from 'academy to community', from 'classroom to clinic', 'from archive to walking tour' as demonstrated above. When considering 'cultural geographies of education' (Mills and Kraftl 2016), we are conscious of how 'formal and informal curricula within education spaces can be important in the (re)production of … social differences' (Holloway and Jons 2012: 484). By teaching inside and outside the classroom, we intend a spatially reflexive response to asylum space in order to 'humanise' and 'relate' its histories and present. Our approach privileges the experience of being 'in the field', a tried and tested learning medium for geographers. We do not want Gartnavel to be simply a record on the page, a sketch, a critique of asylum as a 'ruined idea'. As others have pointed out, 'fieldwork can give us direct experiences that challenge our preconceptions. … This 'enhanced affective response' helps deepen our understanding of the wider world and our place within it. It is for this reason that fieldwork remains a valuable mode of learning for human geography students' (Hope 2009: 170). While the physical experience of walking around, observing and questioning Gartnavel's built and natural environments are a mainstay of the teaching, we also constitute 'the field' elsewhere, in visual and textual archives, relationally with other spaces and times. Our teaching unrolls Gartnavel as a particular site, a place of specific local meaning and history, but also as a representational space through which wider cultures of care and medical regulation can be read and 'set against'.

> Staff: How can we read Gartnavel as a distinctive place where disruptive minds and bodies have been 'treated'?
> How does Gartnavel relate to other times and spaces of psychiatric intervention?

N – Nature

Teaching between the page and 'the field' involves debating the significance of nature in asylum landscapes. Ideas about therapeutic nature were integral to locational debates surrounding the early asylums, and the walking tour traverses the grounds (Fig. 12.4). Gartnavel's original site and situation were influenced by Tukes' famous York Retreat from the 1790s, itself deliberately fashioned as a safe space away from damaging urban industrial society and as one where the 'curative power of pure romantic nature itself … was called into service' (Philo 2004: 483; see also Edginton 1997). As Digby (1985: 55) also argues, the pleasant order of the gardens was assumed to encourage calm, rational reflection in the patients: a space where 'patients could regain their serenity through contact with beautiful, natural surroundings'. Pictures of the York Retreat and its landscapes feature in some of Gartnavel's annual reports. In archives, the working of nature is referenced, and similar plaudits advanced, with 'outdoor labour'being seen to keep patients in good physical condition as well as helping to 'tranquilise their nervous system' (Glasgow Royal Lunatic Asylum 1861: 14). Nature assumes magical properties in asylum histories (Fig. 12.5):

Fig. 12.4 Extensive grounds near the West Wing of the Old Building. (Photograph reproduced with permission of Ebba Högström)

Fig. 12.5 Farm work in Nottingham Asylum circa the 1930s: Cross-Referencing Body Practice. (Reproduced from Parr, 2007)

'It seemed as if the vapours of the brain exhaled with the sweat of the brow', which thus discouraged incoherent thought (Glasgow Royal Lunatic Asylum 1817: 20, cited in Sneddon, 1993: 29).

Agricultural and garden work, which could employ up to 60 males in 1 day in the asylum, was 'highly beneficial' in terms of 'diverting patients from their particular hallucinations' (Glasgow Royal Lunatic Asylum 1873: 9). Such discourses are suggestive of the supposedly inherent healthy powers of nature, a belief entirely in keeping with later eighteenth- and nineteenth-century debates about rurality and the physical geography of asylum location as found in the pages of contemporaneous professional medical journals (Philo 1987, 2004). These are the same landscapes through which we move on the walking tour during our field visit. The labours of the land above are now at far remove from the trails, as the asylum landscape become instead a 'tranquil walk' (Fig. 12.6). While we might critique such commodification of asylum roots and routes, contemporary staff and patient voices bare witness to the *effect* of asylum land:

It's lovely and light and bright. And, because of the trees and the greenery surrounding, you really don't feel like you're in hospital. It feels quite calm and peaceful and separate here. We also have this garden area out the back which people are free to use if they would like to. So folk can go outside too: there's quite a lot of wildlife, you see squirrels and birds and things. So it's all quite calming and quite relaxing … and I think the clients over time that stay with us longer term, they do come to feel quite comfortable within this space. (Staff member, via Philo and Hogstrom's Gartnavel Project)

Fig. 12.6 Trails on asylum grounds. (Photograph: Hester Parr)

Students reflect on the move from narrative evidence of restorative nature to walk-manicured contemporary pathways.

> Staff: What is the significance of asylum landscape for the management of patient bodies in the past and the present?

A – Archives

The management of bodies once again comes to light as we work through documentary sources relating to the histories of psychiatric care, in order to penetrate the inner spaces of the asylum, in keeping with new waves of historical exposure in asylum studies (McGeachan 2018; Coleborne 2006; Piddock 2001). The wide-ranging ambition in the geohumanities to forge new connections with past lives and places by illuminating lost, forgotten and neglected histories (Hawkins et al. 2015) connects well with geography's creative turn (Hawkins 2018) and together opens up pathways for creative archival encounters. One such encounter is found in geo-biographical narratives (Patchett 2017). Biographical and geo-biographical approaches have become increasingly useful tools for geohumanities scholars seeking to tell stories about lives and afterlives of people, objects and places (Lorimer, 2015). Our desire to take the students on a journey through the *embodied geographies* of the asylum leads us to consider the possibilities of drawing on geographical biography in the classroom, because, as Harmione Lee (2005: 3) notes:

> What makes biography so curious and endlessly absorbing is that through all the documents and the letters, the context and the witness, the conflicting opinions and the evidence of the work, we keep catching sight of a real body, a physical life.

Group Exercise: Historical Geographies of Bodies in the Asylum

Task

Focussing on RD Laing's biographical experiences of being in the refractory room of Gartnavel's East Wing during the 1950s, describe and discuss his embodied encounters with the patients and the ward spaces.

Materials

Extracts from RD Laing's notes written during his time at Gartnavel

Extracts from RD Laing's own autobiography *Wisdom, Madness and Folly* (1985)

Fig. 12.7 Example of Asylum Week workshop exercise relating to biography

While these 'real bodies' are endlessly absorbing, they are also frustratingly difficult to capture from the historical records left behind. Lee (2005: 5) argues that, instead of a body, all that can be found are 'body parts … relics, legends and fragments, with the parts and bits and gaps which are left over after the life has ended'. And so the class turns to investigate archival fragments through the figure of RD Laing, an influential psychiatrist (Fig. 12.7), who worked at Gartnavel for a period in the 1950s.

Laing described women patients as 'introjected objects' in his ward notes, although he wished he could find another phrase to describe those he encountered in the spaces of Gartnavel during the mid-1950s: 'ghosts, maybe' (Laing, 1967: 1). Ronald David Laing was one of Gartnavel's most (in)famous of staff members, taking up the position of consultant psychiatrist during one of the hospital's most stretched periods (Beveridge 2011). A revolutionary figure in psychiatry and a controversial advocate for humanistic mental health-care reform, Laing's reflections become, for us, a key educational tool in highlighting the complexities of experimental practice in the asylum. A selection of Laing's archival materials, including notes written while in Gartnavel and his own autobiography, allows students access to historic psychiatric care *in action*.

On entering the locked doors of the 'Women's East Wing', Laing was confronted by an incredible soundscape that reverberated through the spaces of the ward. Describing it as 'like an out-of-tune orchestra endlessly tuning up' (Laing 1985: 114), Laing was forced to retune himself to his new environment, stating that, 'with some acclimatisation, it began to dawn on me that the autism of each patient, although autistic, was interwoven with that of the others' (Laing 1985: 115). Upon looking, Laing noticed almost 60 women in cotton day dresses slipping in and out of view down the narrow corridors of the ward. Describing them as 'strange creatures' (Laing 1955), they reminded him of Homer's description of the ghosts in Hades: 'separated on their side from the living by the width of the Ocean, and, on the part of the living, by the Rivers of Fear' (Laing 1985: 112). They appeared to him on first encounter to exist in a disembodied state of near-death, and yet he wished to know them better. Students read Laing's personal notes about his uneasy feelings at the physicality of encounters with these women:

> My first impressions therefore were mainly of my own reactionist anxiety at raw physical advances, aggressive and/or sexual on my body [they ruffled my hair, pulled my tie, ripped off my trouser buttons], and my horror at the general dilapidation of many of the patients, their obscenity and so on. Sometimes I was so physically shaken that I had to leave after half an hour, to find myself trembling like a leaf. (Laing 1955: 3)

In small groups, students are asked to reflect upon the embodied nature of these interactions, an exercise designed to enliven the ruin spaces of the wards where Laing worked and to animate the medical practices and relations in Gartnavel's history (and see 'Experiment', below).

V – Violence

In making sense of this local example of psychiatric medicine and its history, we hear of medical superintendent lives on the walking tour. The current Gartnavel clinical psychiatrist, Dr. Ian Smith, is our guide, and he feels compelled to defend the life-saving interventions of the asylum, past and present, even if students appear uneasy about the apparent violence of some treatments. Towards the end of Asylum Week, we retreat from the physical site of Gartnavel and start to relate its geographies to wider understandings and practices of mental health care in other times and spaces. As such, we use screened representations of other practices of medical power (Fig. 12.8).

Thursday 11–1 pm in Room 416: 'Hurry Tomorrow' (US Documentary Film, 1976)
This is a screening of 80 minutes of rare documentary footage of a US state mental hospital secure ward. The film was made by Richard Cohen who has just distributed new DVD copies in the United States, and we secured one for the course: See: http://richardcohenfilms.com/.

Reviews of *Hurry Tomorrow* speak to the intentions of Asylum Week: 'The emotional impact of the documentary and its thought provoking scenes remain an effective introduction to the concept of total institutions' (http://richardcohenfilms.com/hurry_tomorrow.htm). Through documentary film, we explore the effect of medical power on human being. *Hurry Tomorrow* allows us in, behind asylum walls, into

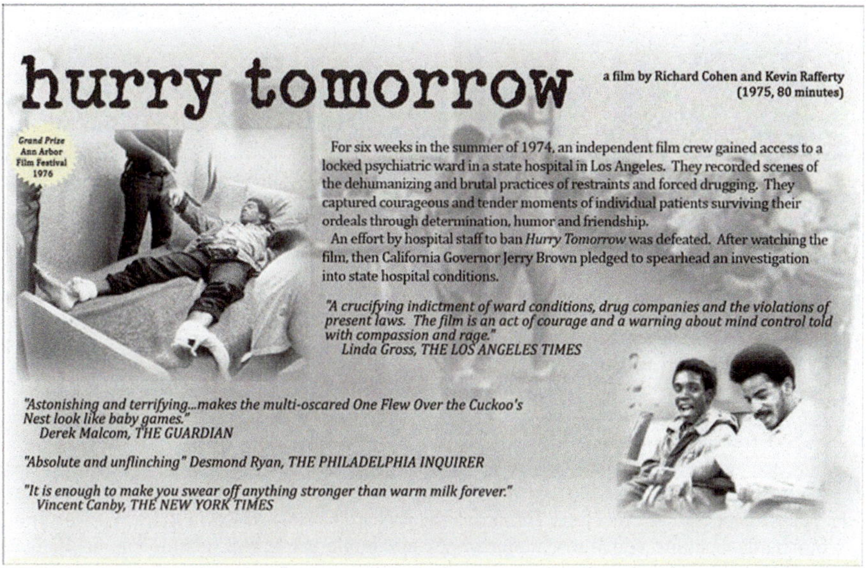

Fig. 12.8 'Hurry Tomorrow' by Richard Cohen and Kevin Rafferty. (www.richardcohenfilms.com; used with permission of Richard Cohen Films)

corridors where the floors are scraped by the shoes of tranquillised feet that drag themselves in and out of the injection room. In 'The way of Jack's Day', a sequence from the film, Jack's pain is evident as he expresses his anxiety about being injected. 'It makes me so stiff I can't walk', he complains. 'I'm taking this under duress'. The camera films Jack as he gets the injection in his left buttock amongst the chatter of the injecting room (Fig. 12.9).

Through discussion, we thus explore asylums as sites of violence where bodies are exposed to strong medication. Tellingly, Hurry Tomorrow has been reviewed as follows:

> As frightening and affecting a documentary as you may ever see … I recommend it highly. Documentary review via http://www.richardcohenfilms.com/hurry_tomorrow_Scenes.html.

Post-screening, the class discusses Goffman's (1961) accounts of hospital life and Foucualt's (1975) panoptic power. We consider the violence of medical power and also bodies becoming docile or resistive in clinical settings.

The documentary itself can also be considered as a kind of violence, even while we acknowledge how 'documentary films have often taken a pivotal role in strategies to internationalize (geography) curricula and classrooms, being used as a method of bringing the world to the classroom' (Hay 2017: 562). By this, we mean the class becomes aware of the power of the visual gaze inside asylum walls. Alongside Hay, we are wary of the representational powers of screen space but value the mode and medium of knowing the asylum via visual documentation as a reflexive 'border pedagogy' (Ansell 2002).

Fig. 12.9 Injection scene in 'Hurry Tomorrow' by Richard Cohen and Kevin Rafferty. (Stills from www.richardcohenfilms.com/hurry_tomorrow_Scenes.html; used with permission of Richard Cohen Films)

E – Experiment

For the geohumanities, experimental modes and practices have become key, signalling its drive to build new (transdisciplinary) relations and to break down and challenge boundaries and conventions (Hawkins et al. 2015). For us, Asylum Week is not only an experimental way of thinking and doing geography but also offers the potential to develop new insights into the very nature of 'experiment' as this has happened *within* the asylum setting. We thus design workshop materials on Gartnavel's experimental practices and attendant spatial dimensions, deliberately drawing attention to how Superintendent Henderson envisaged Gartnavel as both a 'pathological laboratory' (Morrison 2014: 73) and a place of therapeutic potential. One of Laing's most famous experiments, labelled the 'Rumpus Room' (McGeachan 2017), took place at Gartnavel, and through this example, we further unpick and critique experimental psychiatric practice.

Through a medical journal article from *The Lancet* and extracts of Laing's personal diaries, students take a tour inside of Gartnavel's 'East Wing' in the 1950s and trace the development of experimental forms of psychiatric care. We reflect upon the architecture of the experiment and consider the embodied experiences of being in such spaces (Fig. 12.10). Through observing the patients in the 'East Wing', Laing had begun to question the role of the hospital environment in the lives of the patients. Laing considered that the environment was not significantly therapeutic and that there had to be a change in order to foster meaningful relationships between patients and staff in their shared space. Laing took over a separate room in the ward for his experiment, one that was large, bright and newly decorated (Cameron et al.

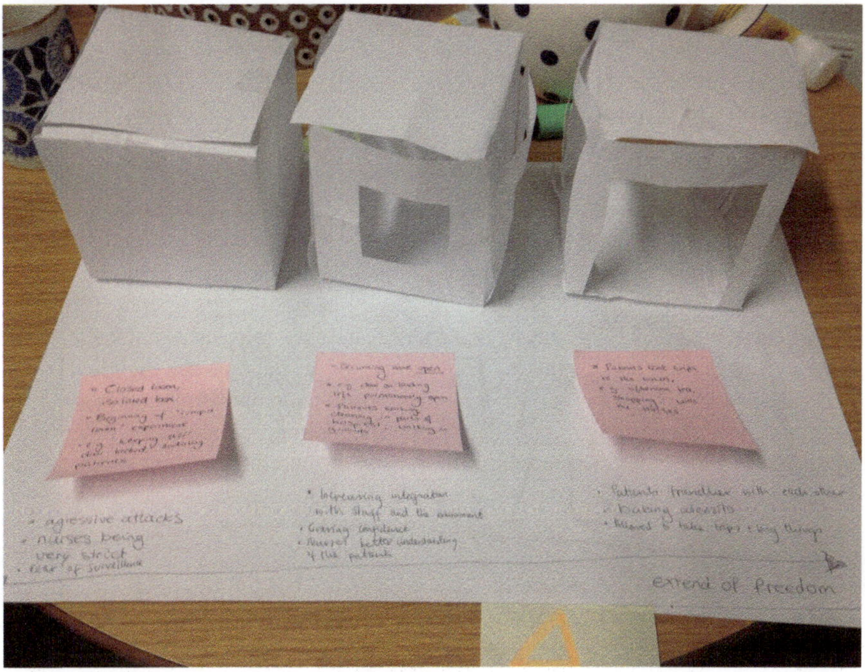

Fig. 12.10 An example of group-work mapping the 'Rumpus Room' experiment. (Photograph: Hester Parr)

1955: 1384). Inside the room were materials for knitting and sewing, magazines and comfortable furnishings, all of which created a visual contrast from the minimal aesthetics of the host ward.

Laing chose 11 women to be part of his experiment, the criteria for selection centred upon their level of isolation from both the world outside the hospital and the actions inside the ward. None of the 11 patients was able to leave the ward unless accompanied by their visitors, and the majority of the women selected were considered too 'disturbed' to take part in any of the ward activities (Cameron et al. 1955: 1384). Laing and his colleagues revealed that the events occurring on the first day became the standard procedure for the entire experiment:

> The nurses appeared for duty on the ward at 7.30 A.M. They shared general ward duties until 9 A.M. when they took their patients over to "the room" where they stayed with them until they took them back to the ward for lunch at noon, leaving them there, while they had their own lunch. At 2 P.M. they collected them again from the ward and took them back to "the room" until 5 P.M. The nurses then returned to "the room" for about half an hour to write up the day's happenings. (Cameron et al. 1955: 1384)

At first, the patients used the space inside the room as nothing more 'than an extension of the ward' (Cameron et al. 1955: 1385), taking up positions within the room reminiscent of their use of space in the ward. The nurses attempted to utilise the material in the room and strove to encourage the patients to knit or draw, but they noted that they initially did not respond to it in any visible way. When the nurses took the patients back to the main ward, they described the way in which they all seemed to become absorbed once again in their long-standing customary roles, taking up the same positions in the ward that they had left that morning; and it appeared that nothing had really changed.

On the second day of the experiment, it was noted that the women were all waiting at the ward door, half an hour earlier than scheduled, to be taken across to the room, and Laing recalled how this was 'one of the most moving experiences of my life' (Clay 1996: 55). Eventually, patients began to use the space of the hospital in different ways from before and created a wholly new micro-geography of 'dwelling' in these spaces. Laing argued that these responses were also indicative of the increased awareness that the nurses were gaining of the patients and their behaviour. The biggest change, Laing believed, lay not in the patients themselves but in how the nurses were beginning to understand them, not simply as another patient in the ward, but as individual people wrapped up in their own differing social situations. After hearing this material, students were asked to reflect upon the architectural and embodied geographies of this experiment, drawing out the complex relations between people and spaces. Using paper, pens, glue and tape (Fig. 12.10), we creatively encounter the transformational experiment of the Rumpus Rooms in our own geohumanities experiment. We play with models of three-dimensional space in order to find new ways to witness experimental space. The asylum and the classroom are duly joined in creative encounter.

L – Learning

In learning from Asylum Week, we ask that all participants engage in a variety of encounters with the asylum, past and present, and we work in an experimental mode. In a course about geographies of bodies, we ask that students walk the ruins of an asylum and around a new build psychiatric facility. We access processes of embodiment behind asylum walls via archives, biographical notes and film. Asylum week juxtaposes temporally and spatially variant forms of psychiatric care in order to ignite geographical imaginations. We ask students to tell us what mattered and the resulting student evaluations included:

> The field class to Gartnavel Royal Hospital was of real educational value. Getting to hear about the institution from an expert [Smith], who had spent many of his working years there, offered another element to the learning process.
>
> Asylum Week was good at getting us to experience the course through our own bodies.
>
> (Asylum week) provides a different analytical framework for the application of knowledge!
>
> They (the workshops) provided a real insight into how embodied social geographies shape real people's lives

Relating the asylum via a field class walking tour, archive and film workshops brought into focus the difficulty of bringing the past and the present together in overwhelming juxtapositions. The geohumanities of mental health that require an experimental 'thinking with' asylum in its multiple forms can involve surprise, discomfort, confusion, as well as learning:

> I was very surprised to learn about the different treatments of the mentally ill, especially in regards to the methods used in the asylums and how the people were categorised.
> When we visited Gartnavel I was most overwhelmed by its contemporary state, the ideas about mental health care that were presented and practiced, and therefore wasn't very well able to focus on the historical significance of the site.
> I was very surprised and I found it shocking. I learnt a lot about the differences between current mental health care and historical mental health care. It was really an interesting and rare insight.

We designed Asylum Week as a teaching experiment that brings the past and the present together via archive records, workshop chatter, footfall on site, questions, reading, voices, watching film, making models and sharing thoughts and feelings. We argue here that there is value in thinking through such asylum pedagogies, and we posit that a geohumanities of mental health might well consider further the shape and purpose of such endeavours. The creation of site-specific geo-histories and geo-futures allows the use of particular institutions, like Gartnavel, to act as the basis of educational encounter that centralises transformations. These transformations are both about the sites themselves – how medical thinking and practice have transformed from 1843 to 2018, for example – and how these very transitions can hold potential for changing our own ideas about minds and bodies and about mental illness and madness. Asylum Week is underpinned by a vision of a 'hopeful asylum pedagogy' in which *patients* are understood as *people* caught up in changing and contingent forms of medical power. We argue that this 'hopeful pedagogy' is one where people and places that have previously been stigmatised and unknown can be 'rescripted' (Parr 2008) via educational practice. As such, we disrupt taken-for-granted understandings and ways of knowing asylum geographies and enlighten students about a local mental health-care institution while also relating to *other* institutions and places in time and space.

Asylum Week is an experiment in geohumanities and mental health. In asking '*Why the geohumanities and mental health?*', we might answer by referencing the future possibilities of a transformative education that seeks to unsettle assumptions about mental health amongst a younger generation. We hope that such an ambition might contribute to a wider body of knowledge, practice and writing on teaching geohumanities.

Acknowledgements Thanks to Iain Smith and Eileen Ross for facilitating our local psychiatric histories and geographies. Thanks to Hazel Morrison for allowing us to use her research in the classroom.

References

Agnew, J. (2012). A review of 'Envisioning landscapes, making worlds: Geography and the humanities'; 'GeoHumanities: Art, history, text at the edge of place'. *Annals of the Association of American Geographers, 102*, 514–516.

Andrews, J., & Smith, I. (Eds.). (1993). *'Let there be light again': A history of Gartnavel Royal Hospital from its beginnings to the present day*. Glasgow: Gartnavel Royal.

Andrews, J., Moran, J. E., & Topp, L. E. (2007). *Madness, architecture, and the built environment: Psychiatric spaces in historical context*. New York: Routledge.

Ansell, N. (2002). Using films in teaching about Africa. *Journal of Geography in Higher Education, 26*, 355–368.

Beveridge, A. (2011). *Portrait of the psychiatrist as a young man: The early writing and work of R.D. Laing, 1927–1960*. Oxford: Oxford University Press.

Cameron, J. L., Laing, R. D., & McGhie, A. (1955). Patient and nurse: Effects of environmental changes in the care of chronic schizophrenics. *The Lancet, 266*(6905), 1384–1386.

Clay, J. (1996). *R.D. Laing: A divided self*. Great Britain: Hodder & Stoughton.

Coleborne, C. (2006). Families, patients and emotions: Asylums for the insane in colonial Australia and New Zealand, *c*. 1880–1910. *Social History of Medicine, 19*, 425–422.

de Leeuw, S., Donovan, C., Schafenacker, N., Kearns, R., Neuwelt, P., Squier, S. M., et al. (2018). Geographies of medical and health humanities: A cross-disciplinary conversation. *GeoHumanities, 4*, 285–334.

Dear, M. (2015). Practicing GeoHumanities. *GeoHumanities, 1*, 20–35.

DeSilvey, C., & Edensor, T. (2013). Reckoning with ruins. *Progress in Human Geography, 37*, 465–485.

Digby, A. (1985). *Madness, morality and medicine: A study of the York Retreat 1796–1914*. Cambridge: Cambridge University Press.

Edensor, T. (2011). Entangled agencies, material networks and repair in a building assemblage: The mutable stone of St Ann's Church, Manchester. *Transactions of the Institute of British Geographers, 36*, 238–252.

Edginton, B. (1997). Moral architecture: The influence of the York Retreat on asylum design. *Health & Place, 3*, 91–100.

Foucualt, M. (1975). *Discipline and punish: The birth of the prison*. New York: Random House.

Galviz, C. L., Bartolini, N., Pendleton, M., & Stock, A. (2017). Reconfiguring ruins: Beyond *ruinenlust*. *GeoHumanities, 3*, 531–553.

Glasgow Royal Asylum for Lunatics. (1861). *Annual Report.*

Glasgow Royal Asylum for Lunatics. (1873). *Annual Report.*

Goffman, E. (1961). *Asylums: Essays on the social situations of mental patients and other inmates*. Oxford, England: Doubleday (Anchor).

Hawkins, H. (2018). Geography's creative (re)turn: Toward a critical framework. *Progress in Human Geography*. https://doi.org/10.1177/0309132518804341.

Hawkins, H., Cabeen, L., Callard, F., Castree, N., Daniels, S., DeLyser, D., et al. (2015). What might geohumanities do? Possibilities, practices, publics, and politics. *GeoHumanities, 1*, 211–232.

Hay, I. (2017). Moving pictures: From ethnographic to autoethnographic documentary in the internationalization of the geography curriculum. *Journal of Geography in Higher Education, 41*, 562–573.

Högström, E., Philo, C., & Smith, I. (2017–18). *Psychiatric spaces in transition: Discourse, dwelling and doing*, funded under EC Research Call H2020-MSCA-IF-2015 (Project No. 706026).

Holloway, S. L., & Jons, H. (2012). Geographies of education and learning. *Transactions of the Institute of British Geographers, 37*, 482–488.

Hope, M. (2009). The importance of direct experience: A philosophical defence of fieldwork in human geography. *Journal of Geography in Higher Education, 33*, 169–182.

Laing, R. D. (1955). *The rumpus room*. University of Glasgow: Papers of R.D. Laing, Special Collections, MS Laing A153.

Laing, R. D. (1967). Seminar at the William Alanson White Institute, New York. University of Glasgow: Papers of R.D. Laing, Special Collections, MS Laing A262.

Laing, R. D. (1985). *Wisdom, madness and folly: The making of a psychiatrist 1927–1957*. London: Macmillan.

Last, A. (2012). Experimental geographies. *Geography Compass, 6*, 706–724.

Last, A. (2018). An experiment in teaching GeoHumanities. Mutable Matter. https://mutablematter.wordpress.com/page/1/. Accessed 21 Mar 2019.

Lee, H. (2005). *Body parts: Essays in life-writing*. London: Chatto & Windus.

Lorimer, H. (2015). Standards of beauty: Considering the lives of W. A. Poucher. *GeoHumanities, 1*, 51–79.

McGeachan, C. (2017). The ghosts of the refractory ward: R.D. Laing and (re)configuring psychiatric spaces of care. In C. Nord & E. Högström (Eds.), *Caring architecture: Institutions and relational practices* (pp. 111–126). Cambridge: Cambridge Scholars Publishing.

McGeachan, C. (2018). Historical geography II: traces remain. *Progress in Human Geography, 42*, 134–147.

Mills, S., & Kraftl, P. (2016). Cultural geographies of education. *Cultural Geographies, 23*, 19–27.

Moon, G., Kearns, R., & Joseph, A. (2015). *The afterlives of the psychiatric asylum: recycling concepts, sites and memories*. Burlington and Farnham: Ashgate.

Morrison, H. (2014). Unearthing the 'clinical encounter': Gartnavel Mental Hospital, 1921–1932. Exploring the intersection of scientific and social discourses which negotiated the boundaries of psychiatric diagnoses. University of Glasgow: Doctoral Thesis.

Morrison, H. (2016). Constructing Patient Stories: 'Dynamic' case notes and clinical encounters at Glasgow's Gartnavel Mental Hospital, 1921–32. *Medical History, 60*, 67–86.

Murray, S., & Lorimer, H. (2015). The ruin in question. *Performance Research, 20*, 58–66.

Olden, R. (2017). Lost world of the nearby: Minimal ethics on Govan Graving Docks. *GeoHumanities, 3*, 510–530.

Parr. (1999). Bodies and psychiatric medicine. In R. Butler & H. Parr (Eds.), *Mind and body spaces: Geographies of illness, impairment and disability* (Critical geographies) (pp. 181–202). London: Routledge

Parr, H. (2007). Mental health, nature work and social inclusion. *Environment and Planning D, 25*, 537–561.

Parr, H. (2008). *Mental health and social space: towards inclusionary geographies?* Oxford: Blackwell Publishing/RGS-IBG book series.

Patchett, M. (2017). Historical geographies of apprenticeship: Rethinking and retracing craft conveyance over time and place. *Journal of Historical Geography, 55*, 30–43.

Philo, C. (1987). Fit localities for the asylum: The historical geography of the nineteenth century 'mad-business' in England as viewed through the pages of the Asylum Journal. *Journal of Historical Geography, 13*, 398–415.

Philo, C. (2004). *A geographical history of institutional provision for the insane from medieval times to the 1860s in England and Wales: The space reserved for insanity*. Lewiston, NY: Edwin Mellon Press.

Philo, C. (2014). 'Encountering asylum geographies', asylum and post-asylum spaces. University of Glasgow: School of Geographical & Earth Sciences.

Philo, C., & Andrews, J. (2017). Introduction: Histories of asylums, insanity and psychiatry in Scotland. *History of Psychiatry, 28*, 3–14.

Piddock, S. (2001). "An irregular and inconvenient pile of buildings": The destitute asylum of Adelaide, South Australia and the English workhouse. *International Journal of Historical Archaeology, 5*, 73–95.

Sneddon, A. (1993). Environment and architecture. In J. M. Andrews & I. Smith (Eds.), *Let there be light again: A history of Gartnavel Royal Hospital from its beginnings to the present day*. University of Glasgow: Wellcome Unit for the History of Medicine.

Wallis, J. (2017). *Investigating the body in the Victorian asylum: Doctors, patients, and practices*. New York: Springer.

Wylie, J. (2005). A single day's walking: Narrating self and landscape on the South West Coast Path. *Transactions of the Institute of British Geographers, 30*, 234–247.

Wynter, R., & Smith, L. (2017). Introduction: historical contexts to communicating mental health. *Medical Humanities, 43*, 73–80.

Chapter 13
Zones of Dissonance and Deceit: Nuclear Emergency Planning Zones

Neil Overy

Abstract Throughout the world nuclear power stations are surrounded by Emergency Planning Zones (EPZs) which are designed to protect people from the hardful effects of radiation in the event of serious accidents. Via the medium of photography, my work explores the scientific and techopolitical deceit of these zones. Because of the nature of 'risk society', the inevitability of accidents and the particular ontology of radiation, EPZs play an illusionary, rhetorical role which attempts to mask the inherent dangers of nuclear power to all life by imposing a flawed 'health and safety' narrative on dangers which cannot be controlled or contained.

> What's it like, radiation? Maybe they show it in the movies? Have you seen it? Is it white, or what? Some people say it has no colour and no smell, other people say that it's black. Like earth. But if it's colourless, then it's like God. God is everywhere, but you can't see Him. –(Monologue about Lies and Truths, *Voices from Chernobyl*. Alexievich 2006: 52)

> At first the question was, who's to blame? But then, when we learned more, we started thinking. What should we do? How do we save ourselves? After realising that this would not be for one year or for two, but we for many generations, we began to look back, turning the pages. –(Monologue about What We Didn't Know: Death Can Be So Beautiful, *Voices from Chernobyl*. Alexievich 2006: 151)

Lies and Truths

While the Fukushima Daiichi nuclear disaster of March 2011 dampened much of the early 2000s' fervour which posited a significant worldwide renaissance in nuclear power, the ongoing climate crisis has witnessed renewed calls for more electricity to be generated by nuclear power because of its alleged low-carbon footprint (Nuttall 2004). Some of these calls have come, unexpectedly, from environmentalists such as George Monbiot (Monbiot 2016) and James Lovelock (Nelles 2014). Others have come, more predictably, from so-called Ecomodernists who see

N. Overy (✉)
Research Associate, University of Cape Town, and Photographer,
Cape Town Area, South Africa
e-mail: neilovery@imaginet.co.za; http://neilovery.com/

S. Atkinson, R. Hunt (eds.), *GeoHumanities and Health*, Global Perspectives on Health Geography, https://doi.org/10.1007/978-3-030-21406-7_13

technology as the only solution to climate change. (Ecomodernist Manifesto 2015). The primary drivers are, however, governments under pressure to meet their respective Kyoto, and now Paris, climate change commitments.

In 2018, the Intergovernmental Panel on Climate Change (IPCC) confirmed the need for more nuclear power in each of its four scenarios which aim to limit the overall rise in temperatures to 1.5 °C by 2100. Even its most conservative assessment for nuclear power, scenario P3, suggests a doubling of energy derived from nuclear power by 2050 (IPCC 2018, 16).

In early 2019, the International Atomic Energy Association (IAEA) claimed that new reactors are under construction in 18 different countries. In total, 55 new reactors are currently being built, of which 24 are being constructed in the so-called Global South (IAEA 2019).[1] According to the World Nuclear Association (WNA), an additional 150 reactors are 'on order or planned', while 300 more 'are being proposed'. The WNA states that 30 countries are considering nuclear power for the first time (World Nuclear Association 2016). For example, South Korea's Kepco is promoting nuclear power in the Middle East, while it completes construction of four new reactors in the United Arab Emirates (Asaba 2018). Rosatom, Russia's nuclear agency, claims it has signed agreements for the 'use of nuclear energy for peaceful purposes' with nine African countries (ROMATOM 2019; Winkler 2018). While we should be cautious of taking these claims at face value because nuclear power 'plans' and 'proposals' are often not realised, there is a renaissance of sorts taking place in nuclear power (Jostow and Parson 2012).

This 'renaissance' is having an impact in South Africa. In 2010, the government stated that it would build nuclear power stations that would generate 9.6 GW of electricity as a low emission alternative to coal (South African Department of Energy 2010: 25). In October 2017, the then Minister of Energy, David Mahlobo, stated that 'South Africa recognises the role of nuclear power in ensuring security of energy supply and meeting the challenge of climate change' and promised to push ahead with procurement processes (Le Cordeur 2017). While the specific plan for 9.6 GW of new nuclear power has stalled under a cloud of corruption, public resistance and reductions in South Africa's predicted energy needs, there are still plans to add 5.6 GW of new nuclear power from 2030 (South African Department of Energy 2010, 25; Winkler 2016).

As a photographer and environmental activist who is opposed to nuclear power on health grounds, this 'renaissance' troubles me, as does the fact that I live close to Africa's only commercially operational nuclear power station at Koeberg, just 25 km north of the City of Cape Town. Koeberg Nuclear Power Station (KNPS), which produces 1.8 GW of power, was built by a French-led consortium between

[1] The Global South refers to the regions of Africa, Asia, Oceania and South America which are largely peripheral to the geo-political centres of power. These regions also tend to share, to differing degrees, experiences of being colonised. See Dados and Connell (2012). I include those being built in Bangladesh (2), Brazil (1), China (11), India (7) and Pakistan (2) as being from the 'Global South'.

1976 and 1985 and was paid for with loans from French banks unmoved by calls to boycott the apartheid regime (Fig 2005: 56).

While completing an interdisciplinary MPhil in Environmental Humanities South at the University of Cape Town, I began to think about the spatial and temporal health-related implications of so-called Emergency Planning Zones (EPZs) that surround nuclear power stations.[2] Nuclear power stations have emergency plans detailing the types of 'actions' that need to take place depending on an assessment of the accident that has occurred or is occurring. The purpose of an EPZ is to designate an area around a nuclear power station within which these pre-planned 'actions' are to occur in the event of an accident (Adalja et al. 2014). Pre-planned actions may include evacuation, sheltering, radiation monitoring, food screening, or the distribution of potassium iodide tablets.[3]

While geographers, historians, anthropologists, artists and sociologists, amongst others, have written about the spatial and temporal meaning of 'zones' as they relate to nuclear power, this attention has focused on zones that have been created after spectacular nuclear events, such as the exclusion zones established after the accidents at Chernobyl and Fukushima (Alexis-Martin and Davies 2017; Pitkanen and Farish 2018). Little work has been done exploring the meaning of these zones in their 'normal', 'everyday' state. The zones are presented as solid lines on maps, and you can physically locate the exact place where a zone ends or begins, but as a representational form, what does this boundary actually tell us? Is it proven scientifically that the boundary represents the border between safe and unsafe – in both physical and temporal terms? Or are the boundaries merely rhetorical, social constructs serving a political end? Whose interests do they serve? The residents who live in them or the technopolitical state and private actors who insist on the necessity, and safety, of nuclear power?

The emerging academic discourse of GeoHumanities is ably placed to begin to answer these questions which explicitly call for a critical dialogue to take place between health, geography, the humanities and the arts. To begin to answer these questions about health and safety, we need a greater understanding of the health implications of nuclear accidents. This is, however, a difficult task, because current understandings are mediated by a scandalous lack of research into the health implications of exposure to low doses of radiation (Brown 2017; Kuchinskaya 2014). Therefore, we need health and environmental scientists to delve deeper into the impact on the body of exposure to low doses of radiation. We need the disciplines of geography and history to enable us to think about the spatial and temporal implications of nuclear power and EPZs. We also need poets, writers, performers, musicians and visual artists to reveal new ways of understanding nuclear power. In short, we need a multidisciplinary approach to enable us to challenge entrenched and dominant discourses that continue to reassure us, despite the evidence to the

[2] See http://www.envhumsouth.uct.ac.za/. Accessed 6 Feb 2019.

[3] During serious nuclear accidents, radioactive isotopes of iodide are released. Iodide prophylaxis tablets can reduce the accumulation of radioactive iodide in the thyroid, thus reducing the chances of developing thyroid cancer.

contrary, that nuclear power is safe and is not a threat to public health (Hilgartner et al. 1983). In this chapter, I want to provide an example of how the visual art of photography has been used to critically reveal the scandal of EPZs.

Of specific interest to this project, but generalizable to all EPZs, is the meaning of the EPZs that surround the KNPS. Given that much of the scholarship on nuclear power is focused on the United States, this analysis makes a useful contribution to work going on elsewhere, especially that coming from the so-called Global South (Alex-Martin, 2018: 863).

Before continuing, it is necessary to say something about the health dangers of nuclear power plants so as to avoid being accused of 'Radiophobia', that is suffering from an irrational fear of nuclear power. This is an accusation typically levelled by many scientists and those in positions of political power against non-scientists who question its safety (Mueller 2009; Ropeik 2016; Stawkowski 2017; Vegter 2018). Research into the immediate and longer-term health impacts of the Chernobyl disaster reveals how both international and national regulatory agencies 'expended a great deal of effort *not* to know about the health effects of the Chernobyl accident, to limit research and to contain judgements' (Brown 2017: 413; Kuchinskaya 2014; Goldstein 2017). Evidence shows that agencies from the United Nations and the World Health Organisation carried out what Kate Brown calls 'the science of political containment' which led to the 'social construction of ignorance' (Brown 2017: 414; Kuchinskaya 2014: 2). This 'containment' meant that only after overwhelming evidence was presented to them, did these agencies reluctantly accept the significant increase in childhood thyroid cancers that took place in Ukraine and Belarus after the disaster (Brown 2017: 432).[4]

They have not to this day, however, accepted that the millions of people exposed to low levels of ionising radiation have suffered, or will suffer, any negative health consequences because of the accident. This is despite the assertion by these agencies that 'it is impossible to assess reliably, with any precision, numbers of fatal cancers caused by radiation exposure due to the Chernobyl accident' (International Atomic Energy Agency Division of Public Information 2006: 7). This is also despite extensive evidence from Soviet doctors who diagnosed what has been described as 'chronic radiation syndrome' in the months and years after the accident (Brown 2017: 417–420).[5]

---

[4] In 2002, the Chernobyl Forum made up of agencies from the UN and WHO and the World Bank acknowledged that 4000 childhood thyroid cancer cases were the result of the accident at Chernobyl. See *Chernobyl's Legacy: Health, Environmental and Socio-Economic Impacts and Recommendations to the Governments of Belarus, the Russian Federation and Ukraine*, The Chernobyl Forum 2003–2005, International Atomic Energy Agency Division of Public Information, 2006, p. 7, http://hps.org/documents/chernobyl_legacy_booklet.pdf. Accessed 20 Jan 2019.

[5] The syndrome is 'defined as a complex of unspecific symptoms that could include malaise, headaches, lower work capacity, loss of appetite, sleepiness during the day, insomnia at night, bleeding gums, disorders of the liver, kidneys, thyroid, menstruation cycles, tonsillitis and chronic gastritis'. Gabrielle Hecht draws attention to how the 'permissible level' of exposure to radiation has constantly been reduced as evidence of the lack of understanding of the effect of low doses. Hecht (2006).

There is very little knowledge about what effect low-level doses of ionising radiation have on the body because no long-term systematic research has ever been undertaken. Donna Goldstein argues that this is a classic case of 'undone science', that is science that is too political or methodologically difficult or that fundamentally challenges established ways of knowing, to be undertaken (Goldstein 2017: 402).[6] There is no doubt that undertaking a large-scale epidemiological study of the impact of low-dose ionising radiation on a significant portion of a population that has been subjected to a nuclear accident would be an enormously complex and difficult undertaking. The level of knowledge production necessary would be significant, because radiation causes no unique illnesses that can be directly attributable to it and because of the potential time delay between exposure and illness (Brown 2017: 420).[7] That said, evidence indicates that political pressure was exerted after Chernobyl to prevent such studies from being undertaken (Brown 2017; Kuchinskaya 2014). If a definitive link were to be established between exposure to low-dose ionising radiation and ill-health, then this would bring into question the entire nuclear power industry and could burden governments with unlimited liability issues if an accident were to occur.

It is, sadly, safe to say that another nuclear accident like Chernobyl or Fukushima will occur, because as Charles Perrow argued in his seminal piece on the accident at Three Mile Island, they are 'normal accidents' which cannot be prevented since they are unanticipated or are the product of what has been deemed acceptable risk. Perrow states that it is 'not feasible to train, design or build in such a way as to anticipate all eventualities in complex systems where the parts are tightly coupled' (Perrow 1981: 17).[8] As such, when serious problems occur, they can become uncontrollable because they either exceed design parameters or operator knowledge. Perrow calls these types of accidents, 'discrete accidents' (Perrow 1981: 18). Evidence from the Three Mile Island and Chernobyl accidents indicate a mixture of operator error and design limitations. The response from the nuclear industry to these accidents is to assert that they will not happen again, because operators will be better trained and designs will improve (Perrow 1981; Plokhy 2018). This response, however, is a form of 'technophilia' – the belief that 'technology is best characterised by its potential for improvement rather than by its limits' – and is a classic example of the 'dogma of infallibility' (Goldstein 2017: 395; Beck 1995: 85). While we can learn from accidents that occur, we can improve designs and operator training; these types of accidents will happen again because they are the result of gaps in

6 She notes that 'scientists' and 'experts' 'appear to simply not have been looking for Chernobyl health effects beyond strictly delimited methodological and theoretical frameworks', p. 161. For a fuller understanding of the term 'undone science', see Hess (2016) .

7 The dominant medico-scientific discourse around the long-term health effects of Chernobyl remains one in which 'lifestyle' factors, such as poor nutrition or alcoholism, are given more credence in explaining ill-health in those exposed to low-dose ionising radiation. Brown (2017), p. 420.

8 He continues, 'it is the novel interactions, the unexpected, the unimagined, that form the basis of a normal accident', p. 25.

knowledge (ignorance). These accidents cannot be predicted as 'we don't know what we don't know' (Wynne 1992: 114).

Perrow draws attention to another form of accident that he calls the 'calculated risk' accident. As it is impossible to create a perfectly safe nuclear power plant, a 'calculated risk' accident is one that is considered so unlikely that it is a risk worth taking (Perrow 1981: 18). The disaster at Fukushima was a 'calculated risk' accident because designers and operators must have known that there was a risk that both reactor cooling systems could fail at the same time. The investigation committee established by the Japanese government after the accident stated that 'risk reduction should be tackled in a drastically different approach. A new approach to safety measures and disaster preparedness should be established for a disaster' (Investigation Committee on the Accident at the Fukushima Nuclear Power Stations of Tokyo Electric Power Company 2012: 3).[9] Despite this call for drastic changes, the response by authorities in Japan has been entirely predictable and entirely inadequate. Rather than question the wisdom of nuclear power, let alone the siting of nuclear power stations on the 'Ring of Fire' where 90% of the world's earthquakes occur, the response has been to build higher walls around Fukushima and other nuclear power stations located on Japan's east coast (Kawakhara 2018).[10]

The existence of EPZs is an acknowledgement that accidents do happen at nuclear power plants, accidents that, to varying degrees, will result in the release of radioactivity and subsequent radioactive contamination. The following section describes how this radioactive contamination is likely to 'behave' after a serious nuclear accident.

Radiation

Chernobyl and Fukushima demonstrate that radiation does not behave predictably; there is no linear relationship between the level of contamination and distance from the accident. The spread of radiation is influenced by the unpredictability of winds, the flow of rivers, the movement of groundwater, the tides of the sea, the absorption rate of animals and plants, the types of soil and a whole host of other variable factors. This results in entirely unpredictable and chaotic patterns of contamination. How else do we explain that some Welsh farmers could not farm their land for 26 years after the Chernobyl disaster, despite being located 2750 km from the

[9] In his 1981 piece, Perrow predicted the accident at Fukushima when he noted that 'It is easy to imagine the loss of both the main and backup power systems during an earthquake or even a storm'. Perrow (1981), p. 18.

[10] 'Sea wall to guard against tsunami completed at Hamaoka nuclear plant', no author, *Japan Times*, 31 March 2016, https://www.japantimes.co.jp/news/2016/03/31/national/sea-wall-guard-tsunami-completed-hamaoka-nuclear-plant/#.XFqOsKBS-Uk, 'Earthquake Glossary', *United States Geological Survey*, https://earthquake.usgs.gov/learn/glossary/?termID=150. Accessed 6 Feb 2019.

explosion? (Wynne 1992: 114; O'Brien 2012).[11] The relationship between dose and effect is also nonlinear; each plant, animal and person is affected differently (Brown 2013: 62–65; Adams 1998: 197–202). Radiation unbundles both the spatial and temporal relationships between exposure and harm because there is no ability to predict where harms will occur, what will to be harmed and when that harm will manifest itself. There is, for example, still no clear idea of how long the contaminated Chernobyl exclusion zone will have to remain in place; scientific estimates vary from 300 to 20,000 years (Plokhy 2018: 343; Saunders 2011). Radiation is, therefore, a form of environmental 'slow violence' in which 'casualties are postponed, often for generations' because 'the nuclear cannot be undone' (Nixon 2011: 2; Carpenter 2016a: 3).

Barbara Adams argues that in 'perforating spatial and temporal boundaries', radiation has 'outgrown' the 'limiting framework' of classical science rooted in the linear equations of cause and effect, expectations of proportionality, the habits of extraction and reduction and the search for simplicity (Adams 1998: 209). It is a phenomenon that cannot be contained in any traditional sense. Radiation is, in fact, a 'hyperobject'. A 'hyperobject', according to Tom Morton, is an object that exists, that we can think about, but that we cannot see or touch or properly comprehend or know all at once. It is an object that is globally distributed both spatially and temporally, which we can only experience locally as pieces of the larger whole (Morton 2014). It is, above all else, an object that challenges the assumptions of human mastery over all things in this age of the Capitalocene.[12]

Given that accidents will happen, and that the spread and impacts of radiation cannot be predicted in any meaningful way, what exactly are EPZs for? Specifically, what is the EPZ around the KNPS for?

KNPS Emergency Planning Zones

The KNPS has three different EPZs. There is the Precautionary Action Zone (PAZ) that extends 5 km around the station within which 'urgent protective actions in the event of a nuclear or radiological emergency to reduce the risk of severe deterministic health effects off the site' are supposed to take place. Then there is the Urgent Protective Action Planning Zone (UPZ) which extends in a circle from the boundary of the PAZ to 16 km from the plant where 'urgent protective actions in the event of a nuclear or radiological emergency to avert doses off the site in accordance with

[11] Scientists in the United Kingdom predicated that the radiation that fell onto Welsh and Cumbrian farms would only be a problem for a number of days as restrictions were imposed on sheep farmers. The last of these restrictions was removed in 2012. See, O'Brien, L. (2012). After 26 years, farms emerge from the cloud of Chernobyl. Independent, 1 June. https://www.independent.co.uk/news/uk/home-news/after-26-years-farms-emerge-from-the-cloud-of-chernobyl-7811615.html. Accessed 24 Jan 2019.

[12] Following Jason Moore's argument, I reject the idea of an Anthropocene, preferring instead his term, the Capitalocene. See Moore (2015).

international safety standards' will allegedly take place. Lastly, there is the Longer Term Protective Action Zone (LPZ) which extends around the plant to a distance of 80 km 'in which contingency plans and procedure are in place for taking effective protective actions to reduce the long term exposure due to deposited radionuclides in the event of a nuclear accident'.[13] These three EPZs form part of the Integrated Koeberg Nuclear Emergency Plan, which sets out exactly what is supposed to happen within the three zones if an accident is declared by KNPS managers. It includes information relating to the 'prompt notification of the public and for implementation of protective measures'.[14]

Implicit in the demarcation of EPZs is the assumption that KNPS staff will be able to identify and exert control over areas that are deemed unsafe or safe, clean or contaminated, that they will be able to 'avert doses off the site' and that they will be able to know the future. It also assumes that by taking appropriate action, mistakes can be undone and corrected. So it is that the language of 'health and safety' is cooperated by the technocratic managerial and scientific elites who run nuclear power stations. However, if we consider the unique qualities of radiation, the fallacy of this type of positivist thinking becomes apparent.

This scientific 'unknowability' of radiation clearly calls into question the purpose of EPZs. If the science on the utility of EPZs was settled, then we would expect there to be uniformity in their size and shape around nuclear power stations and in the 'actions' to be taken within them. This is, however, not the case. Research undertaken by the European Union after the Fukushima disaster indicates that there is no uniformity when it comes to the size of EPZs (including both PAZs and UPZs) amongst European Union member states or what actions should be taken within them. In terms of sheltering or evacuating citizens, the zones range in size from 1 to 30 km, while the issuing of iodide prophylaxis ranges from 1 to 100 km (European Commission 2013: 83–89). The same confusion prevails in the United States. Nuclear power stations in the United States are covered by two zones, one at 16 km and a second at 80 km, but nuclear power station managers differ on what actions they would recommend in the event of a serious accident. For example, in terms of evacuation, some managers would evacuate for 3 km around the plant, while others would do so for 5 km in a keyhole shape (called the 'plume exposure pathway') corresponding to prevailing wind conditions. Other managers reported that they would automatically evacuate for 16 km all around their plants (Adalja et al. 2014: 6–7). Some states routinely issue residents who live within the 16 km zone with potassium iodide tablets; other states only distribute tablets after an accident, while some do not issue tablets at all (Adalja et al. 2014: 9–10).[15]

For any EPZ, and the 'protective actions' to be taken within them, to be 'useful', it is clear that citizens who live within them need to be aware of emergency plans and what to do in the event of an accident. Research demonstrates that, in the United

---

[13] The Integrated Koeberg Nuclear Emergency Plan, KAA-811, Eskom, 21 April 2008, pp. 23–24.

[14] The Integrated Koeberg Nuclear Emergency Plan, KAA-811, Eskom, 21 April 2008, p. 22.

[15] It could be argued that the relative age and different designs of nuclear power stations explain these differences, but this argument is countered by Perrow's definition of a 'discrete accident'.

States, the Nuclear Regulatory Commission (NRC) has 'systematically' refused the public access to nuclear power plant emergency plans (Kyne and Bolin 2016: 13). In South Africa, a local non-profit organisation fought with relevant authorities for nearly 2 years using promotion of access to information legislation before it was granted access to the Integrated Koeberg Nuclear Emergency Plan (Koeberg Alert Alliance 2012). In initially refusing to release the plan Eskom, the state-owned operator of the KPNS, argued:

> We believe that the disclosure of this information could potentially endanger the life or physical safety of individuals. If emergency and evacuation plans fall into the wrong hands, it is possible that someone acting out of malice could take detrimental steps during an emergency. Such steps could possibly circumvent the effective implementation of the plan.[16]

This bizarre response considers the highly unlikely actions of 'someone acting out of malice' as more of a danger to residents living near the KNPS than those same residents not having access to emergency plans. The plan that was eventually released was supposed to be revised in 2014, but no updated plan has been made available to the public. In the absence of the full plan, the residents of Cape Town can access, assuming they are privileged enough to have an Internet connection, a copy of a ten-page leaflet called 'What to do in case of an Emergency at Koeberg?' This leaflet sets out the basic steps that residents need to take in the event of an emergency.[17]

In 2015, research was undertaken within the KNPS UPZ to see how prepared residents were for a serious accident. The research found 'that the majority of interviewed residents had very poor knowledge concerning emergency procedures, and few had made any effort to prepare themselves' (Francioli 2016: 1). Almost half of the residents interviewed were unaware that 'general emergency-related information' was available (Francioli 2016: 7). Research from Canada, Brazil and the United States has found similar low levels of public awareness when it comes to what to do in the event of an emergency (McClenaghan and Blaise 2017; Blando et al. 2007; Adalja et al. 2014; Goldstein 2017).

Given the 'unknowability' of radiation and the inconsistencies in the size of EPZs and the actions to be taken within them, it would be prudent to minimise the number of people who live within them or on their boundaries. IAEA guidelines include three key issues when considering the siting of a nuclear power station: the vulnerability of the site to natural disasters; the characteristics of the site in terms of the transfer of radioactive material during an accident; and the population density

[16] Letter from E. Laubscher, National Deputy Information Officer, Eskom to J. Kruuse, 22 Nov. 2011, letter available on the Koeberg Alert Alliance website. See https://koebergalert.files.wordpress.com/2012/01/response-paia-10238.pdf. Accessed 26 Jan 2019.

[17] 'What to do in case of an Emergency at Koeberg?, *Eskom* www.eskom.co.za/Whatweredoing/ElectricityGeneration/KoebergNuclearPowerStation/EmergencyPlanning/Pages/Emergency_Planning.aspx. Accessed 26 Jan 2019. The City of Cape Town has its own plan, the Koeberg Nuclear Power Station Radiological Release Hazard Disaster Risk Management Plan, in the event of a serious accident at KNPS. This document is, however, not available to the public. Author email communication with Dr. Elmien Steyn Head: Special Planning & Critical Infrastructure, City of Cape Town, 21 July 2017.

and distribution around a site. The guidelines state 'the surrounding demographic setting and dispersion characteristics should be such as to limit the exposure of the population for any plant state to as low as reasonably achievable'. By 'plant state' the guidelines refer to either the normal day-to-day operations of the plant or the plant during an accident. The guidelines note that population density and distribution around any specific site should not get to the point where they may hamper the implementation of emergency plans. Lastly, they indicate that population projections should be used to inform the suitability of a site (IAEA 2015: 10–22).

What is revealing about these guidelines is that they do not specify what a permissible population density or distribution around a nuclear power station is. In the United States, so-called Low Population Zones are supposed to exist around nuclear power stations. However, much like the IAEA guidelines, the US NRC does not define what constitutes a low population (Kyne and Bolin 2016, 6). Between 2000 and 2010, the population of Americans living within 16 km zones around nuclear power stations rose by 17%, significantly more than the national population rise of 10%. In 2010, just over four million lived within 16 km, while nearly one million Americans lived within 8 km of a nuclear power station. According to 2010 census data, some 88 million Americans, nearly 30% of the total population, now live within an 'exposure pathway' (80 km) of at least one nuclear power station (Kyne and Bolin 2016: 4).[18]

In regard to the KNPS, the South African Atomic Energy Board (AEB) proposed in 1970 that no densely populated areas should be within 15 miles (24 km) of the proposed plant. This suggestion was met with consternation from both Western Cape Provincial government and City of Cape Town officials who argued that a significant part of Northern Cape Town was already within 15 miles of the proposed Koeberg site. In 1972, the AEB and provincial and city authorities reached a compromise, and a 10 mile (16 km) zone of restricted development was agreed upon.[19] Despite supposedly being an area of restricted development, in 2011, some 132,000 people lived within the 16 km EPZ, and the most recent information indicates that this number has now grown to 160,000.[20] The closest municipal ward to the KNPS, Ward 23, has seen its population grow from 19,676 in 2001 to 33,448 in 2011, an increase of 70% (City of Cape Town 2011). Development continues apace in the area. In late 2017, the largest regional shopping mall in the Western Cape Province opened within the 16 km EPZ, and significant residential development is planned within the vicinity of the mall.[21] In 1972, when the population of Cape Town was

[18] 'Nuclear neighbours: Population rises near US reactors', no author, *MSNBC*, 14 April 2011, http://www.nbcnews.com/id/42555888/ns/us_news-life/t/nuclear-neighbors-population-rises-near-us-reactors/#.XEmgqmlS-Uk. Accessed 22 Jan 2019.

[19] Report of Committee on Proposed Development Control, 21 May 1970. National Archives of South Africa – Cape Town Archives Repository (KAB) – PAA/LEER/1640 – AF20/7/679 Cape. Cape Town Municipality – Town Planning Scheme Applications – Proposed Development Control around Nuclear Power Station at Duynefontein, 1970–1976.

[20] Email correspondence between author and City of Cape Town officials, 1 & 2 August 2017.

[21] 'Zenprop's Flagship Table Bay Mall Opens', no author, *Propertywheel.co.za*, 28 Sept. 2019, https://propertywheel.co.za/2017/09/zenprops-flagship-table-bay-mall-opens/. Accessed 24 Jan

approximately 1.2 million, the then Chief Medical Officer of Health for the City, Dr. R. Coogan, accused the AEB of 'absolute naivety' in siting a nuclear power station so close to the city (Bamford 2006). Approximately 4.2 million now live in the city.[22]

Research amongst residents living within the KNPS 16 km EPZ revealed that most residents felt either 'safe' or 'neutral' about living near to the power station. However, these residents shared 'an almost fatalistic perception' that if a serious accident occurred, they would not be able to evacuate in time due to the perceived inadequacy of transport routes given the size of the population now living in the EPZ. At least a quarter of respondents erroneously believed that a serious accident at Koeberg would result in a nuclear bomb type of explosion and there would be no need to evacuate because everyone would instantly die in the EPZ (Francioli 2016, 6 and 8). What this research demonstrates is that even those residents who feel that KNPS is 'safe', have little faith in the ability of the EPZ to offer a meaningful way to escape an accident if one were to occur.[23]

Risk

While radiation is something that we can utilise (we can extract energy from it), it is clear that we are not 'in control of the possible relations that arise from it' (Volkmar 2016, 50). This is what is unwittingly revealed in the inconsistencies in the size of EPZs, the actions to be taken, or not, within them and in the public's lack of knowledge of what to do in the case of a serious accident. It is clear that the zones and the 'actions' that may take place within them are not designed to offer genuine solutions to the dangers of a serious nuclear accident. If this were the case, no one

2019. Regulations state that new developments in the 16 km zone can only be approved if they do not compromise the realisation of the Integrated Koeberg Nuclear Emergency Plan. This plan states that residents should be evacuated from the 5 km zone within 4 hours and from the 16 km zone within 16 hours. A Traffic Evaluation Model has to be produced to demonstrate that these evacuation times can be met. The last time the City of Cape Town undertook such an evaluation was in 2007 and was based on population data from 2000, clearly making such an evaluation out-of-date. It is not known that additional evaluations have taken place since 2007. See Draft regulations on development in the formal emergency planning zone of the Koeberg nuclear power station to ensure effective implementation of the Koeberg Nuclear Emergency Plan', *Department of Energy*, 29 Oct 2010. It is important to note that these regulations remain in draft form & Jones et al. (2007).

[22] World Population Review, Cape Town, 22 Oct. 2018, http://worldpopulationreview.com/world-cities/cape-town-population/ & Socio-Economic Profile, City of Cape Town, Western Cape Government, 2017, https://www.westerncape.gov.za/assets/departments/treasury/Documents/Socio-economic-profiles/2017/city_of_cape_town_2017_socio-economic_profile_sep-lg_-_26_january_2018.pdf. Accessed 27 Jan 2019.

[23] The ignorance concerning how a nuclear accident would unfold further demonstrates that Eskom's emergency preparedness information is not reaching the very people it is supposed to assist.

would be living within at least 30 km of a nuclear power station which is the extent of the exclusion zone (or 'zone of alienation' as it is more appropriately known in Russian) around Chernobyl. EPZs appear to be no more than a form of necessary social ritual designed to create the illusion of something being done about the dangers inherent in locating nuclear power stations and residential populations so close to each other (Bloomfield and Vurudubakis 2005: 740). EPZs deceive by falsely promising certainty and order, 'sanitary beginnings and endings', despite the uncertain and messy spatial and temporal impacts that will 'elude rhetorical clean-up operations' if a serious accident occurs (Nixon 2011: 8).

Nuclear power stations located close to population centres are tolerated because we live in a 'risk society' where the scale of the repercussions of industrial accidents have become 'large-scale hazards of civilisation' (Beck 1995: 76). These are hazards which are insusceptible to compensation or remediation but are allowed to exist by authorities and policy-makers who claim that the risks themselves are worth enduring because of the benefits which are claimed to flow from them (Beck 1995: 11). In this sense, people become the externalities of nuclear power, the nuclear industry and energy consumption.

Ironically, the very 'unknowability' of radiation facilitates the acceptance of the risks of nuclear power. While radiation exists as a phenomenon, Gabrielle Hecht asserts that its visibility is dependent upon 'a technopolitical phenomenon' (Hecht 2012: 15). She argues that 'places, spaces, objects or hazards get designated as 'nuclear' due to the complex interplay of history and geography, science and technology, bodies and politics, radiation and race, states and capitalism'. This she refers to as the 'nuclearity' of things or places (Hecht 2012: 14–17; Hecht 2016: 127).

'Nuclearity' has serious consequences for how we understand the role of EPZs, because 'scientists' and 'experts' (those with the equipment necessary to detect radiation) are the only ones able to declare what is unsafe/contaminated or safe/uncontaminated, to declare the 'nuclearity' of an EPZ. This 'power to detect' is professionalised and state-driven, leaving ordinary people vulnerable to manipulation by those looking to minimise the dangers of serious nuclear accidents (Davies 2013: 129).[24] Thus, the creation and stated purposes of an EPZ at any time, before, during or after an accidents, emerge from the social relations where knowledge gets produced. As Ele Carpenter argues, 'the invisibility of radiation is a political project as well as an optical one' (Carpenter 2016b: 15). Evidence from both Chernobyl and Fukushima show how political decisions get taken to reinscribe landscapes from 'unsafe' to 'safe' by expediently changing the definitions of what constituted 'safe' exposure (Kuchinskaya 2012: 91).[25] Thus, by a simple trick, the problem of radioactive contamination can disappear.

[24] Adams notes 'this technologically constituted reality affect the sovereignty of the self, family, nation, species, earth' (Adams 1998: 199).

[25] By changing such definitions, you therefore 'control' the scale of the hazard and any associated liabilities.

The question of the existence of nuclear power plants in populated areas is inexorably linked to the location of political power in society. Who are the 'experts' and 'scientists', shielded by state bureaucracies and private corporations, who calculate what risks are tolerable and which are not? Are they open and transparent in their calculations? What are their motivations? More importantly, how do we, as concerned residents, create new narratives that challenge the assumptions of those who tell us that nuclear power is safe? How do we engage the representational challenges of the relative invisibility of slow violence and the actual invisibility of radiation? (Nixon 2011: 2) I have chosen to respond to Nixon's challenge to 'devise arresting stories, images and symbols adequate to the pervasive but elusive violence of delayed effect' by undertaking a photographic project in and around the KNPS (Nixon 2011: 3).

Photography

The relationship between photography and nuclear power cannot be divorced from the relationship between photography and nuclear weapons, for it was with the invention of nuclear weapons that photography had its first entanglement with the splitting of the atom, with 'nuclearity'. Since that first engagement in 1945, three distinct phases in the relationship can be identified. In brief, phase one, the 'atomic sublime', from 1945 to 1963, is characterised by an aesthetic ecstasy that accompanied the photographic reproduction of atomic detonations (Hales 1991; Kirsch 1997).[26] Susan Schuppli describes this relationship as a period of 'image rapture' as observers and photographers became mesmerised by the spectacle of the explosions (Schuppli 2015: 280). Phase two is characterised by a 'toxic sublime' aesthetic where photographers made pilgrimages to previous test sites and recorded the devastated landscapes they found there (Goin 1991; Misrach 1992). Lastly, there is the phase of the 'apocalyptic sublime', which is characterised by accidents and contemplation as photographers have responded to the disasters at Chernobyl and Fukushima.[27]

Much of the work in this latter phase is characterised by a morbid fascination with landscapes devoid of human life that are being reclaimed by 'nature'. As such, it reflects the current zeitgeist of 'nature' unravelling the conceit of modernity (White 2012). Not all photographers, however, have adopted this particular trope. Roland Barthes stated that 'photography is subversive not when it frightens, repels or even stigmatises, but when it is pensive, when it thinks' (Barthes 2000: 38). Since the 1980s, a number of photographers have produced more 'pensive' thoughtful work. Amongst this work are Carol Gallagher's and Ed Ou's photographs of bodies as landscapes which have been scarred by radiation exposure; Miyako Ishiuchi's

[26] After the 1963 Partial Test Ban Treaty, it was no longer permitted to conduct atmospheric nuclear weapons tests.

[27] For a more detailed account of these three phases, see Overy (2018).

photographs of clothes and personal items worn by victims of the Hiroshima attack; Jeremy Bolen's work which exposes film to radioactive soil; and Tomoki Imai's images of the boundary of the Fukushima exclusion zone (Cotter 2014; Gallagher 1993; Kawashima 2013; Ou 2009; Picard 2013).

The strength of these works is that they clearly and unequivocally incriminate other events, because of their obvious narrative power. They oblige viewers to ask: why are these bodies scarred? Why is there radiation in the soil? John Berger argued that the appearances of an event being photographed implicate other events (Berger 2013: 91).[28] This ability of a photograph to implicate other events is central to my photographic practice within the KNPS EPZ.[29] Through photography, I attempt to compel those viewing my images to make sense of what they are seeing by finding additional information, to seek explanations, however redundant they may be, for what they are seeing. The project is hampered, of course, by the inability to photograph radiation. My solution to this problem is to photograph objects and places of 'nuclearity' within the KNPS EPZ. It is these manifestations of 'nuclearity' that implicate the radiation that would inhabit them were an accident to occur.

Part of my practice has been to locate the spatial boundaries of the 16 km EPZ and photograph these spaces to the north, east and south of the KNPS.[30] In showing the extent of the EPZ on a map in comparison to the exclusion zones around Fukushima and Chernobyl, the ineffectiveness of the EPZ is implicated if an accident similar to those that took place at Fukushima or Chernobyl were to occur (Fig. 13.1).

In visualising the boundaries of the EPZ, sometimes intersecting blocks of residential flats, the work implicates the completely arbitrary nature of the EPZ (Photo No. 2) (Fig. 13.2).

In photographing the KNPS evacuation sirens in heavily populated areas, I am implicating the decisions around risk that have been taken by relevant authorities. In this instance, the 'unknowability' of radiation is also implicated by the same image, by labelling it 'before' and 'after' (Figs. 13.3 and 13.4).[31]

In addition to using a modern digital SLR camera, I made use of two different types of cameras to engage with the temporal aspects of nuclear power. I used an instant camera to implicate how quickly the 'safe' can become 'unsafe' and the 'clean' can become 'contaminated' (Fig. 13.5).

I made use of a pinhole camera, one of the oldest forms of photography, to implicate the unknowability of how long contamination will last after a serious nuclear accident. The imprecise, unclear and uncertain nature of pinhole photography, as seen in the photograph below, mirrors the imprecision with which predications are made about how long contamination will last (Fig. 13.6).

---

[28] Sontag argued that 'the camera record incriminates' (Sontag 2008, 5).

[29] This chapter contains only a fraction of the photographic archive I have created of the KNPS, the EPZ and 'nuclearity' in and around the City of Cape Town.

[30] The western edge of the EPZ is 16 km into the southern Atlantic Ocean, which I hope to one day be able to photograph.

[31] There are 80 evacuation sirens situated within the 16 EPZs.

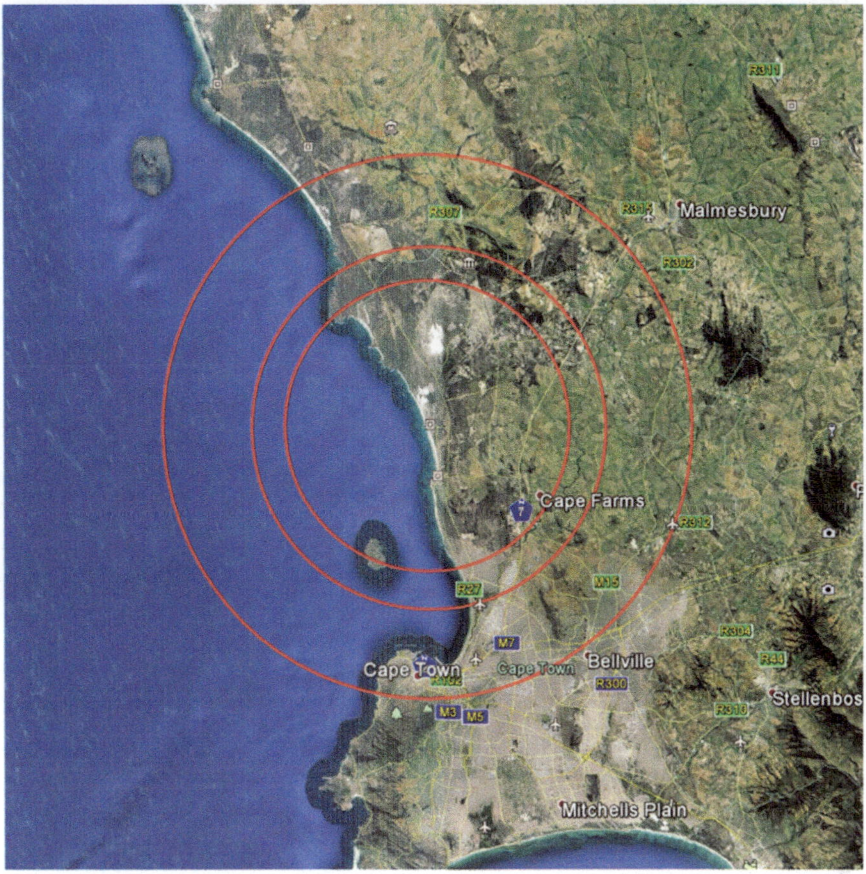

Fig. 13.1 The 16 km 'Urgent Protective Action Planning Zone' compared to 20 km exclusion zone placed around the Fukushima Nuclear Power Station after the 2011 accident and the 30 km exclusion zone placed around the Chernobyl Nuclear Power Station after the 1986 accident. (Google Earth Pro. 2017)

A number of images were also taken with a pair of old lenses from the 1960s that contain a radioactive coating made of thorium oxide, including the image below (Fig. 13.7).[32] By photographing through a radioactive lens, I implicate the pervasiveness of 'nuclearity' in our lives. Despite knowing that thorium oxide is radioactive, the extent of scientific knowledge in the 1960s coupled with calculations of risk decreed that it was safe to use in everyday objects like camera lenses and even eyepieces.

A number of images have text overlaid onto them. In the example below (Fig. 13.8), texts from minutes from City of Cape Town planning meetings in 1970

[32] Thorium oxide was used to improve the optical performance of the lenses. Its use was halted in the 1970s due to fears over health concerns and because the thorium oxide deteriorated over time negatively affecting image quality. Walker (2018).

Fig. 13.2 Boundary of 16 km 'Urgent Protective Action Planning Zone' south of Koeberg (Instant camera. 2017)

Fig. 13.3 Cape Town suburb within the 16 km 'Urgent Protective Action Planning Zone' *before* a major release of radioactive material following an accident at Koeberg Nuclear Power Station, Otto De Plessis Drive, Milnerton (Digital SLR camera. 2017)

Fig. 13.4 Cape Town suburb within the 16 km 'Urgent Protective Action Planning Zone' *after* a major release of radioactive material following an accident at Koeberg Nuclear Power Station, Otto De Plessis Drive, Milnerton (Digital SLR camera. 2017)

are placed over recent residential developments. The minutes relate to a meeting that specifically discussed the need to restrict development within the EPZ. These particular images implicate the risks that continue to be taken when it comes to residential development near nuclear power stations.

Multi-, inter- and transdisciplinary fields like Environmental Humanities, Medical Humanities and GeoHumanities create opportunities to interrogate, research and represent topics in new ways. A multidisciplinary approach enabled me to combine a number of disciplines to provide a written and visual account of the largely rhetorical role that EPZs play. Having the freedom to draw on the analytical and explanatory power of a number of different disciplines and practices, I was able to create not only an academic text but a collection of images which will find their way beyond the academy and published texts. At least two exhibitions of the photographs that formed part of this project have already been confirmed, and tentative discussions have begun for the work to be displayed in other parts of the Global South grappling with the problem of nuclear power. It is through the dialogue afforded by collaborative, challenging fields like Environmental Humanities and GeoHumanities that wider debates about the very real and urgent perils of nuclear power can be deepened. This chapter is offered as a contribution to conversations that may take place about nuclear power, methodologies, collaborations, creative process and other discussion points in GeoHumanities.

Fig. 13.5 Road within the 16 km 'Urgent Protective Action Planning Zone' (Instant camera. 2017)

Conclusion

Radiation has undone rational science. It exposes the inadequacy of the traditional dualisms of cause and effect, safe or unsafe, both spatially and temporally. Radioactive contamination has no neat narrative arc of accident, clean-up and return to normality because it is complex, messy and chaotic. Radiation does not spread out from accident sites in neat circles as implied by EPZs. This reduces the science of radiation 'management' to the science of modelling, predictions, estimates, theories and probabilities; in short, it becomes a science of guesswork. While the dangers of nuclear power are real, the risks are socially constructed and, as such, the

Fig. 13.6 Koeberg Nuclear Power Station (Pinhole camera. 2017)

Fig. 13.7 Boundary fence, Koeberg Nuclear Power Station (35 mm Thorium lens. 2017)

arts and humanities are critically important as a dialogic tool.[33] The GeoHumanities are ably placed to be able to bring specific understandings of spatial and temporal considerations into wider humanities, social science, science, health sciences and artistic discourses. However, while the interdisciplinary spaces that are being

[33] Beck calls them 'constructions of non-hazardousness' (Beck 1995, 86).

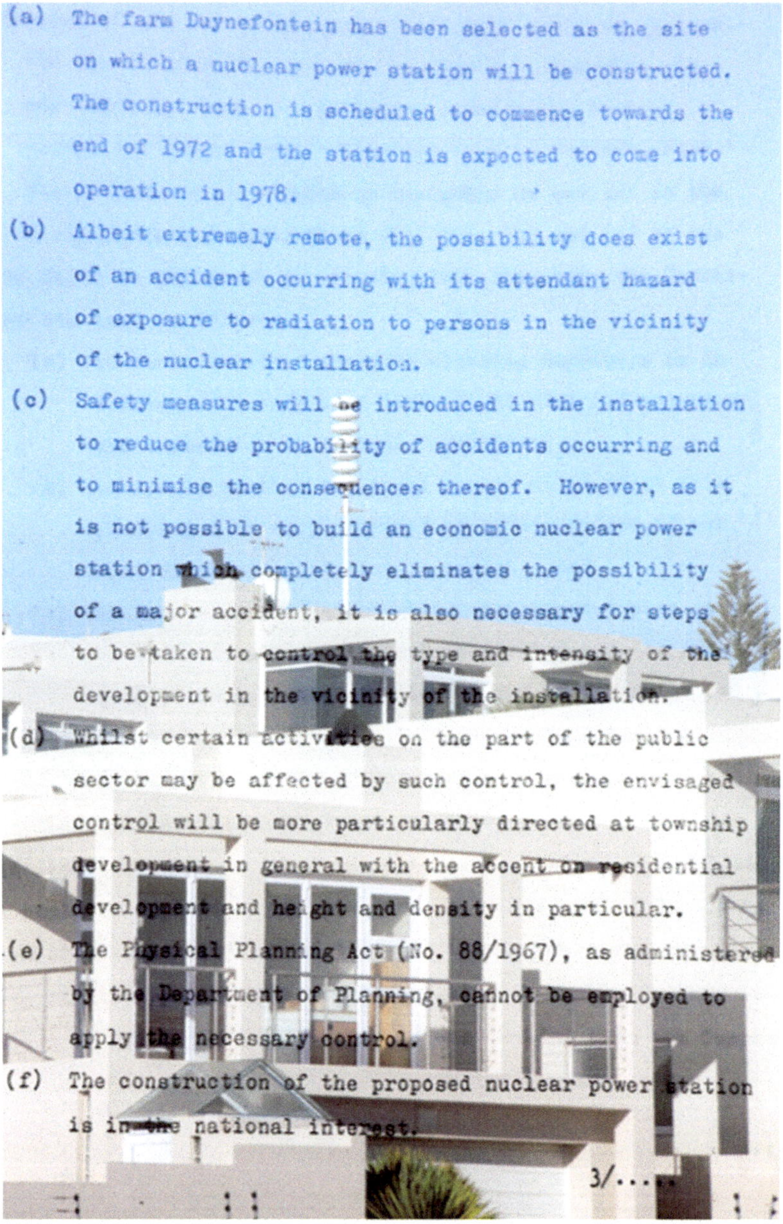

(a) The farm Duynefontein has been selected as the site on which a nuclear power station will be constructed. The construction is scheduled to commence towards the end of 1972 and the station is expected to come into operation in 1978.

(b) Albeit extremely remote, the possibility does exist of an accident occurring with its attendant hazard of exposure to radiation to persons in the vicinity of the nuclear installation.

(c) Safety measures will be introduced in the installation to reduce the probability of accidents occurring and to minimise the consequences thereof. However, as it is not possible to build an economic nuclear power station which completely eliminates the possibility of a major accident, it is also necessary for steps to be taken to control the type and intensity of the development in the vicinity of the installation.

(d) Whilst certain activities on the part of the public sector may be affected by such control, the envisaged control will be more particularly directed at township development in general with the accent on residential development and height and density in particular.

(e) The Physical Planning Act (No. 88/1967), as administered by the Department of Planning, cannot be employed to apply the necessary control.

(f) The construction of the proposed nuclear power station is in the national interest.

3/....

Fig. 13.8 Extensive suburban development within the 16 km ‘Urgent Protective Action Planning Zone’ with 1970 declaration to restrict residential development inside the 16 km zone (Digital SLR camera. 2017)

opened up should be welcomed, the danger of working in slightly expanded silos remains, and constant attention needs to be paid to ensuring communication between, for example, Medical and Health Humanities, Environmental Humanities and GeoHumanities. These broader fields need also to engage with sub-fields and discipline, in respect of the concerns of this chapter, for example, these would include the emergent fields of Nuclear Humanities and Nuclear Geography. Caution must be exercised in creating these new fields of study to ensure that disciplinary boundaries and prejudices do not re-emerge and that epistemic generosity is fostered.[34]

That said, there is a host of future research undertakings that GeoHumanities could seek to answer in regard to EPZs and nuclear power more generally:

How do we think about the spatial and temporal implications and meanings of EPZs around high-level nuclear waste depositories that are being built, or are being planned, around the world that must keep high-level radiation waste 'safe' for hundreds of thousands of years?

Are such repositories nothing more than grand experiments with no known end? (Bloomfield and Vurudubakis, 2005: 740).

While scholarship has contemplated the return of 'nature' to the nuclear testing wastelands of the American West and within the Chernobyl exclusion zone, little consideration has been given to the meaning of the nature reserves that typically surround nuclear power plants?[35] What spatial, political and psychological purposes do they serve?

Are the health impacts of a nuclear accident simply too significant, both spatially and temporally, for any government to properly address? Are they so great that they must be denied? (Kuchinskaya 2014:, 162).

Considering the animal turn, what can we say about animals, both 'wild' and domesticated that live within EPZs?

To break the 'technopolitical' monopoly on 'seeing' radiation, is there a role for residents within EPZs to become 'citizen scientists', copying the work of citizen volunteers who take regular radiation readings as part of the Safecast programme after the Fukushima disaster?[36]

What is the role of residents who live in and around EPZs to counter the dominant medico-scientific discourse about the health impacts of radiation with differing, experiential 'regimes of truth'?

Exactly who is it that lives within and around EPZs? Research undertaken recently in the United States shows that a disproportionate number of 'people of colour' live in EPZs rather than outside of them (Kyne and Bolin 2016: 4). Is this the case elsewhere in the world? What are the 'raced', classed and gendered aspects of EPZs?

---

[34] For more on epistemic generosity, see Tsampiras and Müller (2018).

[35] See Cram (2016). The KNPS is surrounded by the Koeberg Nature Reserve. See Koeberg Nature Reserve, Eskom, http://www.eskom.co.za/Whatweredoing/ElectricityGeneration/KoebergNuclearPowerStation/Pages/Nature_Reserves.aspx. Accessed 29 Jan 2019.

[36] See 'About Safecast', https://blog.safecast.org/about/. Accessed 29 Jan 2019.

Are EPZs 'national sacrifice zones' in waiting?

Scholarship has not considered the spatial and temporal health impacts of naval nuclear accidents to both humans and marine life. Several nuclear powered submarines remain at the bottom of the ocean, with nuclear fuel and weapons on board (Eriksen 1990).

Little has been written about the creation of diasporas after major nuclear accidents. What unique spatial and temporal insights can GeoHumanities bring to the study of these diasporas?

Considerations of the ways in which electricity is generated (nuclear, renewables, coal, etc.) offer a useful point from which to ask a set of broader questions concerned with how geographical, economic and political factors influence each other to determine who it is that has access to electricity. Questions relating to energy consumption, energy justice and climate justice can only be properly answered through multidisciplinary analysis.

In Sontag's 'Imagination of Disaster', she states that we now live in a world of 'unremitting banality and inconceivable terror' (Sontag 1965, 42). She argued that the solution to these terrors was 'fantasy' which acted 'to normalize what is psychologically unbearable' (Sontag 1965: 42). This seems to be an apt description of an EPZ. A banal, imaginary landscape brought to life by lines drawn on maps, enforced by road signs, and described in emergency plans designed to 'distract us from terrors' (Sontag 1965: 42). A place of predictability and tedium, but one that is always 'in a state of constant becoming' as 'the probability of improbable accidents increases' within the zones of dissonance and deceit that surround nuclear power stations (Alex-Martin, 2017: 4; Beck 2008).[37]

References

Adalja, A., Kirk Sell, T., Ravi, S., Minton, K., & Morhard, R. (2014). Emergency preparedness in the 10-mile emergency planning zone surrounding nuclear power plants. *Journal of Homeland Security and Emergency Management, 12*(1). https://doi.org/10.1515/jhsem-2013-0106. Accessed 22 Mar 2019.

Adams, B. (1998). *Timescapes of modernity: The environment and invisible hazards*. London: Routledge.

Alexievich, S. (2006). *Voices from Chernobyl: The oral history of a nuclear disaster*. New York: Picador.

Alexis-Martin, B., & Davies, T. (2017). Towards nuclear geography: Zones, bodies, and communities. *Geography Compass, 11*. https://doi.org/10.1111/gec3.12325. Accessed 22 Mar 2019.

Asaba, B. (2018). Kepco holds nuclear roadshow in Saudi Arabia. *Arabianindustry.com*, 22 Oct. https://www.arabianindustry.com/utilities/news/2018/oct/22/kepco-holds-nuclear-roadshow-in-saudi-arabia-5993221/. Accessed 19 Jan 2019.

[37] Hecht notes that the nuclear industry has always asserted the 'banality' of nuclear power (Hecht 2012, 10).

Bamford, H. (2006). Koeberg: SA's ill-stared nuclear power plant. *Independent Online*, 11 Mar. https://www.iol.co.za/news/politics/koeberg-sas-ill-starred-nuclear-power-plant-269096. Accessed 26 Jan 2019.

Barthes, R. (2000). *Camera lucida*. London: Vintage.

Beck, U. (1995). *Ecological politics in the age of risk*. Cambridge: Polity Press.

Beck, U. (2008). All aboard the nuclear power superjet. Just don't ask about the landing strip. *The Guardian*, 17 July. https://www.theguardian.com/commentisfree/2008/jul/17/nuclearpower.climatechange. Accessed 21 Jan 2019.

Berger, J. (2013). *Understanding a photograph*. London: Penguin.

Blando, J., Robertson, C., & Bresnitz, E. (2007). Evaluation of potassium iodide prophylaxis knowledge and nuclear emergency preparedness: New Jersey 2005. *Journal of American Public Health, 97*, S100–S102.

Bloomfield, B., & Vurudubakis, T. (2005). The secret of Yucca Mountain: Reflections on an object in extremis. *Environment and Planning D, 23*, 735–756.

Brown, K. (2013). *Plutopia: Nuclear families, atomic cities and the great soviet and American plutonium disasters*. Oxford: Oxford University Press.

Brown, K. (2017). Blinkered science: Why we know so little about Chernobyl's health effects. *Culture, Theory and Critique, 58*, 413–434.

Carpenter, E. (2016a). The smoke of nuclear modernity drifts through the Anthropocene. In H. Grove-White (Ed.), *Power in the land* (pp. 17–29). Wales: X-10.

Carpenter, E. (Ed.). (2016b). *The nuclear culture source book*. London: Black Dog Publishing.

City of Cape Town. (2011). City of Cape Town – 2011 Census – Ward 023, January 2013. http://resource.capetown.gov.za/documentcentre/Documents/Maps%20and%20statistics/2011_Census_CT_Ward_023_Profile.pdf. Accessed 26 Jan 2019. The populations of Wards 29 and 32 to north and northeast of the KNPS grew by 28% and 20% respectively over the same period from a combined total of 65 207 in 2001 to 80 699 in 2011. http://resource.capetown.gov.za/documentcentre/Documents/Maps%20and%20statistics/2011_Census_CT_Ward_029_Profile.pdf & http://resource.capetown.gov.za/documentcentre/Documents/Maps%20and%20statistics/2011_Census_CT_Ward_032_Profile.pdf. Accessed 27 Jan 2019.

Cotter, H. (2014). Ishiuchi Miyako: Here and now: atomic bomb artefacts, Hiroshima 1945/2007. *New York Times*, 16 Oct. https://www.nytimes.com/2014/10/17/arts/design/ishiuchi-miyako-here-and-now-atomic-bomb-artifacts-hiroshima-1945-2007.html. Accessed 18 Jan 2019.

Cram, S. (2016). Wild and scenic wasteland: Conservation politics in the nuclear wilderness. *Environmental Humanities, 7*, 89–105.

Dados, N., & Connell, R. (2012). The Global South. *Contexts, 11*(1)., Winter.

Davies, T. (2013). A visual geography of chernobyl: Double exposure. *International Labour and Working-Class History, 84*, 116–139.

Ecomodernist Manifesto. (2015). http://www.ecomodernism.org/. Accessed 18 Jan 2019.

Eriksen, V. (1990). *Sunken nuclear submarines*. Oxford: Oxford University Press.

European Commission. (2013). Review of the current off-site nuclear emergency preparedness and response arrangements in EU member states and neighbouring countries. https://ec.europa.eu/energy/sites/ener/files/documents/2014_nep_epr_review_2012-474_main_0.pdf. Accessed 21 Jan 2019.

Fig, D. (2005). *Uranium road: Questioning South Africa's nuclear direction*. Johannesburg: Jacana.

Francioli, A. P. (2016). Investigating the knowledge and preparedness of proximal residents to a general-emergency event at the Koeberg Nuclear Power Station. *Jamba: Journal of Disaster Risk Studies, 8*(2). https://doi.org/10.4102/jamba.v8i2.208.

Gallagher, C. (1993). *American ground zero: The secret nuclear war*. Cambridge: MIT Press.

Goin, P. (1991). *Nuclear landscapes: Creating the north American landscape*. Baltimore: Johns Hopkins University Press.

Goldstein, D. (2017). Fukushima in Brazil: Undone science, technophilia and epistemic murk. *Culture, Theory and Society, 58*, 391–412.

Hales, P. (1991). The atomic sublime. *American Studies, 32*, 5–31.
Hecht, G. (2006). Nuclear ontologies. *Constellations, 13*(2), 328.
Hecht, G. (2012). *Being nuclear: Africans and the global uranium trade*. Johannesburg: Wits University Press.
Hecht, G. (2016). Nuclearity. In E. Carpenter (Ed.), *The nuclear culture source book*. London: Black Dog Publishing.
Hess, D. (2016). *Undone science – social movements, mobilized publics, and industrial transitions*. Cambridge: MIT Press.
Hilgartner, S., Bell, R., & O'Conner, R. (1983). *Nukespeak: The selling of nuclear technology in America*. New York: Penguin.
IAEA. (2006). *Chernobyl's legacy: health, environmental and socio-economic impacts and recommendations to the governments of Belarus, the Russian Federation and Ukraine*. The Chernobyl Forum 2003–2005, Division of Public Information. http://hps.org/documents/chernobyl_legacy_booklet.pdf. Accessed 20 Jan 2019.
IAEA. (2015). *Site survey and site selection for nuclear installations. Specific safety guide no. SSG-35*. International Atomic Energy Agency.
IAEA. (2019). Under construction reactors. International Atomic Energy Agency. https://pris.iaea.org/pris/worldstatistics/underconstructionreactorsbycountry.aspx. Accessed 20 Jan 2019.
Investigation Committee on the Accident at the Fukushima Nuclear Power Stations of Tokyo Electric Power Company. (2012). *Final report on the accident at the Fukushima nuclear power stations of Tokyo Electric Power Company: Recommendations*. 23 July. https://www.nirs.org/wp-content/uploads/fukushima/SaishyuRecommendation.pdf. Accessed 21 Jan 2019.
IPCC. (2018). *Global warning of 1.5%*. Intergovernmental Panel on Climate Change. https://www.ipcc.ch/sr15/. Accessed 21 Jan 2019.
Jones, J., Naude, S., van Wyngaardt, G., & Marks, A. (2007) *Koeberg Nuclear Emergency Plan: Traffic Evaluation Model*, p. 574. https://repository.up.ac.za/bitstream/handle/2263/5908/021.pdf. Accessed 26 Jan 2019.
Jostow, P., & Parson, J. (2012). The future of nuclear power after Fukushima. *Economics of Energy & Environmental Policy, 1*(2).
Kawakhara, C. (2018). In shift in stance, TEPCO to build extra sea wall at Fukushima. *The Asahi Shimbun*, 15 September. http://www.asahi.com/ajw/articles/AJ201809150033.html. Accessed 21 Jan 2019.
Kawashima, K. (2013). Fukushima photographs that focus on what can't be seen. *The Japan Times*, 19 May. https://www.japantimes.co.jp/life/2013/05/19/general/fukushima-photos-focus-on-what-cant-be-seen/#.XFK2_GlS-Uk. Accessed 31 Jan 2019.
Kirsch, S. (1997). Watching the bombs go off: Photography, nuclear landscapes and spectator democracy. *Antipode, 29*, 227–255.
Koeberg Alert Alliance. (2012). Eskom finally releases Koeberg Emergency Plan. 27 January. https://koebergalert.org/2012/01/27/eskom-finally-releases-koeberg-emergency-plan. Accessed 26 Jan 2019.
Kuchinskaya, O. (2012). Twice invisible: Formal representations of radiation danger. *Social Studies of Science, 43*, 78–96.
Kuchinskaya, O. (2014). *The politics of invisibility: Public knowledge about radiation health effects after Chernobyl*. Cambridge: MIT Press.
Kyne, D., & Bolin, B. (2016). Emerging environmental justice issues in nuclear power and radioactive contamination. *International Journal of Environmental Research and Health, 13*(7). https://doi.org/10.3390/ijerph13070700. Accessed 22 Mar 2019.
Le Cordeur, M. (2017). New Energy Minister Mahlobo's first words on his nuclear vision for SA', *Fin24*, 19 Oct 2017. https://www.fin24.com/Economy/new-energy-minister-mahlobos-first-words-on-his-nuclear-vision-for-sa-20171019. Accessed 18 Jan 2019.
McClenaghan, T. & Blaise, F. (2017). A review of Canada's nuclear emergency preparedness and response plans. *Canadian Environmental Law Association*, July. https://www.cela.ca/sites/cela.ca/files/CELA%20Submission%20to%20the%20CNSC%20-%202016%20Regulatory%20Oversight%20Report%20on%20Nuclear%20Power%20Plants.pdf. Accessed 29 Jan 2019.

Misrach, R. (1992). *Violent legacies: Three cantos*. Manchester: Cornerhouse.
Monbiot, G. (2016). Nuclear power - yes please. Hinkley Point - no thanks. *The Guardian*, 15 September. https://www.theguardian.com/commentisfree/2016/sep/15/nuclear-power-no-hinkley-point-yes-atomic-energy. Accessed 18 Jan 2019.
Moore, J. (2015). *Capitalism in the Web of Life: Ecology and the Accumulation of Capital*. London: Verso.
Morton, T. (2014). *Hyperobjects - philosophy and ecology after the end of the world*. Minneapolis: University of Minnesota Press.
Mueller, J. (2009). *Atomic obsession: Nuclear alarmism from Hiroshima to Al-Qaeda*. Oxford: Oxford University Press.
Nelles, D. (2014). No longer the darling of the green movement. Lovelock explains himself. *The Globe and Mail*, 11 July. https://www.theglobeandmail.com/news/world/no-longer-the-darling-of-the-green-movement-lovelock-explains-himself/article19571394/. Accessed 18 Jan 2019.
Nixon, R. (2011). *Slow violence and the environmentalism of the poor*. Cambridge: Harvard University Press.
Nuttall, W. J. (2004). *Nuclear renaissance: Technologies and policies for the future of nuclear power*. London: Taylor & Francis.
O'Brian, J. (Ed.). (2015). *Camera atomica*. London: Black Dog Publishing.
O'Brien, L. (2012). After 26 years, farms emerge from the cloud of Chernobyl. *Independent*, 1 June. https://www.independent.co.uk/news/uk/home-news/after-26-years-farms-emerge-from-the-cloud-of-chernobyl-7811615.html. Accessed 24 Jan 2019.
Ou, E. (2009). Under a nuclear cloud. http://www.edouphoto.com/index.php?/project/under-a-nuclear-cloud/. Accessed 22 Jan 2019.
Overy, N. (2018). *Zones of Dissonance: Cape Town, Koeberg and Nuclearity, MPhil thesis*. University of Cape Town.
Perrow, C. (1981). Normal accident at three Mile Island. *Society*, (5), 17–26.
Picard, C. (2013). Center field: the stage of scientific reproduction: an interview with Jeremy Bolen. *Art21 Magazine*, 26 February. http://magazine.art21.org/2013/02/26/center-field-the-stage-of-scientific-reproduction-an-interview-with-jeremy-bolen/#.XFK1x2lS-Uk. Accessed 15 Jan 2019.
Pitkanen, L., & Farish, M. (2018). Nuclear landscapes. *Progress in Human Geography, 42*, 862–880.
Plokhy, S. (2018). *Chernobyl: History of a tragedy*. London: Penguin.
Ropeik, D. (2016). The dangers of radiophobia. *Bulletin of the Atomic Scientists*, 2, 311–317.
ROSATOM. (2019). *History of cooperation*. Africa. https://rosatomafrica.com/en/rosatom-in-country/history-of-cooperation. Accessed 22 Mar 2019.
Saunders, D. (2011). Area around Chernobyl remains uninhabitable 25 years later. *Globe and Mail*, 15 March. https://www.theglobeandmail.com/news/world/area-around-chernobyl-remains-uninhabitable-25-years-later/article4266317. Accessed 5 Feb 2019.
Schuppli, S. (2015). Radical contact prints. In J. O'Brian (Ed.), *Camera atomica* (pp. 278–291). London: Black Dog Publishing.
Sontag, S. (1965). The imagination of disaster. In R. Latham (Ed.), *Science fiction criticism: An anthology of essential writings* (pp. 188–199). London: Bloomsbury.
Sontag, S. (2008). *On photography*. London: Penguin.
South African Department of Energy. (2010). *Integrated Resource Plan for Electricity, 2010 (Revision 2)*, 8 Oct.
Stawkowski, M. (2017). Radiophobia had to be reinvented. *Culture, Theory and Critique, 58*, 357–374.
Tsampiras, C., & Müller, A. (2018). Overcoming 'Minimal Objectivity' and 'Inherent Bias': Ethics and Understandings of Feminist Research in a Health Sciences Faculty in South Africa. *Feminist Encounters: A Journal of Critical Studies in Culture and Politics, 2*, 1–16.
Vegter, I. (2018). Nuclear power is extraordinarily safe. *Daily Maverick*, 19 November. https://www.dailymaverick.co.za/opinionista/2018-11-19-nuclear-energy-is-extraordinarily-safe/. Accessed 21 Jan 2019.

Volkmar, A. (2016). Ironic encounters in the Anthropocene: Jürgen Nefzger's nuclear landscape photography. *Frame*, 29 (2). http://www.tijdschriftframe.nl/29-2-perspectives-on-the-anthropocene/1451/. Accessed 22 Mar 2019.

Walker, J. (2018). 'A Radioactive Lens', *PetaPixel*, 7 June. https://petapixel.com/2018/06/07/a-radioactive-lens/. Accessed 31 Jan 2019.

White, L. (2012). Nature, metropolis and the apocalyptic sublime. Paper given at *Landscape and Eschatology*, Tate Modern, 13 January. http://eprints.mdx.ac.uk/8507/. Accessed 22 Mar 2019.

Winkler, H. (2016). How the state capture controversy has influenced South Africa's nuclear build. *The Conversation*, 26 May. https://theconversation.com/how-the-state-capture-controversy-has-influenced-south-africas-nuclear-build-58879. Accessed 19 Jan 2019.

Winkler, H. (2018). Why nuclear power for African countries doesn't make sense. *The Conversation*, 15 May. https://theconversation.com/why-nuclear-power-for-african-countries-doesnt-make-sense-96031. Accessed 19 Jan 2019.

World Nuclear Association. (2016). Plans for new reactors worldwide. http://www.world-nuclear.org/information-library/current-and-future-generation/plans-for-new-reactors-worldwide.aspx. Accessed 20 Jan 2019.

Wynne, B. (1992). Uncertainty and environmental learning: Reconceiving science and policy in the preventive paradigm. *Global and Environmental Change, 2*, 111–127.

Chapter 14
Multiplicity and Encounters of Cultures of Care in Advanced Ageing

Michael Koon Boon Tan and Sarah Atkinson

Abstract The demographic of an ageing population in many countries is increasing the numbers of elderly who are resident in care homes especially in parts of South East Asia. The investments made into care-related activities in residential homes for the elderly largely reflect a medical approach in which priority is given to physical care through bodily maintenance and limited physical exercise, and only limited resources are allocated to other intellectual or imaginative engagements. The study introduces a very different culture of caring practice into a Singaporean nursing home through an arts-based programme in which the medical benefits are secondary to an assertion that the imagination, creativity and self-expression should be intrinsic to how we conceive of human thriving, emancipation and vitality. This difference of purpose draws our attention to the specific practice of the arts practitioner in how they care for and manage the activities, the space and the atmosphere of the arts sessions in the nursing home. In contrast to the dominant containment culture of care of the nursing home, we introduce the notion of the caring artist to capture this culture of careful practice that is attentive to the participants' own choices, supportive to the actions they wish to engage and reflexive of their own practice.

> I have never created work of such quality. I am amazed how I managed to produce this…. Yes, I think I am quite capable. [Alice, nursing home resident]

M. K. B. Tan (✉)
Nanyang Technological University, Singapore, Singapore
e-mail: michaeltan@ntu.edu.org

S. Atkinson
Durham University, Durham, UK
e-mail: s.j.atkinson@durham.ac.uk

S. Atkinson, R. Hunt (eds.), *GeoHumanities and Health*, Global Perspectives on Health Geography, https://doi.org/10.1007/978-3-030-21406-7_14

Introduction

The contemporary and future trends of ageing populations include the growing numbers of those with advanced ageing who require institutionalised support. Caring for those with advanced ageing confronts debates around the meanings of care, from the concerns of medicine and social welfare with issues of quality and human rights through feminist engagements with labour and practice, value and ethics. Guidance for the provision of elder care has largely emerged from the medical and social sector in which care is framed as a potentially universal practice that enables compliance with a universal ethic of how the elderly are to be treated. This is evident, for example, within the WHO's Healthy Ageing Agenda in which long-term care is enshrined within a framework of universal rights, freedoms and dignity (WHO 2016). This universalising approach to care is, however, predominantly based on discussions of how an ethic of care is understood and practised in the global north, and feminist scholars have called for greater attention within care-focused research to the diversity, dynamism and setting of the practice and ethics of care (Raghuram 2016). This critique is complemented by more practical efforts undertaken to recover some of the identities, memories and places that often become reduced through physical and mental constraints associated with advanced ageing. Such efforts aim to challenge and expand dominant perceptions of what ageing bodies can, and should, do and feel (Milligan et al. 2005), by drawing on the creative arts within explicitly interdisciplinary projects. Those exploring the benefits for seniors from participation in the creative arts with the spatiality of an interdisciplinary geohumanities approach include Rose and Lonsdale (2016) who seek to recover and reimagine those landscapes of particular emotional significance and security through painting; Luján Escalante et al. (2017) who demonstrate the potential of creative practices to facilitate the inclusion of people with dementia as co-designers in digital innovations for their support; and Herron et al. in this volume who engage a wider set of benefits from dance than are often considered. These explorations all involve collaborations across not only different academic disciplines but also with arts practitioners and providers of care. Here, we continue in this approach of a transdisciplinary geohumanities (Dear et al. 2011) through a collaboration between an artist working as both an arts-and-health practitioner and an academic, a social scientist working in both geography and the medical humanities, and the managers, carers and residents of a nursing home in Singapore. Despite this attention to practice, however, there has been, to date, relatively limited development of how a creative arts-based practice may relate to care and to the dominant universalising practice of care. The chapter, therefore, seeks to explore and conceptualise what we understand to be different cultures of care evident in the practices of the nursing home team and of the arts practitioner.

Culture is a definitionally complex term with myriad of definitions to be found across the spectrum of academic research. We favour the brief elegance of Redfield, following the Lancet Commission on Culture and Health who defines culture as, 'conventional understandings manifest as art and artefact' (cited in Napier et al. 2014: 1608). This definition captures not only the 'the various complex systems of meaning that constitute everyday life' (Gilleard and Higgs 2014: 2) and the shared and taken-for-grantedness of those meanings but also stresses their inseparability from practice and materiality. Engaging care through a concept of culture then demands 'digging deep' (Killett et al. 2016) to reveal those taken-for-granted assumptions, which comprise ethos and underpin practice and which, by definition, are rarely explicitly articulated. We combine two rather different entry-points for 'digging deep' in order to respond to Raghuram's call for attention to diversity, dynamism and setting. While Raghuram (2016) proposed encounters between geographically differentiated cultures, we seek to reveal co-existing orientations to the practice and ethic of care within the same shared historical and geographical setting. In this, we favour the approach of Mol (2010) who focuses on the mixing of values and practices, and how they interact and shape one another, within the specificities of time and place. In particular, Mol teases out the diverse logics (Mol 2008) and processes of 'valuing' (Mol 2010; Heuts and Mol 2013) that adhere to different understandings of notions such as care. By ethnographically documenting the various meanings given to what makes something good or bad, and the practices undertaken to make it so, for example, in relation to the terms independence and dependence (Struhkamp et al. 2009), she and her colleagues reveal a multiplicity of values and logics to care being expressed and entangled within a single setting.

Our study builds on elements of both Raghuram's differentiated cultures of care and Mol's co-existing values of care to better comprehend multiplicity and encounter in cultures of care in advanced ageing. We draw on a single case study of a long-term care home for the advanced elderly in Singapore. The practices of care in institutionalised homes for those who are particularly vulnerable, sometimes physically, sometimes mentally and often both, afford a space in which we encounter several constituencies with different sets of priorities, desires and experiences of the home. We therefore engaged these constituencies in discussions of expectation, intentions and experiences of working and living in the home. Alongside this exploration of the everyday variations in the cultures of care across different constituencies within the home, we also intentionally staged an exceptional encounter between cultures of care associated with explicitly differentiated traditions of practice. The project introduced into the largely medicalised ethos and practice, a programme of creative arts participation predicated on rather different practices and assumptions. The chapter, thus, describes multiple values and practices of care, sometimes contrasting, sometimes complementary, in a residential home for advanced ageing that emerge through both everyday and exceptional spaces of care and from differently positioned constituencies involved in the home. We conclude by exploring the relationships between these, their groundings in different logics and the implications of differentiated normative schema for variously 'tinkering with' or 'reforming' care provision.

Advanced Ageing and Residential Care

Residential care homes have become well-established in Europe, but countries that traditionally emphasised caring for the elderly in the family setting, including Singapore, are now also witnessing a rapid growth of institutionalised elder care.

Singapore has one of the world's highest rates of ageing, along with several other countries in South East Asia, for example, Japan, and while the numbers of residents in institutionalised facilities are still relatively low, these are increasing rapidly with the expectation of continued growth of the care home sector (Wong et al. 2014). Nursing homes in Singapore reflect trends in the global labour market, with the hands-on care labour undertaken largely by foreign migrant workers, particularly from the Philippines or Myanmar, while Singaporeans occupy the higher-level management roles (Huang et al. 2012; Wong et al. 2014). In this, Singapore mirrors other countries in which caring for the elderly is devalued, usually feminised, often coloured, 'because the elderly are regarded as a marginal group whose bodies are associated with images of decay, disease and disgust (deriving from the loss of control over basic functions)' (Huang et al. 2012: 198; Twigg 2006). Associated, at least in part, with this view of elder care as low grade or dirty work and with the rapid growth in the market for elder care provision, a large percentage of countries have developed regulatory frameworks to mitigate the risk of poor standards of care (Wong et al. 2014).

The importance of providing a good standard of care in relation to bodily hygiene for those no longer able to do this themselves privileges the body and the bodily deterioration with age. Other care-related activities in residential homes for the elderly likewise largely reflect a medical approach in which priority is given to physical care through bodily maintenance and limited physical exercise. The disadvantage of viewing the elderly in residential care as primarily in need of care for physical maintenance is that this becomes blurred with viewing the elderly as largely passive recipients of the care provided. As a result, the provision of activities to engage or entertain often has an instrumental purpose in promoting movement or in occupying the residents' time rather than promoting leisure, enrichment or quality of life (Theurer et al. 2015). While physical activity benefits the health and wellbeing of people in later life, an exclusive attention to this overlooks the concomitant benefits that derive from an engaged lifestyle (Lennartsson and Silverstein 2001; Son et al. 2007; Wang et al. 2002). Medicalised care that attends only to the bodily needs of a 'condition', in this case ageing, implicitly limits the identity of the person to that condition (Mclaughlin 2006). Typically, residential homes tend to be associated with top-down governance and with little opportunity given for input from the residents about how things are organised, what activities they might prefer or how they themselves might be active participants in what has become their home. In effectively limiting the opportunities and access to the resources and technologies for an engaged lifestyle, the care for the residents in the nursing home produces a leisure-scarce space with implications for health and wellbeing outcomes (Hutchison and Nimrod 2012; Iwasaki 2007). There has been critique and calls for new approaches to providing residential care, particularly in terms of the inclusion of the resident voice and building connections between the residential home and other organisations in the neighbourhood. The culture-change movement in the United States, for example, has campaigned since at least the 1990s for resident-centred care that promotes a more home-like environment through smaller residential subunits, continuity in the care teams and greater involvement in decision-making for both residents and the care workers (Koren 2010; Rahman and Schnelle 2008). In particular, the culture-change movement responded to the three 'plagues' of nursing home life: boredom, loneliness and helplessness (Thomas 1996), and these resonate with three of the basic needs articulated by self-determination theory: autonomy, relatedness and competence (Deci and Ryan 2008; Shura et al. 2010). While calls to reform care provision in residential homes have repeatedly emphasised the need to improve autonomy and social interaction, unfortunately, boredom, helplessness and competence have received relatively little attention (Shura et al. 2010). The provision of activities in residential care homes is frequently identified and programmed with very little input, if any, from residents. Moreover, the activities are often little more than passive forms of entertainment that demand nothing from the residents and are subsequently rather childlike or patronising in their simplicity. Rather than offer depth of experience, these forms of leisure appear designed primarily to keep the residents occupied, to pass the time and mask the lack of purpose or meaning in residential home life (Theurer et al. 2015). Certainly, there appears to be little attention to occupational activities which are explicitly designed to care for the residents' continued feelings of competency.

This brief and partial review of the literature related to care provision in residential homes supports the claims made by Raghuram (2016) that understandings of care are dominated by those of the global north and that we need a more geographically differentiated accounts. The approach in much of the literature on care homes aims and claims to offer generalisable, perhaps universally so, observations and insights. The offerings are premised on an often implicit, normative framework of what residential care provision needs to entail. Those arguing for various forms of cultural change explicitly position themselves as promoting justice for those least powerful within the homes and as seeking universal principles for how care should be provided. The ethic of care espoused within feminist social theory does something rather different from this; rather than seek universal principles grounded in an ethics of justice, it argues for awareness of the multiplicity of norms or goods on which care may be and could be defined and argues, '[i]nstead, care implies a negotiation about how different goods may co-exist in a given, specific, local practice' (Mol et al. 2010: 13). As such, cultures of care are to be comprehended not as distinct entities, but as multiple, relational and dynamic, as both mutually complementary and conflicting and as situated entanglements of the material, affective, normative labour and practices of doing care, as Puig de la Bellacasa states in relation to thinking with care: '…it is important to insist that knowledge-making based on care, love and attachment is not incompatible with conflict; that care should not be reduced to smoothing out of differences' (Puig de la Bellacasa 2012: 204).

Living at Evercare

Everyday living in our case study residential home, which we are calling Evercare, showed many of the features typically criticised within the reform movements. At present, the public image of nursing homes in Singapore is both highly negative and

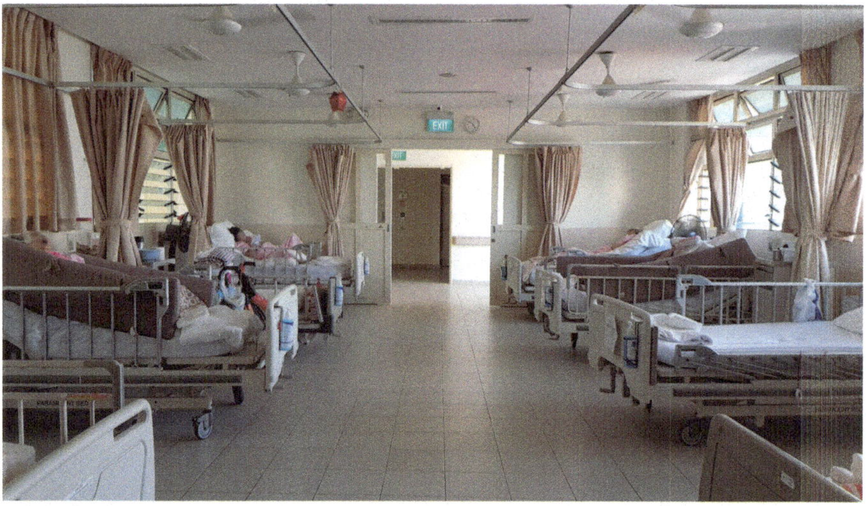

Fig. 14.1 A ward at Evercare. (Photograph: Michael Tan, used with permission of Evercare)

stigmatised with new developments having been met with public protest (AsiaOne 2012). These are pervasive, and residents of Evercare admitted that they too had similarly cynical impressions of residential living, prior to admission.

In part, these attitudes both reflect and inform the ways in which elder care is conceived and practised. Our case study, Evercare, exemplified the rolling out of a medicalised model of elder care (Kane 2005; Ronch 2004). The home was arranged very much as a hospital, with the 400 residents housed over several floors, each of which comprised 4 eight-bed wards and a shared communal area for eating and sitting (see Fig. 14.1).

The home catered for residents with a range of care needs. The majority are non-ambulant, requiring varying degrees of assistance from the staff with their daily needs such as toileting, showering and feeding, while a small number are ambulant and largely able to manage their self-care. There are also residents who are fully bed-bound in late-stage decline for whom the home effectively offers palliative care. It is due to these residents that the terminology used for Evercare as a nursing rather than care, home, is justified. It is however also these bed-ridden residents who are perhaps those who furnish the popular image in Singapore of the nursing home as a depressing place to inhabit:

> In actuality, those bed-ridden residents are simply left with a mouthful of breath. Have you see those that are in that state? Their eyes are always closed and they just keep breathing. Their eyes may open when you nudge them but will close shortly after. What can you expect from those residents? What do you want them to do? [Joe]

Another factor in creating a negative impression is that the ward structure offered almost no privacy and only little space for personal possessions. Residents were further depersonalised in being obliged to wear the nursing home's uniform pyjama-style clothes (Tan 2017). The nursing home was, however, very clean and fresh-smelling at all times, and the environment bright compared with others in the area (Tan 2017), a point acknowledged by those residents who had visited several others which they described as, 'very dark and dirty'. The nursing home, thus, delivered a high quality of care in relation to the physical concerns of environmental and bodily hygiene. Evercare was also attentive to the physical wellbeing of the residents, evident in the provision of activities for exercise and movement. The association of physical movement with continued health, in the sense of the absence of disease, of mental capacity and of capacity to carry out some functions appeared to be recognised and given primary emphasis in the nursing home through the employment of several physiotherapists. Additional activities, however, that hold the potential to enrich residents' cognitive and imaginative capacities and engender a sense of control and development in their life (see Edvardsson 2008; de Macedo et al. 2015; de Oliveira et al. 2014) were rather limited. Even when these were available, as in the provision of puzzles or colouring sheets, staff were quick to highlight that these activities represented an additional strategy for maintaining motor coordination skills. While such activities might pass the time, their provision was nevertheless underpinned by medical rather than pleasure-based reasoning. As Jess highlights:

> Our main focus is on the patients' physical hand movement. Creating an art piece is not our priority. We can give them different things to feel with their hands….All these activities provide tactile stimulation and encourage them to use their hands…. When using art, the occupational therapist is interested in stretching and strengthening and mobilising activities that accompany art making. [Jess, Therapist Aide]

An occupational therapist was contracted to provide reminiscence activities, bingo, art and crafts and cooking sessions, but these were held erratically and irregularly. On rare occasions, voluntary groups would visit the home to offer one-off cultural performance, a sing-a-long session or some kind of craft activity with the residents (Tan 2017). However, the opportunity to participate in these was uneven given that staff were not always available to help those with limited or no mobility access the event:

> There's singing on level two that they always asked me to join. But no one pushes me there. Whenever I request the ward staff to bring me, they would brush me off, telling me that I don't know how to sing, that I don't know the song. So I didn't go. They don't want to bring me there. [Mag]

Each floor has a communal space where residents could sit and interact informally, although few seem to do so. Taking up the chance to interact informally depends again on residents having sufficiently independent mobility as well as there being residents with interests in common on any specific floor:

> Sometimes, it's like making friends, if one doesn't share commonality on a topic of conversation, it is useless to speak another word…. you see, now we speak in different languages….. [Francis]

There seems to be little opportunity or encouragement for informal socialising beyond the formally organised physical exercise classes, a situation that resonates with the description of life in a care home as being 'alone together' (Turkle 2012):

> When the session is on, the others will gather around to chitchat, that is good. But when you stop coming, the residents no longer come together. They don't come up. [Alice]

Residents themselves recognised that the care they were receiving was very medicalised and focused on their bodily maintenance. Several expressed concerns about the lack of mental, sensory or imaginative stimulation, particularly in contrast to the creative arts intervention:

> The nurse doesn't. They attend to things that concern our body. They don't take care of the ability of our mind and visual acuity. They don't. [Clare]

> You help residents pass the time, provide some interest for them to play. We don't get this from the nurses. They help out in taking care of our daily need, such as meals, medication. They are concerned about our safety, to ensure that you don't fall. [Joe]

The nursing home management team was aware that there could be better provision of stimulating and learning opportunities for the residents. But, as in many countries, the staff were already working at full capacity with little time, inclination or skills to explore additional forms of care:

> ..because the health care staff will say that 'we don't have enough staff, how to carry out the activities.' How much the staff want to participate in the activity is a concern. [Agnes, Evercare manager]

> They don't have the kind of skills to facilitate the residents in to the art session kind of thing. [Agnes, Evercare manager]

Time management has also been flagged as an issue within residential care homes, with a medical model drawn from hospital structuring a tight schedule of maintenance tasks and with little spare capacity room for including additional tasks, such as taking a resident to join an activity:

> There is time management, due to the maintenance situation we have, we also have to communicate with the maintenance people, because we are back clashing by time. Because we are also facing a lot of the time we get clashes. Here the job is heavy duty. The problem is the time. [Jess, Therapist Aide]

This 'medical value of time' which puts a premium on physical health, disease and curing has been subject to critique for some 20 years, as it can result in neglected or suboptimal psychosocial care (see Henderson 1995). It also underpins an argument for excluding those who may need additional support to participate in activity programmes, including those already displaying disengagement and who might most benefit (Kolanowski et al. 2006).

This medicalised approach to care reflects an underlying shared assumption of ageing bodies as incapable and passively dependent, a view recognised as compromising the wellbeing of residents (Meyer and Owen 2008; Vladececk 2003; Kane 2001; Ronch 2004; Gerritsen et al. 2004; Ryvicker 2009) through generating processes such as learned helplessness, lower motivation and loneliness (Coudin and Alexopoulos 2010).

> They are thinking that all the people here are already incapable. These things are not required. Daily leisure needs are not required. [Joe]

Moreover, the popular stigma attached to ageing bodies may be seen in ageist attitudes amongst staff in the nursing home with consequences on how care is conceptualised and delivered for the elderly in the home and expressed through staff behaviours that may be at best paternalistic and patronising, thereby diminishing dignity and self-esteem (Giles and Gasiorek 2011) or worse, resulting in physical and emotional neglect (Angus and Reeve 2006; Band-Winterstein 2015). Residents are dependent on the staff for help with their physical needs and are very aware of their vulnerability to staff attitudes in respect of this:

> Showers are carried out very quickly, in a slapdash manner. For non-communicative residents, showers may last less than two minutes while for the verbal they will take about five minutes- they are able to complain. [Joe]

> If they are in a good mood, they will perform their task well. But when they are not in a good mood….. [tails off, others describe rough handling] [Mag]

Acting to compound the vulnerability of the residents was the lack of structure within the management of Evercare through which residents could express their views and voice. Residents did offer suggestions during the research interviews, thereby showing their ability to articulate ways to improve the nursing home. These included improvements to the environment through having more flowering plants and creating shady and cool spaces outside in which residents could relax. The model of care, however, echoed that of a hospital in which the residents are passive and subject to the expert care of the facility staff.

Making Art

Michael, a Singaporean artist, viewed the nursing homes he visited as leisure-scarce environments and responded by opening dialogue with the Evercare nursing home to set up an experimental participatory visual arts programme. This comprised 12 weekly sessions and introduced the group of 10 participants to 8 different creative projects that involved a mixture of two-dimensional and three-dimensional media (see Figs. 14.2 and 14.3). The participants were recruited through a purposive sample (Patton 2005) with the assistance of staff to identify potential and interested residents. The only criteria for participation were a willingness to be interviewed and a facility in one of the study languages (Mandarin, Teochew, Hokkien or English); there was no requirement for any prior experience with any visual arts approach. Michael treated the arts sessions as a form of creative assemblage (Fox 2013) and documented them through video, a research diary and subsequent interviews and group discussions with the ten participants and seven of the staff at the nursing home (see Tan 2017, 2018a, b). As such, the intervention shared much with an action-research approach in which Michael, as a reflexive practitioner, considered what it meant to work as a caring artist and generated a framework to capture his practice (Tan 2018a, b).

Fig. 14.2 Examples of residents' artwork

Fig. 14.3 Examples of residents' 3-D artwork

The intervention clearly benefited the participants as evidenced from their own descriptions of what they appreciated about the sessions and from observations made by others, particularly the staff, Michael and relatives. The arts engagement corresponded with the findings of many other arts and health projects in providing an opportunity for socialising and enjoyment,

> It is only through your session that we sat together. I became acquainted with them, conversations developed from the initial nodding of the head. [Betty]

and for helping build and rebuild the participants' confidence and self-esteem:

> I doubt myself, not capable of art, in a way it's good for me to learn something new. At least I have tried something new for myself. [Peter]

> I did not expect myself to splash so much. I didn't expect that I would know how to splash these. I see that I am quite clever. [Alice]

Alice's comment, however, points at something that goes further than enjoyment and extends or underpins self-confidence; Alice is talking about a mixture of competence and creativity. Other participants also explicitly referenced the appreciation they had of the arts sessions in terms of learning new skills, gaining knowledge and producing a concrete and aesthetic output (see Fig. 14.2):

> You can help us improve our memory. You can also pass us knowledge. Otherwise, over here, there is not much knowledge. Beside reading newspapers and watching TV to gain knowledge, no-one has the means….. Very happy, very excited that I can produce a thing of such standard. I consider that is not a bad achievement for a disabled person to be doing this. I was challenging creating the goldfish piece, because it was hard to manipulate the form of the fish. [Clare, who only has the use one arm]

The art, moreover, provided a means through which to assert to others their continuing competency, both to family and to the professional staff in the home:

> It is good to let the people outside know that the patient here has done something….Frame one for me so that when my family member comes, I will ask them to bring it home, bring back to home and display it in our, our house, our main hall. [Peter]

> We never expected from Uncle [Joe] who did a perfect piece, we just though in the end it will be a simple piece. We never expected, I never expected that this piece will become like this. [Jess, Therapist Aide]

> Don't underestimate them actually, right? Because some times they can do really impossible things for us to see. Right, so continue to give the encouragement. I believe they can do even more for us to know, I think nothing is impossible. I think by nurturing they can be much better. [Matt, Staff]

In the context of the nursing home, the art sessions enabled participants to leave their ward, to meet other people from the home as well as engage in the arts practice. Some participants were reliant on staff bringing them to the sessions, and Michael did himself fetch one or two residents on occasions to enable their continued participation. While the opportunity to relieve the boredom of life in the nursing home might be offered by any range of activity sessions, the participants recognised the arts sessions as offering a particular benefit to their mental capacities. And this, again in the context of the nursing home, is of high significance for the participants:

> When you teach us things to do, we are alert. When you guys come and teach us things to do, our spirits are better. [Alice]

> They can keep their mind going, instead of spending most of our time sleeping. I am afraid I will become demented. At least this is something that keeps the mind running. [Betty]

> We can use our brain to think how we are going to accomplish the work. It can also train the eye. That is very good. You have to see as you create. [Clare]

The creative practice offered in this project differed explicitly with the kinds of art and craft activities more commonly available. They offered particularly stark contrast to sets of colouring-in sheets. These reflect a format that is prestructured and the output predetermined with, as seen before, a primary focus on motor-coordination skills and time-passing.

> Your art activity is different. You lead them to discover and show their talents with using the colour, the clay, the paintings. [Jess, Therapist Aide]

> What the OT did wasn't something very dynamic. What you have done is something very different. Because, at the end of the project itself, erm, even with the watercolour, you don't have any base, outline for them to colour. You just let them fill the papers with colours ….. there was some beautiful work that is being done… and you allow the elderly to express it in their own way. You are not fixated, eh, you coloured out of line already, you know, we don't do that to the patient. So whereas your art is very free flowing in itself, it allows the patients or the resident to express their own sense of art work. [Agnes, Evercare Manager]

The nursing home manager and the physiotherapist fully understand the difference that these creative arts sessions offer compared with the highly structured occupational therapy sessions. The participants unpick this further to indicate how the openness of the creative arts sessions affords direct mental stimuli, exercise of the imagination, opportunities for expression and exploration, and aesthetic satisfactions (see Fig. 14.3):

> I can realise my imagination. I didn't think what I was going to create. I simply drew. [Freddy]

> It is not the same. That you have to use your eyes, your mind, to figure out how to do the thing, how to pursue, how to do things to avoid damaging it. Colouring [sheets given by staff] is different. You simply colour. Even when it is not good, they [the staff] will still say it is good. That is not good. [Clare]

> I find myself using my brain, you make a decision about choice of colour. [Francis]

> I never imagined myself to be creating this, it is all arbitrary and they do look adorable….. Even if it was an arbitrary effort, they came out to be quite adorable. [Alice]

This acquisition of skills, the opportunity to exercise the imagination, to be expressive and creative and to produce an aesthetic product suggest that the arts and health intervention effectively valued and supported differently conceived goods for care. Here, the participants felt engaged, competent and able to continue forms of personal growth:

> You taught us how to draw, how to create. To appreciate the point of creation, the expression of it. This is something we don't know at all previously. It was from you that we learn and know what to do. [Betty]

This stimulation of mental capacities, engagement and creativity and the support of competency, aesthetics and learning, we argue, constitute caring for vitality. We use the term vitality in the sense of aliveness, in which feeling alive differs from being alive, as in Bergson's *élan vital* (Bergson 1911) or Stern's *vital element* (Stern 2010). For Stern, vitality is emergent from the connections and flows between five elements of movement, time, force, space and intentionality with its 'basis in physical action and traceable mental operations' (Stern 2010: 9). He gives particular emphasis to the body and its own movements: '[t]he experience of vitality is inherent in the act of movement. Movement, and its proprioception, is the primary manifestation of being animate and provide the primary sense of aliveness' (Stern 2010: 9). The descriptions given by participants of their experiences of the creative arts programme detail the physical action and traceable mental operations of aliveness and vitality and thereby intimate the significance of offering an assemblage of movement, time, force, space and intentionality in the care home.

Multiplicity and Encounters in Cultures of Care

In interpreting the positive changes in mood and behaviours displayed by some of the participants, Agnes, the nursing home manager, relates these almost explicitly to a sense of empowerment that producing the art works may have engendered:

> The problem with the patient here is, they don't have the chance to gain ownership to an item very much. In this art programme, you can feel that, hey I made this, this is mine. So it gives them a sense of ownership….. I have never seen them, so proud in their lives before. I can see the happiness on the patients' faces, which is wonderful. Wonderful. [Agnes, Evercare Manager]

Other staff intimated their recognition through the creative arts project of the potential of older bodies to become flourishing and productive bodies with the right support and care:

> They can blossom, they just need the motivation and encouragement for them each and everybody, they can come out from their lives. [Jess, Therapist Aide]

> It will at least let people know that, yes they are old, they are sick but if you care enough, you show them your patience, they can be productive in a way, you know. They can show you the other side of themselves. [Selma, Staff]

> If only we can give them time, give them opportunity, I believe they can do it definitely. [Matt, Staff]

The staff's appreciation of the intervention and the manager's insight into its benefits did not, however, translate into any evident plan to try to continue with any similar creatively directed practice in the home once Michael's intervention had ended. The staff did, nonetheless, feel obliged to offer explanations of why this was not possible, and these explanations, or logics, perhaps more than anything, reveal tensions between different caring 'goods' and practices in deciding on the priorities for investing finite resources.

First, staff treated the success of the intervention as resulting from the exceptional qualities of this particular artist which effectively dismisses the experiment as potentially indicative of what partnership with creative practitioners more widely might bring or how a more creative practice might be engendered in existing staff:

> How you led the group made a big difference to the patients' art and their experience. The freedom they had in their creations, your encouragement, and prompting to get them to try different approaches when things didn't work. You showed versatility to work with different patients. You were sensitive to the residents' needs. [Agnes, Evercare Manager]

The management also invoked the limits on the care that could be provided within the finite resources available, money, time and personnel. Effectively, this reveals a culture of care in which these aspects of care are valued but not prioritised as highly as maintenance and physical care, such that with:

> But here we also, difficult to manage because of the human resources. Then I also have some interest in the art, but .. there's no choice. I cannot manage my physiotherapy side, so I bring this to the occupational therapist side. I'm also thinking how to manage, how to keep this thing going. So we need you, we need you. [Susie, Physiotherapist]

The logic of care in the nursing home returns, by default, to an accountancy approach in which the value of investments in activities has to fit a standardised cost-effectiveness instrument:

> [let out a breath] Cost, will be an issue ! I have to look for finances to fund the programme itself. Especially the material costs. Let's say if I have an artist that comes in and on a volunteer basis, then of course I need to fund the project work being done. I cannot be expecting the artist to come out with money, that kind of thing. That in itself, then, I have to look for funding to create the projects .... Then we will need to see how much it benefits the patient in totality as number per se. If a group is too small, then it may not be very beneficial to pay the artist to do that. It has to be a group that is substantial enough, you know the kind of thing, so that we could actually, um, yeah, then the artist pay could be involved. [Agnes, Evercare Manager]

The explorations of what is valued as care by different constituencies and through different practices in the residential home of Evercare easily lend themselves to interpretation as two rather different, and conflicting, cultures of care. In the everyday routines of the nursing home, a medically informed culture of practice focuses on cleanliness, hygiene, safety and some maintenance of physical mobility and movement. The limited and somewhat irregular care provided for psychosocial dimensions of life often appears to meet descriptions of containment, occupation or entertainment that reflect and reinforce a limited view of the capacities of those in advanced ageing. There is little care given to vitality through supporting a more engaged forms of occupation that provide mental stimulation, opportunities for learning new skills and for developing and exercising competency. There is no care at all given to supporting empowerment and voice in decision-making. These absences, moreover, fit within a logic of care in which the significance of physical deterioration masks the potential for other capacities. Michael, by contrast, as a creative practitioner supported and cared for vitality of these through the open-ended processes of art-making in which the participants themselves directed the process, made the key decisions and were only guided by the artists in developing new skills in using new artistic media. As an outsider coming into the nursing home, Michael himself viewed the culture of care as seriously lacking in leisure and enrichment opportunities which he located within a mode of thinking about universal principles for justice in advanced ageing and for practice as a caring artist (see Tan 2017, 2018a, b).

We can, however, think this apparent neglect of care for vitality in advanced ageing in a different less oppositional and conflicting frame if we think in terms of a multiplicity of caring goods, logics and practices that are differently situated and embedded within caring landscapes and that are in constant and negotiated tension. The staff and the creative practitioner are very differently situated in relation to the residents. The majority of staff are low paid, migrant workers who are employed explicitly to do the hands-on caring work with and for bodies, what Huang et al. (2012: 198) refer to as 'dirty work'. The physiotherapy team is situated within their profession in which enabling particular forms of movement constitutes providing high-quality care. Both the staff caring for bodily hygiene and the physiotherapists have been contracted to deliver these forms of care. Their time is already fully committed, and they have little capacity or evident incentive to add extra caring roles to their workload. Michael, by contrast, has chosen to come to the nursing home as a volunteer to run the participatory arts sessions and, as such, always has the option to leave as his employment is elsewhere. The caring practices, then, come from different positionalities in the home but nonetheless share many values. Everyone, the staff, the artist, the residents, values the attention given to hygiene and cleanliness of the environment and the residents. The erratic and unruly nature of bodies is a defining characteristic of advanced ageing (Mol et al. 2010). The inability of residents to care themselves for their bodies is a primary rationale for moving to a residential home and the need to provide assistance with these tasks is a primary rationale for why residential homes exist. Michael, coming into the home with a set of practices specifically targeting creativity, imagination and productivity, had the benefit of being able to take this bodily care as a given, which he explicitly acknowledged when making his choice of nursing home.

Tensions between caring goods and caring logics become evident beyond the practices of basic maintenance. The creative arts intervention thus extends rather than conflicts with what is already being done by articulating additional practices through caring for vitality, for mental stimulation, social interactions, learning opportunities and expression of competency, as equally central to the understanding of care in advanced ageing. The management team and other staff fully recognise the value of caring for vitality; where the tensions lies is in whether this is valued equally centrally as the routine care of bodily hygiene and physiotherapeutic exercise. The management team appear to only consider whether the kinds of activities that Michael has enacted could be 'added' to existing work practices and conclude that staff are already too busy and unwilling to commit themselves further. Within the existing funding model, additional activities would need to be provided through fundraising, voluntary provision or donation. And in the contemporary dominance of accountancy practices within public sector management, the benefits would need to be what Mol refers to as 'squeezed' into single-dimension scales (2010).

The research, by staging an encounter of the everyday care practices with the exceptional practices of the caring artist, effectively produced a space of negotiation between these different practices and logics of care concerning the position of different goals or 'goods' of care in relation to one another. All constituencies, the management, the residents and the creative practitioner all effectively mobilised a hierarchy of care in which cleanliness of the environment and of the residents was prioritised. Beyond this, however, there were tensions across the different constituencies regarding the relative importance of other care goods. The management invoked a medical approach in which physical care of the body through movement became the next priority in the provision of care. Care for occupying the residents' time was provided through a basic offer of TV, newspapers, colouring sheets or puzzles. As such, the logic of care was premised on the centrality of the physical, material body. The creative practitioner, by contrast, offered a logic of care that was premised on what we have termed vitality. Vitality recognises a distinction between being alive and feeling alive, and, while physical movement is crucial in connecting to mental and emotional vitality, the exercise of mental capacity, imagination and creativity prompted a greater alertness in the participants, as observed by both the practitioner and the staff.

This excursion through the values given to different forms of caring practice for residents in a Singaporean nursing home raises several interrelated issues about how we think through and think with a notion of care. Attending to the practices and logics of care in terms of multiplicity and negotiation allows for a greater sensitivity to how different goods are valued, prioritised or dismissed. This permits a broader spectrum of potential values from different perspectives to inform how care in advanced ageing is conceptualised and delivered. Our experiences in the nursing home nonetheless leave us dismayed at the lack of value given to the residents' vitality, and we find it hard to let go of a wish for a universal principle premised on the right for care in advanced ageing. This prompts us to contemplate the different academic cultures of care, of universal rights and of locally specific ethics and practices and ask whether we can find a way also to better negotiate between these and 'tinker' with different goods mapping onto different dimensions to care.

Acknowledgements The empirical research was supported through a Postgraduate Arts Scholarship from the National Arts Council of Singapore and subsequent conceptual work supported through a Wellcome Trust Development Award to Durham University (WT209513). We especially acknowledge and express our gratitude to all the staff and participants at 'Evercare' nursing home who generously gave their time, care and insights to the work.

References

Angus, J., & Reeve, P. (2006). Ageism: A threat to 'ageing well' in the 21st century. *Journal of Applied Gerontology, 25*, 137–152.
AsiaOne. (2012). No nursing home in our neighbourhood please: Bishan residents. 28th May. Singapore Press Holdings.
Band-Winterstein, T. (2015). Health care provision for older persons: The interplay between ageism and elder neglect. *Journal of Applied Gerontology, 34*, NP113–NP127.
Bergson, H. (1911). *Creative evolution*. [Translated Mitchell, A.]. London: Macmillan.
Coudin, G., & Alexopoulos, T. (2010). 'Help me! I'm old !' How negative aging stereotypes create dependency among older adults. *Aging & Mental Health, 14*, 516–523.

de Macedo, L. D. E. D., de Oliveira, T. C. G., Soares, F. C., Bento-Torres, J., Bento-Torres, N. V. O., Anthony, D. C., & Picanço-Diniz, C. W. (2015). Beneficial effects of multisensory and cognitive stimulation in institutionalized elderly: 12 –months follow-up. *Clinical Interventions in Ageing, 10*, 1351–1360.

de Oliveira, T. C. G., Soares, F. C., de Macedo, L. D. E. D., Diniz, D. L. W., Bento-Torres, N. V. O., & Picanço-Diniz, C. W. (2014). Beneficial effects of multisensory and cognitive stimulation on age-related cognitive decline in long-term care institutions. *Clinical Interventions in Aging, 9*, 309–321.

Dear, M., Ketchum, J., Luria, S., & Richardson, D. (Eds.). (2011). *GeoHumanities: Art, history, text at the edge of place*. London: Routledge.

Deci, E. L., & Ryan, R. M. (2008). Hedonia, eudaimonia, and well-being: An introduction. *Journal of Happiness Studies, 9*, 1–11.

Edvardsson, D. (2008). Therapeutic environments for older adults: Constituents and meanings. *Journal of Gerontological Nursing, 34*, 32–40.

Fox, N. J. (2013). Creativity and health: An anti-humanist reflection. *Health, 17*, 495–511.

Gerritsen, D., Steverink, N., Ooms, M., & Ribbe, M. (2004). Finding a useful conceptual basis for enhancing the quality of life of nursing home residents. *Quality of Life Research, 13*, 611–624.

Giles, H., & Gasiorek, J. (2011). Intergenerational communication practices. *Handbook of the Psychology of Aging, 7*, 231–245.

Gilleard, C., & Higgs, P. (2014). *Cultures of ageing: Self, citizen and the body*. London: Routledge.

Henderson, J. N. (1995). The culture of care in a nursing home: Effects of a medicalised model of long term care. In J. N. Henderson & M. D. Vesperi (Eds.), *The culture of long-term care: Nursing home ethnography*. Westport: Bergin and Garvey.

Heuts, F., & Mol, A. (2013). What is a good tomato? A case of valuing in practice. *Valuation Studies, 1*, 135–146.

Huang, S., Yeoh, B. S. A., & Toyota, M. (2012). Caring for the elderly: The embodied labour of migrant care workers in Singapore. *Global Networks, 12*, 195–215.

Hutchison, S. L., & Nimrod, G. (2012). Leisure as a resource for successful aging by older adults with chronic health conditions. *The International Journal of Aging and Human Development, 74*, 41–65.

Iwasaki, Y. (2007). Leisure and quality of life in an international and multicultural context: What are major pathways linking leisure to quality of life? *Social Indicators Research, 82*, 233–264.

Kane, R. A. (2001). Long-term care and a good quality of life: Bringing them closer together. *The Gerontologist, 41*, 293–304.

Kane, R. A. (2005). Changing the face of long-term care. *Journal of Aging and Social Policy, 17*, 1–18.

Killett, A., Burns, D., Kelly, F., Brooker, D., Bowes, A., La Fontaine, J., et al. (2016). Digging deep: How organisational culture affects care home residents' experiences. *Ageing & Society, 36*, 160–188.

Kolanowski, A., Buettner, L., Litaker, M., & Yu, F. (2006). Factors that relate to activity engagement in nursing home residents. *American Journal of Alzheimer's Disease and Other Dementias, 21*, 15–22.

Koren, M. J. (2010). Person-centred care for nursing home residents: The culture-change movement. *Health Affairs, 29*. https://doi.org/10.1377/hlthaff.2009.0966.

Lennartsson, C., & Silverstein, M. (2001). Does engagement with life enhance survival of elderly people in Sweden? The role of social and leisure activities. *The Journals of Gerontology Series B: Psychological Sciences and Social Sciences, 56*, S335–S342.

Luján Escalante, M. A., Tsekleves, E., Bingley, A., & Gradinar, A. (2017). 'Ageing playfully': A story of forgetting and remembering. *Design for Health, 1*, 134–145.

Mclaughlin, J. (2006). Conceptualising intensive caring activities: The changing lives of families with young disabled children. *Sociological Research Online, 11*(1). http://journals.sagepub.com/doi/pdf/10.5153/sro.1219 [Accessed 10.08.18].

Meyer, J., & Owen, T. (2008). Calling for an international dialogue on quality of life in care homes. *International Journal of Older People Nursing, 3*, 291–294.
Milligan, C., Bingley, A., & Gatrell, A. (2005). Healing and feeling: The place of emotions for older people. In J. Davidson, M. Smith, & L. Bondi (Eds.), *Emotional geographies* (pp. 49–62). Aldershot: Ashgate.
Mol, A. (2008). *The logic of care. Health and the problem of patient choice*. London: Routledge.
Mol, A. (2010). Care and its values: Good food in the nursing home. In A. Mol, I. Moser, & J. Pols (Eds.), *Care in practice: On tinkering in clinics homes, and farms* (pp. 215–234). Bielefeld: Verlag.
Mol, A., Moser, I., & Pols, J. (2010). Care: Putting practice into theory. In A. Mol, I. Moser, & J. Pols (Eds.), *Care in practice: On tinkering in clinics homes, and farms* (pp. 7–25). Bielefeld: Verlag.
Napier, A. D., Ancarno, C., Butler, B., Calabrese, J., Chater, A., Chatterjee, H., Guesnet, F., Horne, R., Jacyna, S., Jadhav, S., Macdonald, A., Neuendorf, U., Parkhurst, A., Reynolds, R., Scambler, G., Shamdasani, S., Smith, S. Z., Stougaard-Nielsen, J., Thomson, L., Tyler, N., Volkmann, A. M., Walker, T., Watson, J., Williams, A. C., Willott, C., Wilson, J., & Woolf, K. (2014). Culture and health. *Lancet, 384*(9954):1607–1039. https://doi.org/10.1016/S0140-6736(14)61603-2.
Patton, M. Q. (2005). *Qualitative research*. Wiley Online Library.
Puig de la Bellacasa, M. (2012). Nothing comes without its world. Thinking with care. *The Sociological Review, 60*, 197–216.
Raghuram, P. (2016). Locating care ethics beyond the global north. *ACME: An International Journal for Critical Geographies, 15*, 511–533.
Rahman, A. N., & Schnelle, J. F. (2008). The nursing home culture-change movement: Recent past, present and future directions for research. *The Gerontologist, 48*, 142–148.
Ronch, J. L. (2004). Changing institutional care: Can we re-value the nursing home? *Journal of Gerontological Social Work, 43*, 61–65.
Rose, E., & Lonsdale, S. (2016). Painting place: Re-imagining landscapes for older people's subjective wellbeing. *Health & Place, 40*, 58–65.
Ryvicker, M. (2009). Preservation of self in the nursing home: Contradictory practices within two models of care. *Journal of Aging Studies, 23*, 12–23.
Shura, R., Siders, R. A., & Dannerfer, D. (2010). Culture change in long-term care: Participatory action research and the role of the resident. *The Gerontologist, 51*, 212–225.
Son, J. S., Kerstetter, D. L., Yarnal, C. M., & Baker, B. L. (2007). Promoting older women's health and well-being through social leisure environments: What we have learned from the red hat society. *Journal of Women and Aging, 19*, 89–104.
Stern, D. N. (2010). *Forms of vitality: Exploring dynamic experience in psychology, the arts, psychotherapy, and development*. New York: Oxford University Press.
Struhkamp, R., Mol, A., & Swierstra, T. (2009). Dealing with in/dependence. Doctoring in physical rehabilitation practice. *Science, Technology and Human Values, 34*, 55–76.
Tan, M. K. B. (2017). The caring artist: exploring the role of an arts-health practitioner in a nursing home and a model of arts-health practice. Doctoral Thesis, Durham University.
Tan, M. K. B. (2018a). Towards a caring practice: Reflections on the processes and components of arts-health practice. *Arts and Health*. https://doi.org/10.1080/17533015.2018.1494452.
Tan, M. K. B. (2018b). Connecting reminiscence, art making and cultural heritage: A pilot art-for-dementia care programme. *Journal of Applied Arts and Health, 9*, 25–36.
Theurer, K., Mortenson, W. B., Stone, R., Suto, M., Timonen, V., & Razanova, J. (2015). The need for a social revolution in residential care. *Journal of Aging Studies, 35*, 201–210.
Thomas, W. H. (1996). *Life worth living: How someone you love can still enjoy life in a nursing home*. Acton: Vanderwyk & Burnham.
Turkle, S. (2012). *Alone together: Why we expect more from technology and less form each other*. New York: Basic Books.
Twigg, L. (2006). *The body in health and social care*. Basingstoke: Palgrave Macmillan.

Vladececk, B. C. (2003). Unloving care revisited: The persistence of culture. *Journal of Social Work in Long-term Care, 22*, 1–9.

Wang, H.-X., Karp, A., Winblad, B., & Fratiglioni, L. (2002). Late-life engagement in social and leisure activities is associated with a decreased risk of dementia: A longitudinal study from the Kungsholmen project. *American Journal of Epidemiology, 155*, 1082–1087.

WHO. (2016). *Global action and strategy plan on ageing and health.* Geneva: WHO.

Wong, G. H. Z., Pang, W. S., & Yap, P. (2014). A paradigm shift in regulating and running nursing homes. *Journal of the American Medical Directors Association, 15*, 440–444.

Chapter 15
Cartographies of Health: From Remote to Intimate Sensing

Ronan Foley

Abstract Within geographies of health, mapping remains an important research element, with a complex and mobile history that moves beyond the subject into other branches of the humanities. This paper will critically discuss, under the banner of *cartographies of health*, how mapping, as both process and form, informs health and wellbeing in different ways. In discussing mappings of health in place, I introduce the history of cartography to trace a number of important genealogies of practice. These have included shifts from cartographic to citizen science and from official and remotely generated maps to more intimate mappings of place. The shift from distant to more intimate inscriptions of health in place can also be tracked within health geography in moves towards a more-than-medical focus and a deeper understanding on places and spaces that enable wellbeing and human flourishing. Many newer forms of place mapping – what Casey calls 'somatography' – have a strong focus on health and wellbeing, with experience, embodiment and emotion all significant components. This essay will document empirical examples of citizen mappings of health and wellbeing, drawn from a range of culturally shaped global settings including Cambodia, the USA and Ireland. Finally, such new mappings also utilise creative and artistic approaches that, alongside new forms of sensory technologies, open up possibilities to explore how different spaces, built and natural, operate as co-producers of new knowledge on health practices across societies.

Red-Light Readable
In Rattanakiri the word for basket
Has been spoken for three thousand years
A tiny tribal man sharpens arrows
As we plan an election
For which they have no word.

R. Foley (✉)
Department of Geography, Maynooth University, Maynooth, Ireland
e-mail: ronan.foley@mu.ie

S. Atkinson, R. Hunt (eds.), *GeoHumanities and Health*, Global Perspectives on Health Geography, https://doi.org/10.1007/978-3-030-21406-7_15

I scan the alveoli of jungle
from a helicopter
Searching for villages
along the Ho Chi Minh Trail.
In my hands an American map
from the early seventies.

The Engineers' insignia
surround the map's scale-bar
as the bombing escalates.

The scales of Mao's fish glisten
under scales of justice
hanging by a laddered thread.

A simple stamp declares:
Destroy when no longer needed.

Reading these maps I am
scaled, scalded, white-hot,
ready to plot each crater
each pockmark on the skin.

Red print emerges from the green:
This map is red-light readable.

Blood lines flow from
jungle and ricefield

A kind of foreign policy
in a rubber stamp.

Cambodia a map
destroy when
no longer needed.

[Foley, C. and Foley, R. (2012)]

Introduction and Context

Everyone loves maps. Equally everyone identifies strongly with the places they know or come from. When Google Earth was first launched in 2004, an overwhelming proportion of the initial zooms made by users were to their family home (Patterson 2007). From a health perspective, finding our way and place in the world seems to lend itself quite naturally to finding our healthy place in the world. An entire global industry around location-based services has developed over the past two decades, within which searches for health and wider care facilities feature prominently (Cromley and McLafferty 2012). As a starting point for knowledge, the map has also been an efficient synthesiser of knowledge and data, and an interest in cartography has acted as a key material meeting point between academics and the public. In witnessing the huge popularity of recent books such as Jerry Brotton's *A History of the World in Twelve Maps* (2013) or on place-writing that has a deep cartographic instinct such as Robert Macfarlane's *The Old Ways* (2012), one might argue for a very evident cartographic turn in the humanities. That broad interest in mapping, as process, subject and object, identifies a phenomenon that has a place in both deep science and deep art. In developing a discussion on cartographies of health, it too can be framed against the different positionalities of clinical and emotional health.

The history of cartography itself has seen a shift in how and by whom maps have been made. From the private cartographers of the seventeenth century to the state-led organisations of the nineteenth and twentieth centuries, the twenty-first century has seen a shift towards citizen science and new forms of collective mapping. Nonetheless, the critical contexts for who does the mapping and what they choose to show continue to identify the importance of representation (Harley 1992). From a technical, even aesthetic point of view, there has been a shift from official maps, with their politicised and power-laden singular 'clean' lines of representation to the development of a more challenging and democratic cartography that practices delineations; multiple representations that break down sharp divides and represent space in more fluid and complex ways (Crampton 2009). A specific example of this might be the new interest in cartographics of which Beck's London Underground map is a classic early example (Hadlaw 2003). But there is also value to official representations as the production of a starting point for order and orientation that in turn leaves room for others to disorient that representation. Transgression rarely works in a vacuum and needs something to bounce off, so there is a sort of co-dependency in these various lines of representation. Similarly, access to digital representations of ever-increasing accuracy via tools like LiDAR, drones and GPS brings up contrasting and contested responses. On the one side, new forms of mapping reflect the notion of an invisible technology that potentially serves a neoliberal and technocratic ordering of the world (Wright et al. 1997). Equally, they reflect an ongoing child-like fascination on how to place ourselves and allow us to use new digital resources to represent ourselves, our wishes and needs in much more immediate and visible ways.

A similar continuum emerges within cartographies of health and place, from diagnosis, diffusion and cure to lived experience, narrative and lay perspective. The maps that inform our understandings also emerge from parallel tracks. On the one side maps are used within epidemiology based on the epistemologies of biomedicine, the security and safety of expert/technical/scientific knowledge and as defence against disease/contagion. On the other side, they provide the means to map out in specifically spatial terms, a complementary and holistic understanding of the complex fluid experience of health in and through place. As noted by Dorn and others, medical/health geographers sometimes also like to work with this co-dependency between the two different strands of the subject (Dorn et al. 2009). Across the subject this can be tracked in how health geography emerged from medical geography and the shifts more widely from the quantitative revolution to a new culturally informed and qualitative variant of the sub-discipline. In our approaches to mapping health, we might give space in our cartographic representations to positive dimensions of health/wellness, alongside acute or chronic moments of illness and disease. In cartographic terms, we might envisage diagnosis as delineation and patient experience as disruptor and working with that analogy, suggest the need to consider both epidemiologies and emotional ecologies.

This chapter works from the broad hypothesis that both historical colonial maps and contemporary official versions can be regarded as remotely sensed. They are enumerated and measured, standardised and fixed in a relatively distant way, through censuses, satellite imagery and digital signals. They are also instruments of governance. Yet what modern cartographies of health seem to make visible is a wider public shift from remote to more intimate mappings of place; where markings of presence and belonging are more important. The work of the 'new topographers' like Tim Robinson, Rebecca Solnit and Robert Macfarlane, suggest that mapping may be developing in ways that also show a direct shift from distant (disembodied) to bodily inscriptions. While much of that writing is about honouring place, it also has, I feel, an implicit aim to open us up to the relationships between place and health and wellbeing. That implicit, indeed almost affective sense draws from older and more connected indigenous traditions that always understood that a withdrawal from the connection to land and landscape was an unhealthy act, both for human and more-than-human health.

What is at stake here is the potential for cartographies of health to continue to emphasise the importance of connection and to show that this has had a cyclical and relational history. In tracing how both place and health can be understood together, we might see that historically the line of connection shifted from a song, poem or stick/finger, to a pen, a theodolite, a satellite or Lidar scanner. Yet new mappings of health are beginning to use both together and even shift back to a combined analogue/digital approach. My brother's poem at the start of this chapter was written when he got a job as a UN cartographer mapping electoral boundaries for the first post Khmer Rouge elections in Cambodia in 1992. This was a period when digital mapping was relatively new and the cartographers often had to draw their maps in the field using crayons as the humid environment made all other inks next to useless. But the poem is also a response to the ironic experience of working with communities to update maps for community gain, when those same maps (very early examples of remote sensing) were originally designed to potentially destroy them.

The bulk of the chapter will be built around five vignettes of cartographies of health in action. Each incorporates specific embodied and emotional mapping forms and types I have either been inspired by or have been directly involved in. They range in space from Cambodia to County Kerry; from Minnesota to Maynooth. While the health component is more explicit in some than others, I would argue that each explores a cartography of health that has a strong geo-humanities component. The production of each is in various ways inspired by artistically framed, collective and in situ expressions of mapping that try to blur the traditional and exploratory in cartographic terms, in an attempt to better understand place-based health.

Mapping Health

In thinking about mapping health, the famous John Snow Map of Cholera in London in 1854 is a foundational artefact for medical/health geographers. His decision to map cases of cholera in central London has been credited with being the first epidemiological map that turned representation into knowledge and in turn into a translational health outcome that alerted public authorities to the likely source of the disease (polluted water) and, by extension, a likely cure (restricting access to that water). Nick Lally *(personal communication)* argues that Snow's was the first example of a health-specific relational map, in that its express purpose was to visualise spatial relationships but also of a disease as being representable by a vector-data model. As a core element in subsequent cartographies of health, the mapping of disease diffusion, crucially developing Snow's more fixed model into more fluid representations of time and flow, have been essential tools in recent epidemiological success stories, from measles in Iceland (Cliff et al. 1993) to HIV/AIDS in the US (Gould 1993) to the 2014 Ebola outbreak. Here the maps, often derived from remote data, were crucial in telling part, though not all of the story. An essential breakthrough in the Ebola outbreak was the recognition of a traditional West African tradition of touching the dead body, something that was a highly risky intimate practice given those same post-mortem bodies were still extremely virulent. Once this practice was no longer allowed, transmission rates declined significantly, but this interrelationship between remote (scientific) and intimate (cultural) knowledge has always informed the best epidemiological research.

Another significant strand in the use of cartography within medical geography has been the mapping of health care services and patient accessibility. Joseph and Philips' influential 1984 text, *Accessibility and Utilisation*, identified different examples of how mapping was used to identify inequalities and spatial variations in facility location, patient accessibility and health outcomes. The modelling identified in that text has, in different ways, been updated into a much more computationally powerful digital age using readily available geo-spatial data and GI Science tools; the online *International Journal of Health Geographics* provides many valuable examples (https://ij-healthgeographics.biomedcentral.com/). These quantitative roots give rise to technical cartographic explorations that provide strong associational, cross-sectional and geo-statistical evidence yet continue to struggle with causality. There are wider structural components of accessibility that are aspatial and relate not to the objective placing of 'service dots' on a map, but the wider structures of power that shape the subjective lives of the 'patient dots' that can often get lost in a spatial aggregation or economically framed demand statistic. As a GIS researcher myself, I continue to value the power and potential of taking geo-spatial data and establishing patterns and associations with it, yet it always seems only one part of a wider cartographic assemblage of knowledge.

More recent directions in research in the area of health geography has seen the beginning of more intimate, though still often digital cartographies, that more fully recognise that assemblage approach. One key strand has been the combined areas of therapeutic landscapes and healthy nature research, which have generated an ever-expanding literature across multiple subjects on engaging with nature, being outdoors, enhancing mental health and tackling concerns around obesity and physical activity (PA) (Bell et al. 2015, 2018). With strong connections to health promotion, the development of mapped and mappable evidence has placed cartography as a valuable tool in public health policy. The ways in which maps are generated can be a contrast between objective motion sensors and biomedical markers and subjective (e)motion sensors. They both echo quite different but not unconnected paradigms of 'being moved', crucial given the rising interest in PA and anti-obesity policy, but also social prescriptions that address mental health. There is always a slight sense of a risk of governmentality and responsibalisation inherent in these mappings of health (Brown et al. 2017), especially in relation to individual's risky health behaviours and new examples of online happiness mapping that carry a shadow of neoliberalism and place marketing. But equally, there are community and collective mappings of health in place that emerge as balancing and potentially transgressive counter-cartographies (Young and Gilmore 2013).

While sometimes more evident in geo-humanities than health geography, there is an implicit overlap with new mappings of place and its power to enhance wellbeing. New topographers are developing different types of mapping and especially revisiting older ideas of 'deep mapping', often driven by artistic and poetic interactions with cartography as subject/object/process, supported by a significant public interest and engagement (Biggs 2010; Miller 2014). Echoing Edward Casey's somatography and his idea of a 'trace of a place', such mappings almost always contain significant embodied and emotional components that emerge not as a single trace, but as multiple delineations. The cartographic works of Tim Robinson from the West of Ireland can be described as 'slow maps'. Landscape geographer Paddy Duffy (2016), values Robinson's slow mapping and argues that the aspect of translation that occurs in more traditional and bounded cartographic representation can dry out the lively/pulsing trace of a place and sometimes produces a desiccated or anaemic geography, an aspect echoed in wider writing on emotional geographies (Anderson and Smith 2001). I am not sure I would entirely agree as original markers of boundaries and landscapes of material power remain important, even if only as lines around which to waver, re-lineate and trace over.

In extending the embodied and pulsing connections with place and our own health, being open to the place and the air and checking it via our senses – sound, smell, vision, touch and taste – are all routes in and out of the natural world around us. We often have them switched off – or perhaps more accurately, the dial on our senses is turned down. Nature is always on, we just need to turn up the dial as it were, to connect to what architect Vesely (2004) calls the latent world; a world that is never mute or completely silent but always has a background hum. In mapping out healthy places for example, we are generally trying to tap into the latent power of nature and perhaps too our own latent selves for our health and wellbeing. Equally the places we go and how we engage with them always leave an impression – both figuratively and physically – like little bruises or scars we carry back with us, but good ones. The philosophy of the 'leave no trace' movement is interesting and reflects a shared care not to leave a bad impression on the landscape. But it is very much a living and two-way process and those place impressions are often where memory and even identity are stored. Our local places leave stronger and more lasting impressions, but passing or fleeting visits to other more distant places can also leave a strong impression that can be tapped into as a health-enabling resource.

Recent emergent research on therapeutic landscapes and health enabling spaces, places and practices (Bell et al. 2015) is beginning to provide a very valuable route into these cartographies of health. In particular, they reflect an assemblage approach that combines both digital/analogue and remote/intimate approaches. The work of Duff (2011, 2012) and Bell et al. (2015, 2018) provide excellent examples of individualised health-enabling space mappings using a mix of visual and technical approaches. Duff's exploration of health-enabling places in Australia (see also Duff, Chap. 10, this volume), using disposable cameras and other visual mapping tools with different groups with mental health issues, shows how unexpected geographies, such as previously under-regarded patches of ground and built environments, identify very personal mappings of health in place. The approach of Sarah Bell and her colleagues has in part been built around an interesting combination of new technologies (GPS, accelerometers, video) and in situ methodologies (walk and kayak alongs) to generate rich geo-narratives of how people encounter places and use them for health purposes (Bell et al. 2015, 2018). Bell has also been developing new work on the mapping out of sensescapes; specifically in relation to those who have different routes to healthy enablement, such as people with chronic illness or visual impairment (see Bell, Chap. 6, this volume). Using a healthy hybrid mix of quantitative and qualitative approaches, one is getting closer here to a more intimate and collective citizen sensing, that ironically remains dependent on remote sensing technology to make it happen. Yet geo-narratives also map out the trace of a place in ways that are co-produced – walking and talking while recording – that in turn help to make visible choices, emotional responses, wider stories that 'place' the individual in terms of where they come from and where they are going to. In producing new and dynamic cartographies of health, this work has considerable potential to provide insights into different levels and shades of condition and how these are experienced differently in different types of geographical spaces.

Cartographic Vignettes

Community Mapping, Battambang, Cambodia

I visited my brother Connell in March, 1997, and shadowed him on a work journey to Battambang across deeply rutted roads to meet with a group of people developing community maps in the city. For this group, the mapping was intended to identify a form of community health via the intimate mapping of poverty and need within the town and outlying villages. A large piece of paper and pens were used by the participants (mostly male it must be said), to map out incomes and shared resources in the village. As a process and meeting place it was fascinating to see how they local actors mapped out houses, animals and figures to represent family. While scale and representation were a little vague, there was a broad coherence to the map. However, there was a real and perhaps understandable reluctance to come up with a way of mapping relative income, which was a core intent of the work. Perhaps this should not be surprising. Other research on mapping access to water resources in rural Uganda was equally unsuccessful, as people preferred to keep knowledge of local wells private in case external groups would come in and colonise it (Murphy et al. 2016). In other parts of the country, the development agencies had interestingly (and slightly unethically) identified children as much more open participants, who would happily map the rich and poor people of the village in a way adults were reluctant to do. The production of maps was a slightly unfamiliar process as well, though the group did instinctively include a mix of symbols and text in the mapping. What was noticeably absent, especially outside the town, was any sort of formal medical system or visible supports and indeed getting in to debt to pay informal ad hoc practitioners for irregular medicines was a real issue in Cambodia at that time (Fig. 15.1).

Fig. 15.1 Battambang Community Mapping Project. (Source: Connell Foley)

Mapping Joy and Pain, Minneapolis, USA

Artistic practices have been increasingly using mapping as a source and inspiration in their work. Rebecca Krinke's work in Minnesota is an especially interesting example and her recent offline/online Map of Joy and Pain has an explicit health dimension that echoes recent therapeutic landscapes work noted above. The initial work was based on collective public mapping and has significant experiential and psychotherapeutic geographies dimensions. She set up a large physical map on a giant outdoor table in a public space in downtown Minneapolis and asked passers-by to mark on it sites in the city that represented joy (a sticky gold dot) and pain (a sticky grey dot) on to the map. Many dots were placed and in many cases the gold and grey dots were put in the same place to show what Conradson (2005) identified as differential outcomes in the same place. High schools were especially prominent, but equally several locations where identified where something good/bad had happened to either the participant or a family member. This very relational act, of mapping oneself but also one's family into the map as an extension of self, was telling on how such a relational geography of health emerged across the life course and reflects recent writing in health geography in this area (Cummins et al. 2007; Foley 2017). Yet for Krinke, it was the event itself, the conversations around the map between complete strangers that was as powerful as the dots on it. This specific event of mapping produced in turn very specific collective intimacies that were central to its cartographic power and its ability to capture a shared mapping of community emotional health through place. Krinke has subsequently transferred the idea to an online map (http://www.rebeccakrinke.com/Projects/The-Mapping-of-Joy-and-Pain-Online) that anyone can sign up to and place dots on, but it feels like something quite different to the place-based intervention (Fig. 15.2).

Spatial Video-Based Emotional Mapping, Maynooth University, Ireland

Following on from the discussion above about new directions in therapeutic landscapes research that combine the intimate and remote, researchers at Maynooth University's National Centre for GeoComputation have developed a new spatial video app called Ubipix (http://www.ubipix.com). The technology looks like it creates a normal video on the smartphone, but when uploaded on the website it also shows screens with the track on a background map as well as graphics showing height and speed of movement (Fig. 15.3 below). But the relationship between the video and the map is geo-referenced so that any point on the image can be tagged and uploaded. But equally it is a valuable tool for a walk-along interview. One recent example was a walk I carried out with an Artist-in-Residence, Lisa Cahill, who had been working with students from the Froebel School of Education at Maynooth University for the previous 3 years. She was interested in how place and

Fig. 15.2 Mapping Joy and Pain: Minneapolis Map. (Curated by Rebecca Krinke)

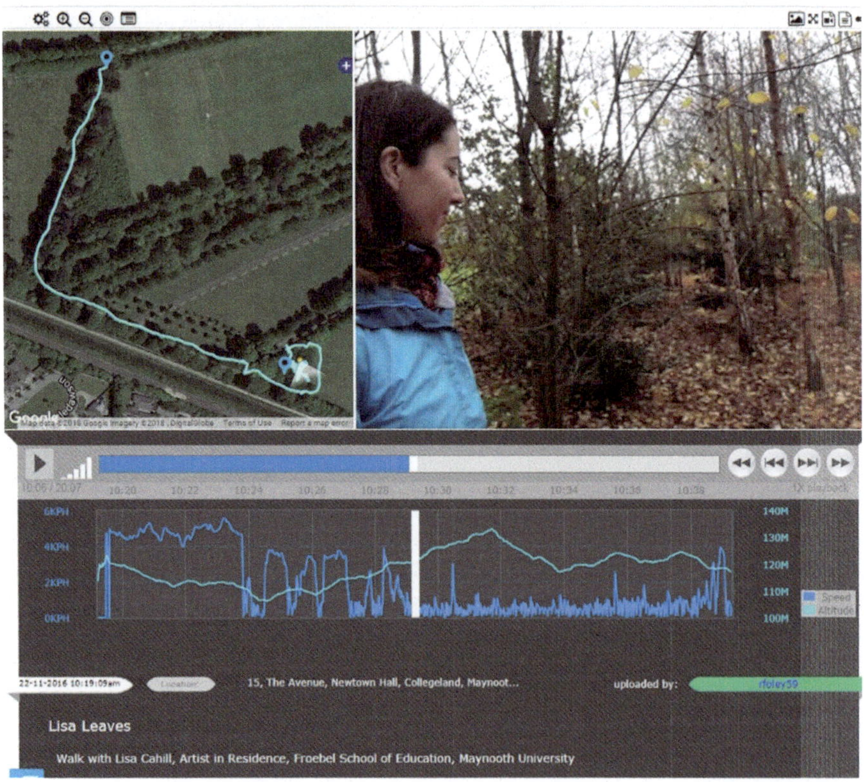

Fig. 15.3 Ubipix Sample Screen. (Source, Author)

nature informed her own dance practice, but equally as a space for students to learn about human-nature interaction. We recorded a walk to a small copse at the back of the Maynooth campus and Lisa talked me through the space and her respectful engagement with it. We talked about how nature expressed itself and as she mentioned the movement of the trees, in that same moment the previously still leaves began to move and whisper above us, as if they had been listening. While that moment had an emotional effect, the technology also allowed us to record it as a precise moment in time and space and as an unexpected but healing outcome from a shared place experience. Indeed Lisa's embodied interaction with the little copse made it both an emotional space for engagement with nature but also a small-scale and intimate setting that was captured through both affective and technological registers. More broadly her dance practice incorporated leaves, bark and paper in an energetic performance that was directly inspired by her arboreal relationships in place. This shared-health vision was visible in both the art/movement practice and through a more phenomenological engagement with and respect for the more-than-human from which emerged an emotional health in and with nature.

Campus Health Mapping, Maynooth University, Ireland

Still on the campus at Maynooth University and very directly engaged with a cartography of health, in April, 2015, I worked with students from my MA module on Landscapes of Healing to create a campus map of healthy places. We followed Rebecca Krinke's model and produced some large printed maps of the campus using OpenStreetMap and GoogleEarth. We set these up in the foyer of the University Library and then invited passing staff and students to pin a location to a place they either found 'happy' or 'unhappy', though it was made clear that these were more like synonyms for healthy/unhealthy. A fuller description is available from the 'Eyes on the World' geography department blog (Department of Geography 2016), which also contains a short time lapse recording of a 2-h event squeezed into 3.23 min/s (https://www.youtube.com/watch?v=9msd8PbIE98&feature=youtu.be). Each student and staff member had different coloured pins with additional colours for happy/unhappy place responses. In addition participants were encouraged to write on sticky notes and also write directly onto the maps themselves about feelings about the campus. A significant clustering emerged for students around spaces of socialisation such as the library and students union building, while uncomfortable lecture theatres but also darker parts of the campus were identified as unhappy/unhealthy. Indeed the wider responses identified a real lack of sitting spaces and benches on campus which were identified as important mental health supports for students, especially during busy and stressful periods like exams. This was a valuable finding given that one of the main ways in which psychologists measure the benefits of the outdoors for human health and wellbeing is through the idea of attention-restoration. Our attention gets swamped in everyday life and work and sort of needs to be re-(at)tuned. But the other part of that term, restoration was also identified by students to map out places to recharge and rebuild their reliance and feelings of wellness. Ironically the mapping exercise coincided with a consultation period associated with a new map of the campus plan and the corporate planners were surprisingly wary of a mapping process that might uncover negative or unhealthy spaces. They suggested the single term 'happy' instead, suggesting a certain reluctance to map a real as opposed to an imagined campus landscape. Yet we would argue it was the more negative findings emerging from the maps that best identified specific actions to which planners might respond (Fig. 15.4).

Healthy Green and Blue Space Mapping, Derrynane, Ireland

One final vignette relates to a similar example to the previous one, but in this case shifted to a different public space and with a more direct participant-led component. I was invited to give a mapping workshop in October 2017 for the Iveragh Learning Landscapes Workshop in County Kerry. Given the recent upsurge in work on green and blue space and how people respond to it for health and wellbeing purposes, I designed a mapping exercise to compare responses to three different settings in close proximity. These included a dark woodland (green), an open sand-dune near the sea (green-blue) and a sandy beach (blue). In each space I asked each participant

Fig. 15.4 Mapping Happy Campus Event, Maynooth, April 2015. (Source, Author)

(working in pairs) to record their sensory response to place with specific question on what I termed sound, emotional and wellbeing mapping, with a mix of text and numerical score responses. At the end of the process the group gathered together for a shared mapping exercise similar to Vignette IV, though time restricted a fuller discussion. In broad terms, the aim was to explore, using maps how people responded to different potentially health enabling space and to also see if one might be able to identify a quantifiable score for that response. It was a small sample, and some respondents got the scoring back to front, but for the specific wellbeing questions, the table below shows that even for a small sample, one can quickly generate interesting evidence and how open/safe/connected people felt (Table 15.1).

It seemed that each place elicited slightly different responses in relation to the question people were asked about them. While the beach scored best (had the lowest score) in relation to openness and safety, the woodland was preferred for the connected question. It was interesting to pick up on the protected sense of that woodland space and perhaps a greater closeness to leaves, trunks and leaf mulch that might explain that higher level of place connection. In general, the Dunes scored least well, in part perhaps because they were an intermediate and very open and relatively featureless space, but also were the middle site that people visited. Yet what was as important and evident in the images and sample from the final shared map was how people physically placed themselves in the landscape during the exercise and expressed their different traces on the final maps. Here, we can see more directly the delineations of an intimate encounter in place and with the more-than-human. This work too has been described a little more fully in a recent blog (Department of Geography 2017) (Fig. 15.5).

Table 15.1 Felt response to different spaces: results

Question/space	Woodland	Dune	Beach
Open	2.7	2.2	2.1
Safe	3.0	3.0	2.5
Connected	1.7	2.4	2.2

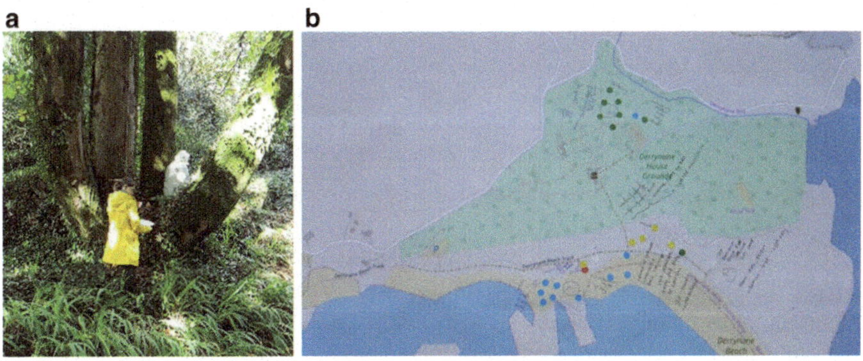

Fig. 15.5 (**a**) Participants Mapping in the Woodland and (**b**) Extract from Shared Map. (Source: Author)

Discussion

In framing cartographies of health in terms of a shift from remote to intimate sensings of place, many of the vignettes above bear witness to the lovely quote, cited by Anna Ryan (2012) that 'the physical body holds a conversation with a topography or landscape through which it moves'. It is not a huge stretch to suggest that cartography can act as a means to record or document – and not just represent – that embodied, mobile and flowing conversation. In technical terms, remote sensing relies in part on 'ground-truthing' to check on the ground what's picked up from outer space. But mapping out health in its many forms suggests a more 'fluid-storying' approach, where geo-narratives can incorporate the remote (via GPS driven location devices and health monitors) with the very intimate and immediate response from place. Equally, it would be impossible in a relational theory sense to suggest we can separate off the remote from the intimate as we have explicitly identified examples that simultaneously employ both. From health geographies, work by Aday and Anderson (1974) and Khan and Bhardwaj (1994) on the complexities of accessibility shows that the consideration of distance in solely geographical terms is only one part of a much wider understanding of more aspatial aspects. A wider socio-relational distance can be emplaced, by virtue of the system and services in which an individual finds themselves, but is equally shaped by the relational place in any individual's life course along which healthcare need emerges in complex and unpredictable ways. Recent research from the CRESH team in Edinburgh and Glasgow is beginning to look at older historical health records to record the kinds of life course-related longitudinal mappings that identify those wider relational geographies across time and space (Pearce et al. 2016).

In addition, new research on relational geographies of health (Cummins et al. 2007) argues that scale and distance should be seen less in terms of global or local but rather in terms of short and long reach; that reach can equally be imaginative. Cartographer and researcher Nick Lally's background in art, maths and sociology had found a natural home in what he terms 'geographical imagination systems' and the use of cartography – once static, now mobile – to reflect the role of art to challenge limits. Within some of the vignettes listed above, there are strong examples from broad art practice that recount the performative and mobile aspects of both academic and artistic work and the importance of public engagement and public shared spaces to marry the two together. While the various vignettes bring in aspects of attention restoration and wider rest and chronic illness management, they are equally about co-produced knowledge and shared spatial practice as a helpful act. Artistic work on nature and borders and embodied health practices reflect this and also represent in both explicit and implicit ways, how cartographies of health emerge through art practice. Francis Alys's border walk along the Israel-Palestine border using a line of green paint is a classic and explicitly political example of the trace of a place for oppressed/excluded communities (http://francisalys.com/the-green-line/). Here a blurred embodied line is created to challenge boundaries and enact an alternative re-lineation. Other examples include Vanessa Daws' psychoswimography work in Ireland and the USA that is partially mapped using GPS but also explore shared active communities that support physical and mental health (http://vanessadaws.com/psychoswimography/).

Several of the vignettes describe strongly engaged public mapping and new forms of citizen sensing. Again this can be a double-edged element and citizen sensing certainly contains strong echoes of concerns around Big Data and its increasing commercial exploitation. Indeed if we think at all about remotely sensed data in the contemporary world it is often in relation to the power/control of trans-national corporations like Google, Facebook and Twitter. Equally in terms of a democratic process, they might be seen as parallel strands to traditional state mapping that are increasingly replacing it. These new forms of mapping can incorporate for example, explicit monitoring of health behaviours by private health insurance companies to potentially exclude access to cover (Lupton 2015). Others argue for the likes of OpenStreetMap as a simultaneous democratic counter-mapping though it too is unregulated and produces quite partial and skewed representations based on privileged knowledge. But equally more indigenous and empowering uses of GIS and big data to map land ownerships and property boundaries for disadvantaged and marginalised groups have a lengthening history in places like Scotland, Canada and Central America (Wightman 1997; Lauriault and Welch 2003; Young and Gilmore 2013). Given that they increasingly gather health-related data, it is equally possible to see a national Census as an innate form of longitudinal intimate sensing exercise. As a '(c)sense-us' it does attempt to get the sense of who we are. It can even be a very intimate questioning at that, hence the high-ish rates of non-response or open answers. Citizens can choose what to say and how to answer it, which can have a strong subjective aspect that filters through into the mapping of self-reported health. Nonetheless, the ability to map that self-reported health data down to neighbourhood level can be a crucial tool and support for health researchers.

Mapping and cartographic metaphors are found across health and wellbeing settings as well. The idea of the body as a mapping template has meant that GIS has been used in medical education, mapping the body through the skin with multiple layers representing multiple systems in the body. The addition of 3-D into the toolkit has meant that new traces inside the body, of endocrine, blood, breathing and lung functions can be visualised, with flow and movement picked up as well. The relationship between visualisation and mapping patterns means that there is a striking similarity in some MRI scans to remotely sensed maps, yet in this context they are explicitly intimate. At a more community-based level, it is also helpful to think about cartographies of shared assets or resources as being a form of collective intimate sensing for those communities; especially in the absence of more formal statutory supports. As my colleague Andy Power (*personal communication*) notes from the voluntary and disability sector, getting or finding 'a place' has a meaning that has significance beyond the spatial and cartographic.

Across geographical and wider humanities writing that has a place component we have seen some important shifts in theoretical positions. From a sort of god-like and remote view above the world, we have moved to ideas of dwelling and being-in-the-world and on to more recent explorations of becoming and sensing-in-the-world.

Cartography has always had some place in these shifts as an always evolving tool that allows us to trace our place and allow us to mark out routes towards health. In turn certain types of places, of many different shades and palettes, have left a healthy mark on us. Online digital mapping has also allowed the public to better locate and identify the presence of healthy places, marked through co-ordinates as locations that provide services and supports. Yet the technological developments in geo-spatial technologies, while they have on one level, greater enhanced remote sensing capacities, have also begun to enable different voices and more intimate accounts to emerge as well. In the vignettes discussed in this chapter, we see a public response to mapping as a means to connect with intimacy as a route to wellbeing and positive effects. In discussing sensing in general, this is a blurred contemporary reality, an assemblage of the simultaneously multiply sensed, from near and afar. Sometimes, it is hard to make sense of it all, so a visual representation helps. Finally, the role and potential for cartographies of health can be linked to the notion of therapeutic accretion (Foley 2017); something that is an active embodied process at all times and across shared and individual life courses. Cartography provides a tool yet also a way of seeing, that emphasises more longitudinal and relational understandings of how health emerges in place; in these more sustained and collective mappings the power and meanings of place will also be recorded. Maps will always be needed.

References

Aday, L. A., & Anderson, R. (1974). A framework for the study of access to medical care. *Health Services Research, 9*, 208–220.

Anderson, K., & Smith, S. (2001). Editorial: Emotional geographies. *Transactions of the Institute of British Geographers, 26*, 7–10.

Bell, S., Phoenix, C., Lovell, R., & Wheeler, B. (2015). Using GPS and geo-narratives: A methodological approach for understanding and situating everyday green space encounters. *Area, 47*, 88–96.

Bell, S., Foley, R., Houghton, F., Maddrell, A., & Williams, A. (2018). From therapeutic landscapes to healthy spaces, places and practices: A scoping review. *Social Science & Medicine, 196*, 123–130.

Biggs, I. (2010). Deep mapping as an 'essaying' of place. Illustrated talk given at the Writing seminar at the Bartlett School of Architecture, Friday 9th July, 2010. http://www.iainbiggs.co.uk/text-deep-mapping-as-an-essaying-of-place/. Accessed 4 Apr 2018.

Brotton, J. (2013). *A history of the world in twelve maps*. London: Penguin.

Brown, T., Andrews, G., Cummins, S., Greenhough, B., Lewis, D., & Power, A. (2017). *Health geographies. A critical introduction*. Chichester: Wiley-Blackwell.

Cliff, A., Haggett, P., & Smallman-Raynor, M. (1993). *Measles: An historical geography of a major human viral disease from global expansion to local retreat, 1840–1990*. Oxford: Blackwell.

Conradson, D. (2005). Landscape, care and the relational self: Therapeutic encounters in rural England. *Health & Place, 11*, 337–378.

Crampton, J. (2009). Cartography: Performative, participatory, political. *Progress in Human Geography, 33*, 840–848.

Cromley, E., & McLafferty, S. (2012). *GIS and public health* (2nd ed.). New York: The Guilford Press.

Cummins, S., Curtis, S., Diez-Roux, A., & Macintyre, S. (2007). Understanding and representing 'place' in health research: A relational approach. *Social Science & Medicine, 65*, 1825–1838.
Department of Geography. (2016). A pin on the map: identifying healthy places at Maynooth University. https://maynoothgeography.wordpress.com/2016/10/26/a-pin-on-the-map-identifying-healthy-places-at-maynooth-university/. Accessed 4 Apr 2018.
Department of Geography. (2017). Emotional mapping at the Iveragh Learning Landscapes Workshop. https://maynoothgeography.wordpress.com/2018/02/19/emotional-mapping-at-the-iveragh-learning-landscapes-workshop-october-2017/. Accessed 4 Apr 2018.
Dorn, M., Kierns, C., & Casino, D., Jr. (2009). Doubting dualisms. In T. Brown, S. McLafferty, & G. Moon (Eds.), *A companion to medical and health geography* (pp. 55–74). New York: Wiley.
Duff, C. (2011). Networks, resources and agencies: On the character and production of enabling places. *Health & Place, 17*, 149–156.
Duff, C. (2012). Exploring the role of 'enabling places' in promoting recovery from mental illness. A qualitative test of a relational model. *Health & Place, 18*, 1388–1395.
Duffy, P. J. (2016). Genius loci: The geographical imagination of Tim Robinson. In D. Gladwin & C. Cusick (Eds.), *Unfolding Irish landscapes: Tim Robinson, culture and environment* (pp. 21–40). Manchester: Manchester University Press.
Foley, R. (2017). Swimming as an accretive practice in healthy blue space. *Emotion, Space & Society, 22*, 43–51.
Foley, C., & Foley, R. (2012). From Rathdowney to Ratanakiri: Honouring Prof. In P. J. Duffy & W. Nolan (Eds.), *At the anvil: Essays in honour of W.J. Smyth* (pp. 43–51). Dublin: Geography Publications.
Gould, P. (1993). *The slow plague: A geography of the AIDS pandemic*. Oxford: Wiley-Blackwell.
Hadlaw, J. (2003). The London underground map: Imagining modern time and space. *Design Issues, 19*, 25–35.
Harley, B. (1992). Deconstructing the map. *Cartographica, 26*, 1–20.
Joseph, A. E., & Phillips, D. R. (1984). *Accessibility and utilization: Geographical perspectives on health care delivery*. London: Harper & Row.
Khan, A. A., & Bhardwaj, S. (1994). Access to health care: A conceptual framework and its relevance to health care planning. *Evaluation and the Health Professions, 17*, 60–76.
Lauriault, T. P., & Welch, D. (2003). Community mapping and new possibilities for building democracy: The changing faces of Ottawa. *Canadian Journal of Social Policy, 52*, 9–32.
Lupton, D. (2015). Health promotion in the digital era: A critical commentary. *Health Promotion International, 30*, 174–183.
Macfarlane, R. (2012). *The old ways. A journey on foot*. London: Penguin.
Miller, K. (2014). *The cartographer tries to map a way to Zion*. Manchester: Carcanet.
Murphy, C., Tembo, M., Phiri, A., Yerokun, O., & Grummell, B. (2016). Adapting to climate change in shifting landscapes of belief. *Climatic Change, 134*, 101–114.
Patterson, T. C. (2007). Google Earth as a (not just) geography education tool. *Journal of Geography, 106*, 145–152.
Pearce, J., Shortt, N., Mitchell, R., & Rind, E. (2016). Life course, green space and health: Incorporating place into life course epidemiology. *International Journal of Environmental Research and Public Health, 13*, 331. https://doi.org/10.3390/ijerph13030331. Accessed 22 Mar 2019.
Ryan, A. (2012). *Where land meets sea: Coastal explorations of landscape, representation and spatial experience*. Farnham: Ashgate.
Vesely, D. (2004). *Architecture in the age of divided representation: The question of creativity in the shadow of production*. Boston: MIT Press.
Wightman, A. (1997). *Who owns Scotland?* Edinburgh: Cannongate Press.
Wright, D., Goodchild, M., & Foster, J. (1997). Demystifying the persistent ambiguity of GIS as 'tool' versus 'science'. *Annals of the Association of American Geographers, 87*, 346–362.
Young, J., & Gilmore, M. (2013). The spatial politics of affect and emotion in participatory GIS. *Annals of the Association of American Geographers, 103*, 808–823.

Index

S. Atkinson, R. Hunt (eds.), *GeoHumanities and Health*, Global Perspectives on Health Geography, https://doi.org/10.1007/978-3-030-21406-7

Printed by Printforce, the Netherlands